Lecture Notes in Artificial Intelligence 12898

Subseries of Lecture Notes in Computer Science

Editors
Vicenç Torra
Umeå University
Umeå, Sweden

Yasuo Narukawa
Tamagawa University
Tokyo, Japan

ISSN 0302-9743 ISSN 1611-3349 (electronic)
Lecture Notes in Artificial Intelligence
ISBN 978-3-030-85528-4 ISBN 978-3-030-85529-1 (eBook)
https://doi.org/10.1007/978-3-030-85529-1

LNCS Sublibrary: SL7 – Artificial Intelligence

This Springer imprint is published by the registered company Springer Nature Switzerland AG
The registered company address is: Gewerbestrasse 11, 6330 Cham, Switzerland

Preface

This volume contains papers presented at the 18th International Conference on Modeling Decisions for Artificial Intelligence (MDAI 2021), celebrated online at Umeå University, Sweden, September 27–30, 2021.

This conference followed MDAI 2004 (Barcelona), MDAI 2005 (Tsukuba), MDAI 2006 (Tarragona), MDAI 2007 (Kitakyushu), MDAI 2008 (Sabadell), MDAI 2009 (Awaji Island), MDAI 2010 (Perpinyà), MDAI 2011 (Changsha), MDAI 2012 (Girona), MDAI 2013 (Barcelona), MDAI 2014 (Tokyo), MDAI 2015 (Skövde), MDAI 2016 (Sant Julià de Lória), MDAI 2017 (Kitakyushu), MDAI 2018 (Mallorca), and MDAI 2019 (Milano).

The aim of MDAI is to provide a forum for researchers to discuss different facets of decision processes in a broad sense. This includes model building and all kinds of mathematical tools for data aggregation, information fusion, and decision-making; tools to help make decisions related to data science problems (including, for example, statistical and machine learning algorithms as well as data visualization tools); and algorithms for data privacy and transparency-aware methods so that data processing procedures and the decisions made from them are fair, transparent, and avoid unnecessary disclosure of sensitive information.

The MDAI conference included tracks on the topics of (a) data science, (b) machine learning, (c) data privacy, (d) aggregation functions, (e) human decision-making, (f) graphs and (social) networks, and (g) recommendation and search.

The organizers received 50 papers from 20 different countries, 24 of which are published in this volume. Each submission received at least three reviews from the Program Committee and a few external reviewers. We would like to express our gratitude to them for their work.

The conference was supported by the Dept. of Computing Sciences, Umeå University, the European Society for Fuzzy Logic and Technology (EUSFLAT), the Catalan Association for Artificial Intelligence (ACIA), the Japan Society for Fuzzy Theory and Intelligent Informatics (SOFT), and the UNESCO Chair in Data Privacy.

July 2021

Vicenç Torra
Yasuo Narukawa

Organization

General Chair

Vicenç Torra	Umeå University, Sweden

Program Chairs

Vicenç Torra	Umeå University, Sweden
Yasuo Narukawa	Tamagawa University, Japan

Advisory Board

Didier Dubois	Institut de Recherche en Informatique de Toulouse, CNRS, France
Jozo Dujmović	San Francisco State University, USA
Lluis Godo	IIIA-CSIC, Spain
Janusz Kacprzyk	Systems Research Institute, Polish Academy of Sciences, Poland
Sadaaki Miyamoto	University of Tsukuba, Japan
Pierangela Samarati	Università degli Studi di Milano, Italy
Sandra Sandri	Instituto Nacional de Pesquisas Espaciais, Brazil
Michio Sugeno	Tokyo Institute of Technology, Japan
Ronald R. Yager	Machine Intelligence Institute, Iona College, USA

Program Committee

Kayode S. Adewole	Umeå University, Sweden
Laya Aliahmadipour	Shahid Bahonar University, Iran
Nuria Agell	ESADE, Spain
Eva Armengol	IIIA-CSIC, Spain
Edurne Barrenechea	Universidad Pública de Navarra, Spain
Gloria Bordogna	Consiglio Nazionale delle Ricerche, Italy
Humberto Bustince	Universidad Pública de Navarra, Spain
Alina Campan	North Kentucky University, USA
Francisco Chiclana	De Montfort University, UK
Susana Díaz	Universidad de Oviedo, Spain
Josep Domingo-Ferrer	Universitat Rovira i Virgili, Spain
Yasunori Endo	University of Tsukuba, Japan
Vladimir Estivill-Castro	Griffth University, Australia

Zoe Falomir	Universität Bremen, Germany
Giorgos Flouris	FORTH-ICS, Greece
Camilo Andres Franco De Los Rios	Universidad de los Andes, Colombia
Katsushige Fujimoto	Fukushima University, Japan
Joaquin Garcia-Alfaro	Institut Mines-Télécom and Institut Polytechnique de Paris, France
Michel Grabisch	Université Paris I Panthéon-Sorbonne, France
Enrique Herrera-Viedma	Universidad de Granada, Spain
Aoi Honda	Kyushu Institute of Technology, Japan
Van-Nam Huynh	JAIST, Japan
Masahiro Inuiguchi	Osaka University, Japan
Simon James	Deakin University, Australia
Marc Juarez	USC Viterbi, USA
Aránzazu Jurío	Universidad Pública de Navarra, Spain
Yuchi Kanzawa	Shibaura Institute of Technology, Japan
Sema Kayapinar Kaya	Munzur University, Turkey
Hiroaki Kikuchi	Meiji University, Japan
Petr Krajča	Palacky University Olomouc, Czech Republic
Marie-Jeanne Lesot	Université Pierre et Marie Curie (Paris VI), France
Giovanni Livraga	Università degli Studi di Milano, Italy
Jun Long	National University of Defense Technology, China
Beatriz López	University of Girona, Catalonia, Spain
Maite Lopez-Sanchez	University of Barcelona, Catalonia, Spain
Jean-Luc Marichal	University of Luxembourg, Luxembourg
Michael Mayo	University of Waikato, New Zealand
Radko Mesiar	Slovak University of Technology, Slovakia
Andrea Mesiarovà-Zemànkovà	Slovak Academy of Sciences, Slovakia
Anna Monreale	University of Pisa, Italy
Toshiaki Murofushi	Tokyo Institute of Technology, Japan
Guillermo Navarro-Arribas	Universitat Autònoma de Barcelona, Spain
Tomaharu Nakashima	Osaka Prefecture University, Japan
Shekhar Negi	Umeå University, Sweden
Jordi Nin	Esade, Universitat Ramon Llull, Spain
Miguel Nunez-del-Prado	Universidad del Pacífico, Peru
Anna Oganyan	National Institute of Statistical Sciences (NISS), USA
Gabriella Pasi	Università di Milano Bicocca, Italy
Oriol Pujol	University of Barcelona, Catalonia, Spain
Maria Riveiro	Jönköping University, Sweden
Julian Salas	Universitat Oberta de Catalunya, Catalonia, Spain
H. Joe Steinhauer	University of Skövde, Sweden
László Szilágyi	Sapientia-Hungarian Science University of Transylvania, Hungary
Aida Valls	Universitat Rovira i Virgili, Spain
Zeshui Xu	Southeast University, China
Yuji Yoshida	University of Kitakyushu, Japan

Local Organizing Committee

Kayode S. Adewole
Aso Bozorgpanah
Saloni Kwatra
Shekhar Negi
Sudipta Paul
Mariam Taha

Additional Referees

Najeeb Jebreel
Thananut Phiboonbanakit
Alberto Blanco-Justicia
Mau Toan Nguyen
Rami Haffar
Sergio Martinez Lluis

Supporting Institutions

Department of Computing Sciences, Umeå University, Sweden
The European Society for Fuzzy Logic and Technology (EUSFLAT)
The Catalan Association for Artificial Intelligence (ACIA)
The Japan Society for Fuzzy Theory and Intelligent Informatics (SOFT)
The UNESCO Chair in Data Privacy

Contents

Data Science and Data Privacy

Invited Papers

Andness-Directed Iterative OWA Aggregators

Jozo Dujmović(✉)

Department of Computer Science, San Francisco State University,
1600 Holloway Avenue, San Francisco, CA 94132, USA
jozo@sfsu.edu

Abstract. In this paper we introduce andness-directed iterative OWA aggregators. Iterative OWA aggregators belong to the family of OWA aggregators, where the aggregated value of additive aggregators is a scalar product of the sorted vector of arguments and the vector of logic weights that determine conjunctive or disjunctive properties of OWA aggregators. The overall logic properties of any OWA aggregator are characterized by andness (a conjunction degree) and orness (a disjunction degree), as well as the presence/absence of support for annihilators 0 and 1. In this paper we present iterative OWA aggregators where all logic weights are explicit functions (simple polynomials) of andness or orness. In such a way, the desired andness of aggregator is explicitly visible and easily adjustable, yielding ultimate simplicity in applications. Iterative OWA aggregators can also be weighted with importance weights.

Keywords: Iterative aggregation · Andness-directedness · Soft aggregators · Hard aggregators · ItOWA · SItOWA · HItOWA

1 Introduction

The simplest logic aggregators of two logic variables $x_1 \in I = [0, 1]$ and $x_2 \in I$ are $y = c(x_1 \wedge x_2) + d(x_1 \vee x_2)$ and $y = (x_1 \wedge x_2)^c (x_1 \vee x_2)^d$, where $c \in I, d \in I$, and $c + d = 1$. If $0 < c < 1$. The additive form is "soft" (it supports no annihilators), and the multiplicative form is "hard" (it is used only for high values of c because it always supports the annihilator 0). The parameters c (the conjunction degree or andness) and d (the disjunction degree or orness) are used to provide a continuous transition in a selected range of andness between conjunction and disjunction. The resulting aggregators are idempotent, i.e. means.

These simple bivariate functions yield the obvious question: how to make the same aggregators in the case of $n > 2$ variables? The first idea, introduced in [1], was to expand the additive form, creating a scalar product of the vector of sorted arguments and the vector of logic weights. The second idea, introduced in [2], was to iteratively apply the elementary bivariate additive form to various pairs of arguments, so that in each iteration the dispersion of the set of n arguments reduces, and after several iterations converges to the resulting aggregated value.

V. Torra and Y. Narukawa (Eds.): MDAI 2021, LNAI 12898, pp. 3–16, 2021.
https://doi.org/10.1007/978-3-030-85529-1_1

Let $\mathbf{X} = (x_1, ..., x_n)$ denote an array of logic variables: $x_i \in I$, $i = 1, ..., n$. According to [3], all integrable aggregators $A(\mathbf{X})$, $\mathbf{X} = (x_1, ..., x_n)$, $n \geq 2$ can be both geometrically and logically characterized using the *global andness* α, or the *global orness* ω, which are defined as follows:
$\alpha = \frac{n-(n+1)V}{n-1} \in [0, 1], \omega = \frac{(n+1)V-1}{n-1} \in [0, 1], V = \int_{I^n} A(\mathbf{X})dx_1 \cdots dx_n, \alpha+\omega = 1.$
These definitions were first introduced in 1974 and used in multi-criteria decision making and decision-support software systems for characterizing all logic aggregators based on Bajraktarević means and their numerous descendants (quasi-arithmetic means, exponential means, power means, Gini means, etc.). Generally, they are applicable to all integrable aggregators.

The OWA aggregator $OWA(\mathbf{X}; \alpha)$, as an idempotent model of simultaneity/substitutability, was proposed by R.R. Yager in [1] as follows:

$$OWA(\mathbf{X}; \alpha) = v_1 x_{(1)} + \cdots + v_n x_{(n)}, \quad \max(\mathbf{X}) = x_{(1)} \geq \cdots \geq x_{(n)} = \min(\mathbf{X}) ,$$
$$\mathbf{v} = (v_1, ..., v_n), \quad v_i \in I, \quad i = 1, ..., n, \quad v_1 + \cdots + v_n = 1 .$$

The arguments are sorted and $x_{(i)}$ denotes the i-th largest argument. The logic properties of the OWA aggregator are characterized by the OWA andness α and the OWA orness ω, as follows:

$$\alpha = [v_2 + 2v_3 + \cdots + (n-2)v_{n-1} + (n-1)v_n]/(n-1)$$
$$\omega = [(n-1)v_1 + (n-2)v_2 + \cdots + v_{n-1}]/(n-1) ; \quad \alpha \in I, \quad \omega \in I, \quad \alpha + \omega = 1.$$

The logic weights $\mathbf{v} = (0, ..., 0, 1)$, $(1/n, ..., 1/n)$, $(1, 0, ..., 0)$ respectively reduce OWA to the pure conjunction $\min(\mathbf{X})$, the arithmetic mean $(x_1 + \cdots + x_n)/n$, and the pure disjunction $\max(\mathbf{X})$. We use the terms "logic weights" or "OWA weights" because they adjust the logic properties of OWA aggregator (modeling simultaneity or substitutability). We use the term "importance weights" for weights that reflect the importance of inputs.

By selecting appropriate values of the OWA weights $\mathbf{v}$, it is possible to make a continuous transition from conjunction to disjunction. The conjunctive properties of OWA aggregator are obtained by adjusting the logic weights to emphasize the impact of small values of arguments, and disjunctive properties are obtained by adjusting the logic weights to emphasize the impact of large values of arguments.

When OWA was introduced in 1988, it was not known that the OWA andness and orness are identical to the previously used global andness and orness. Ten years after the introduction of OWA, Marichal [4] proved that the OWA andness/orness is equivalent to the global andness/orness. Consequently, we can use the same notation α, ω for the global andness/orness and the OWA andness/orness.

In the case of OWA aggregator the andness and orness must be computed from the selected values of OWA weights. Thus,α and ω are not explicitly visible input parameters of the OWA aggregator. On the other hand, in decision engineering practice, both andness and orness are inputs, and consequently it is indispensable to have aggregators that use andness/orness as explicitly visible and easily adjustable input parameters [5, 7, 13]. Such aggregators are called graded andness-directed logic aggregators. Thus, there is a strongly justifiable interest in the following (soft additive, or hard multiplicative) *andness-directed OWA*:

$$Y_n(\mathbf{X};\alpha)=\sum_{i=1}^{n} w_{ni}(\alpha)x_{(i)},\ \text{or}\ Y_n(\mathbf{X};\alpha)=\prod_{i=1}^{n} x_{(i)}^{w_{ni}(\alpha)},$$

$$\sum_{i=1}^{n} w_{ni}(\alpha)=1,\ Y_n(\mathbf{X};0)=x_{(1)}\geq x_{(2)}\geq\cdots\geq x_{(n)}=Y_n(\mathbf{X};1),\quad n>1.$$

If $w_{ni}(\alpha)$ are simple functions (e.g. polynomials) of α, then such aggregators have andness α as a visible and adjustable parameter, and in this paper, our goal is to develop such functions.

All andness-directed aggregators, including $Y_n(\mathbf{X};\alpha)$, satisfy the obvious condition

$$\alpha=\frac{n}{n-1}-\frac{n+1}{n-1}\int_{I^n} Y_n(\mathbf{X};\alpha)dx_1\cdots dx_n,\quad n\geq 2.$$

This condition strictly holds for all exact theoretical forms of andness-directed aggregators. On the other hand, if $Y_n(\mathbf{X};\alpha)$ is an approximation, then the quality of approximation can be characterized using the following error function:

$$E_n(\alpha)=\left|\alpha-\frac{n}{n-1}-\frac{n+1}{n-1}\int_{I^n} Y_n(\mathbf{X};\alpha)dx_1\cdots dx_n\right|,\ n\geq 2.$$

For example, assuming that $w_{ni}(\alpha;\mathbf{p}_{ni})$ is a real polynomial of α, and $\mathbf{p}_{ni}$ is a vector of its coefficients, we could define the following error function:

$$E_n(\alpha;\mathbf{p}_{n1},\ldots,\mathbf{p}_{nn})=\left|\alpha-\frac{n}{n-1}-\frac{n+1}{n-1}\int_{I^n}\left(\sum_{i=1}^{n} w_{ni}(\alpha;\mathbf{p}_{ni})x_{(i)}\right)dx_1\cdots dx_n\right|,\ n\geq 2.$$

Then, we can find the array of optimum coefficients $\mathbf{p}_{n1}^{opt},\ldots,\mathbf{p}_{nn}^{opt}$ so that $E_n(\alpha;\mathbf{p}_{n1}^{opt},\ldots,\mathbf{p}_{nn}^{opt})$ satisfies a given optimization criterion, e.g. the smallest value of $\max_{0\leq\alpha\leq 1} E_n(\alpha;\mathbf{p}_{n1}^{opt},\ldots,\mathbf{p}_{nn}^{opt})$. This brute force optimization problem is not simple and deserves to be avoided.

This paper presents a simple solution of this optimization problem, based on iterative OWA functions (ItOWA) introduced in [2]. It complements the analysis of andness-directedness for classic OWA aggregators presented in [5, 6, 17]. The proposed ItOWA aggregators can be soft or hard and have directly adjustable andness.

The paper is organized as follows. A soft ItOWA aggregator and its weights are presented in Sect. 2. Andness-directedness of soft and hard ItOWA aggregators is studied in Sect. 3. Section 4 shows a simple technique to introduce importance weights in ItOWA aggregators, and Sect. 5 offers conclusions.

2 The Soft Iterative SItOWA Aggregator and a Recursive Method for Computing Its Weights

Similarly to OWA, the soft iterative OWA aggregator (SItOWA) [2, 7] uses the sorted list of arguments ($x_{(i)}$ denotes the ith largest argument) and applies logic weights that affect the andness α: $SItOWA(\mathbf{X};\, c) = v_{n1}(c)x_{(1)} + \cdots + v_{nn}(c)x_{(n)}$. The weights are functions of the parameter c, and consequently the andness α is also a function of c. Since $\alpha = f_n(c)$, where f is an invertible function, it follows that the andness-directed SItOWA (presented in Sect. 3) can be obtained using $c = f_n^{-1}(\alpha)$: $SItOWA(\mathbf{X};\, \alpha) = w_{n1}(\alpha)x_{(1)} + \cdots + w_{nn}(\alpha)x_{(n)}$, $w_{ni}(\alpha) = v_{ni}(f_n^{-1}(\alpha))$, $i = 1, \ldots, n$. The additive structure of both OWA and SItOWA makes them soft for all $0 < \alpha < 1$. In addition, both OWA and SItOWA are means, and consequently idempotent aggregators.

SItOWA is characterized by the conjunction degree $c \in [0, 1]$ and the disjunction degree $d = 1 - c$, used in the bivariate case. For both OWA and SItOWA, in the case of two variables x_1, x_2, the parameters c and d are the global andness and orness, as follows:

$$y_2(x_1, x_2; c) = d \max(x_1, x_2) + c \min(x_1, x_2) = (1 - \alpha)x_{(1)} + \alpha x_{(2)},$$
$$c = \alpha, \quad d = 1 - \alpha = \omega.$$

In the case of $n > 2$ variables we can iteratively apply the basic bivariate aggregator to various pairs of variables x_i, x_j, $i \in \{1, \ldots, n\}$, $j \in \{1, \ldots, n\}$, $i \neq j$, and replace them with the aggregated value $x_{ij} = d \max(x_i, x_j) + c \min(x_i, x_j)$. We will now show that after iteratively applying this procedure, the dispersion of x_{ij} points quickly decreases; the iterative process converges, and yields the resulting aggregated value $y_n(x_1, \ldots, x_n; c)$. Since the aggregation process is fully controlled by the conjunction degree c the conjunctive/disjunctive properties of the bivariate aggregator penetrate and systematically infiltrate in the multivariate aggregation process, yielding unique values of the ItOWA logic weights. We will start with the case $n = 3$ variables.

In the case of three variables $x_{(1)} \geq x_{(2)} \geq x_{(3)}$, the basic bivariate aggregator is iteratively applied to all pairs of inputs as follows:

while $x_{(1)} - x_{(3)} \geq \varepsilon$ **do** // ε = small positive error (e.g. 10^{-6})
$t_{12} = (1 - c)x_{(1)} + cx_{(2)};$ // $x_{(1)} \geq t_{12} \geq x_{(2)}$
$t_{13} = (1 - c)x_{(1)} + cx_{(3)};$ // $x_{(1)} \geq t_{13} \geq x_{(3)}$; $x_{(1)} \geq t_{12} \geq t_{13}$
$t_{23} = (1 - c)x_{(2)} + cx_{(3)};$ // $x_{(2)} \geq t_{23} \geq x_{(3)}$; $t_{13} \geq t_{23} \geq x_{(3)}$
$x_{(1)} = t_{12};\ x_{(2)} = t_{13};\ x_{(3)} = t_{23};$ // $t_{12} - t_{23} \leq x_{(1)} - x_{(3)} \Rightarrow$ convergence
end
$y_3 = x_{(2)};$ //the middle value $x_{(2)}$ is the desired value of function $y_3(x_1, x_2, x_3; c)$

In each iteration, this process keeps the order $x_{(1)} \geq x_{(2)} \geq x_{(3)}$ and reduces the size of triplet $x_{(1)} - x_{(3)}$. The temporary values $t_{12} \geq t_{13} \geq t_{23}$ are closer to the resulting value than the initial triplet $x_{(1)} \geq x_{(2)} \geq x_{(3)}$. When the size of triplet becomes less than an arbitrary small error $\varepsilon \ll 1$ the iterative process terminates and returns the middle value $y_3 = x_{(2)}$. The fast converging of the iterative process for $c = 0.25, 0.5, 0.75$ and $x_{(1)} = 1$, $x_{(2)} = 1/2$, $x_{(3)} = 0$ is shown in Fig. 1, where the precise final answer is obtained in less than 10 iterations.

In the case of n variables, we might apply the convergent pairwise aggregation process to $n(n - 1)/2$ pairs of points but that would produce a quadratic growth of points and inefficient $O(n^2)$ aggregation. However, the presented idea of reducing the aggregation of three variables to three aggregations of two variables can be generalized in a recursive way: the aggregation of n variables can be organized as n aggregations of $n-1$ variables. Below, we present this method and use it to compute andness-directed weights of aggregators of $n = 3, 4, 5$ variables.

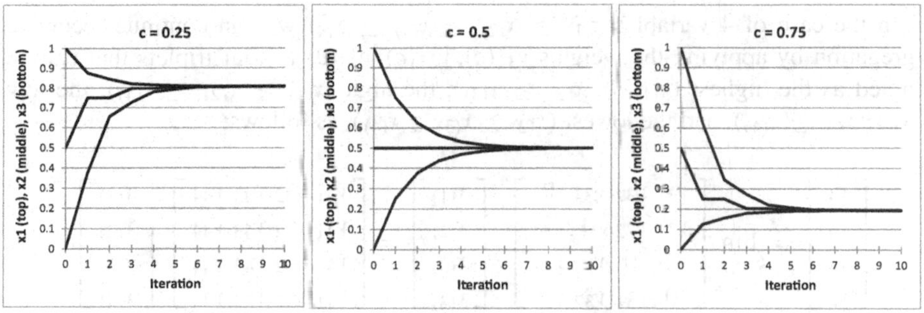

Fig. 1. Iterative aggregation of three variables for three different conjunction degrees

For n = 3, the iterative aggregation process can be written in the following matrix notation:

$$\begin{bmatrix} y_3 \\ y_3 \\ y_3 \end{bmatrix} = \lim_{k\to\infty} \begin{bmatrix} 1-c & c & 0 \\ 1-c & 0 & c \\ 0 & 1-c & c \end{bmatrix}^k \begin{bmatrix} x_{(1)} \\ x_{(2)} \\ x_{(3)} \end{bmatrix} = \begin{bmatrix} v_{31} & v_{32} & v_{33} \\ v_{31} & v_{32} & v_{33} \\ v_{31} & v_{32} & v_{33} \end{bmatrix} \begin{bmatrix} x_{(1)} \\ x_{(2)} \\ x_{(3)} \end{bmatrix}$$

This is a convergent process that returns the following result [2]:

$$y_3 = \frac{(1-c)^2 x_{(1)} + c(1-c)x_{(2)} + c^2 x_{(3)}}{1-c+c^2}.$$

Consequently, SItOWA of three variables can be written as follows:

$$\begin{aligned}
&y_3(x_1, x_2, x_3; c) = v_{31}(c)x_{(1)} + v_{32}(c)x_{(2)} + v_{33}(c)x_{(3)},\\
&v_{31}(c) = (1-c)^2/(1-c+c^2),\\
&v_{32}(c) = c(1-c)/(1-c+c^2),\\
&v_{33}(c) = c^2/(1-c+c^2),\\
&v_{31}(c) + v_{32}(c) + v_{33}(c) = 1,\\
&y_3(x_1, x_2, x_3; 0) = x_{(1)},\\
&y_3(x_1, x_2, x_3; 1) = x_{(3)},\\
&y_3(x_1, x_2, x_3; 1/2) = (x_{(1)} + x_{(2)} + x_{(3)})/3.
\end{aligned}$$

These formulas validate the example shown in Fig. 1 as follows:

$$y_3(1, 1/2, 0; c) = \frac{(1-c)^2 + c(1-c)/2}{1-c+c^2} = \begin{cases} 0.808\,, & c = 0.25 \\ 0.5\,, & c = 0.5 \\ 0.192\,, & c = 0.75 \end{cases}$$

In the case of 4 variables $x_{(1)} \geq x_{(2)} \geq x_{(3)} \geq x_{(4)}$ we can continue recursive aggregation by applying the weights $v_{31}(c), v_{32}(c), v_{33}(c)$ to four triplets that can be denoted as the highest $(x_{(1)} \geq x_{(2)} \geq x_{(3)})$, the high $(x_{(1)} \geq x_{(2)} \geq x_{(4)})$, the low $(x_{(1)} \geq x_{(3)} \geq x_{(4)})$, and the lowest $(x_{(2)} \geq x_{(3)} \geq x_{(4)})$, as follows:

$$\begin{bmatrix} y_4 \\ y_4 \\ y_4 \\ y_4 \end{bmatrix} = \lim_{k\to\infty} \begin{bmatrix} v_{31} & v_{32} & v_{33} & 0 \\ v_{31} & v_{32} & 0 & v_{33} \\ v_{31} & 0 & v_{32} & v_{33} \\ 0 & v_{31} & v_{32} & v_{33} \end{bmatrix}^k \begin{bmatrix} x_{(1)} \\ x_{(2)} \\ x_{(3)} \\ x_{(4)} \end{bmatrix} = \begin{bmatrix} v_{41} & v_{42} & v_{43} & v_{44} \\ v_{41} & v_{42} & v_{43} & v_{44} \\ v_{41} & v_{42} & v_{43} & v_{44} \\ v_{41} & v_{42} & v_{43} & v_{44} \end{bmatrix} \begin{bmatrix} x_{(1)} \\ x_{(2)} \\ x_{(3)} \\ x_{(4)} \end{bmatrix}$$

The results for 4 variables are the following [7]:

$$\begin{aligned} &y_4(x_1, x_2, x_3, x_4; c) = v_{41}(c)x_{(1)} + v_{42}(c)x_{(2)} + v_{43}(c)x_{(3)} + v_{44}(c)x_{(4)}\,, \\ &v_{41}(c) = (1 - c - c^2 - c^3 + 4c^4 - 2c^5)/(1 + 4c^4)\,, \\ &v_{42}(c) = c(1 - c^2 - 2c^3 + 2c^4)/(1 + 4c^4)\,, \\ &v_{43}(c) = c^2(1 + c - 2c^3)/(1 + 4c^4)\,, \\ &v_{44}(c) = c^3(1 + 2c + 2c^2)/(1 + 4c^4)\,. \end{aligned}$$

Similarly, for 5 variables:

$$\begin{aligned} &y_5(x_1, x_2, x_3, x_4, x_5; c) = v_{51}(c)x_{(1)} + v_{52}(c)x_{(2)} + v_{53}(c)x_{(3)} + v_{54}(c)x_{(4)} + v_{55}(c)x_{(5)}, \\ &v_{51}(c) = (1 - c - c^2 - c^3 - c^4 + 11c^5 - 8c^6 - 3c^7 + 2c^8 + 2c^9 - c^{10})/(1 + 11c^5 - c^{10}) \\ &v_{52}(c) = c(1 - c^2 - 2c^3 - 3c^4 + 8c^5 - 3c^7 - c^8 + c^9)/(1 + 11c^5 - c^{10}). \\ &v_{53}(c) = c^2(1 + c - 2c^3 - 5c^4 + 3c^5 + 3c^6 - c^8)/(1 + 11c^5 - c^{10}). \\ &v_{54}(c) = c^3(1 + 2c + 2c^2 - 5c^4 - 2c^5 + c^6 + c^7)/(1 + 11c^5 - c^{10}). \\ &v_{55}(c) = c^4(1 + 3c + 5c^2 + 5c^3 - 2c^5 - c^6)/(1 + 11c^5 - c^{10}). \end{aligned}$$

The coefficients $v_{ij}(c)$ introduced in [2] and presented in [7] were recently verified and expanded to any number of variables in [8] and [9]. Troiano and Díaz [8] proved that $SItOWA(\mathbf{X}; c)$ can generally be written as follows:

$$SItOWA(\mathbf{X}; c) = v_{n1}(c)x_{(1)} + \cdots + v_{nn}(c)x_{(n)} = \frac{\sum_{i=1}^{n} (1-c)^{n-i} c^{i-1} x_{(i)}}{\sum_{i=1}^{n} (1-c)^{n-i} c^{i-1}}\,, \quad n \geq 2.$$

3 Andness-Directed SItOWA and HItOWA Aggregators

The conjunction degree c can be used as a degree of simultaneity or substitutability of the SItOWA aggregator. However, the conjunction degree c is equal to the global andness α only for $n = 2$. For $n > 2$, there are differences between c and α. According to [10] there are nine versions of andness/orness that can be used for adjustment of degrees of simultaneity or substitutability of logic aggregators, and for each of them, evaluators can be trained to adapt to their properties and use them efficiently in various applications. It is possible to use the SItOWA conjunction and disjunction degrees (c and d), but it is more convenient to parameterize SItOWA aggregators using the standard global andness α. Therefore, let us numerically analyze the differences between the SItOWA conjunction/disjunction degrees and the global andness/orness, with intention to express the coefficients $v_{n1}, \ldots, v_{nn},\ \ n \geq 2$ as explicit functions of the global andness α, and create desired andness-directed aggregators.

For the most frequently used values $n = 2, 3, 4, 5$ the global andness of SItOWA $\alpha(c) = [n - (n+1)V_n(c)]/(n-1)$ can be determined using a numerical computation of the volume $V_n(c) = \int_0^1 \cdots \int_0^1 y_n(x_1, \ldots, x_n; c)dx_1 \cdots dx_n$. The corresponding results are shown in Fig. 2. If $n = 2$ we have $\alpha = c$, but for $n > 2$ we have nonlinear relationships $\alpha = f_n(c)$, where $f_n(0) = 0, f_n(1/2) = 1/2$, and $f_n(1) = 1$. The function $c \mapsto f_n(c)$ is strictly increasing and invertible. Since these are numerical relationships we can compute inverse functions $c = f_n^{-1}(\alpha)$ and approximate them with polynomials as follows:

$$c(\alpha) = \begin{cases} 3.044\alpha^5 - 7.6122\alpha^4 + 7.7913\alpha^3 - 4.0731\alpha^2 + 1.8498\alpha + 8 \cdot 10^{-5}, & n = 3 \\ 6.9006\alpha^5 - 17.265\alpha^4 + 16.984\alpha^3 - 8.2002\alpha^2 + 2.5761\alpha + 0.0023, & n = 4 \\ 11.078\alpha^5 - 27.712\alpha^4 + 26.623\alpha^3 - 12.21\alpha^2 + 3.2114\alpha + 0.0045, & n = 5. \end{cases}$$

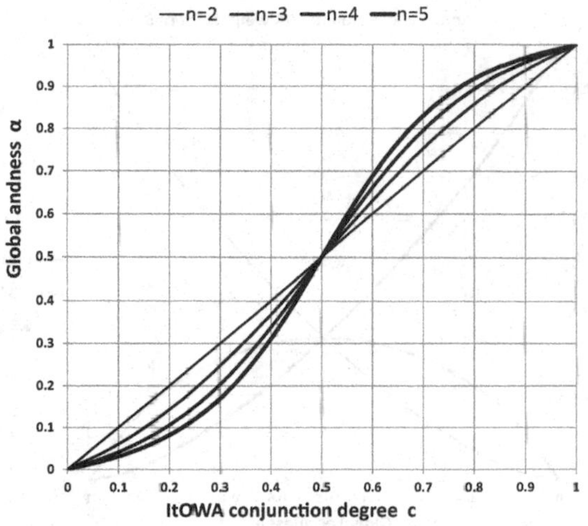

Fig. 2. Global andness of the SItOWA aggregator

After inserting $c(\alpha)$ in weights $v_{ni}(c)$ we get the andness-directed SItOWA with the global andness-directed coefficients. In the case $n = 3$, we have

$$\begin{aligned} y_3(x_1, x_2, x_3; \alpha) &= v_{31}(c(\alpha))x_{(1)} + v_{32}(c(\alpha))x_{(2)} + v_{33}(c(\alpha))x_{(3)} \\ &= \frac{(1 - c(\alpha))^2 x_{(1)} + c(\alpha)(1 - c(\alpha))x_{(2)} + c(\alpha)^2 x_{(3)}}{1 - c(\alpha) + c(\alpha)^2} \\ &= V_{31}(\alpha)x_{(1)} + V_{32}(\alpha)x_{(2)} + V_{33}(\alpha)x_{(3)} \end{aligned}$$

In the case of three variables, the aggregator $y_3(x_1, x_2, x_3; \alpha)$ has the global andness that is sufficiently close to the desired andness α:

$$D_3^{\max} = \max_{0 \le \alpha \le 1} \left| \alpha - \left[1.5 - 2 \int_0^1 \int_0^1 \int_0^1 y_3(x_1, x_2, x_3; \alpha)\, dx_1 dx_2 dx_3 \right] \right| = 0.00158.$$

The characteristic shapes of the andness-directed weights $V_{ni}(\alpha) = v_{ni}(c(\alpha)),\ i = 1, .., n$ for $n = 3$ are shown in Fig. 3. These curves are rather smooth and can be numerically approximated by the following simple polynomials:

$$\begin{aligned} w_{31}(\alpha) &= 0.8916\alpha^4 - 1.7754\alpha^3 + 1.7641\alpha^2 - 1.8798\alpha + 0.9994 \approx V_{31}(\alpha) \\ w_{32}(\alpha) &= -1.7791\alpha^4 + 3.5585\alpha^3 - 3.5471\alpha^2 + 1.7677\alpha + 0.0007 \approx V_{32}(\alpha) \\ w_{33}(\alpha) &= 0.8876\alpha^4 - 1.7832\alpha^3 + 1.783\alpha^2 + 0.1121\alpha - 0.00009 \approx V_{33}(\alpha). \end{aligned}$$

In this way it is possible to reduce computational complexity because both $v_{ni}(c)$ and $c(\alpha)$ can be intricate polynomials, while the coefficients $w_{ni}(\alpha)$ are significantly simpler polynomials and it is possible to show that their use is not reducing the accuracy and applicability of the resulting SItOWA aggregator. Therefore, we can use the following simplest approximate form of the andness-directed SItOWA aggregator for $n = 3$:

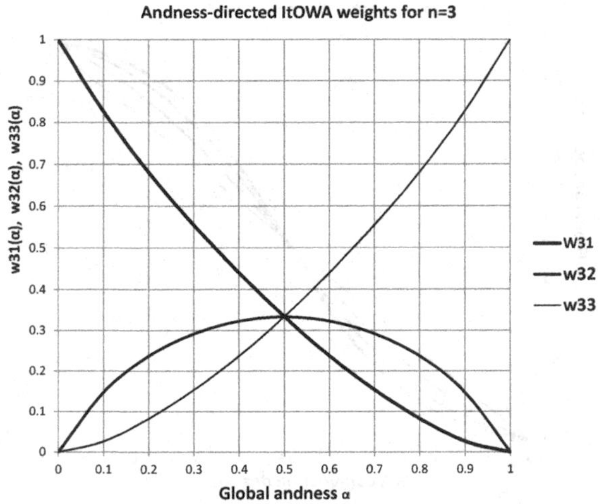

Fig. 3. Andness-directed SItOWA weights in the case of 3 variables

$$
\begin{aligned}
Y_3(x_1,x_2,x_3;\alpha) &= w_{31}(\alpha)x_{(1)} + w_{32}(\alpha)x_{(2)} + w_{33}(\alpha)x_{(3)} \\
&= (0.8916\alpha^4 - 1.7754\alpha^3 + 1.7641\alpha^2 - 1.8798\alpha + 0.9994)\max(x_1,x_2,x_3) \\
&+ (-1.7791\alpha^4 + 3.5585\alpha^3 - 3.5471\alpha^2 + 1.7677\alpha + 0.0007)\mathrm{mid}(x_1,x_2,x_3) \\
&+ (0.8876\alpha^4 - 1.7832\alpha^3 + 1.783\alpha^2 + 0.1121\alpha - 0.00009)\min(x_1,x_2,x_3) \\
&\approx y_3(x_1,x_2,x_3;\alpha)\ .
\end{aligned}
$$

This aggregator provides the explicitly visible and easily adjustable desired andness α. Evaluation of the quality of approximation $Y_n(x_1, ..., x_n; \alpha)$ can be defined similarly as for $y_n(x_1, ..., x_n; \alpha)$:

$$
E_n(\alpha) = \left| \alpha - \left[\frac{n}{n-1} - \frac{n+1}{n-1} \int_0^1 \cdots \int_0^1 Y_n(x_1, ..., x_n; \alpha)\, dx_1 \cdots dx_n \right] \right|
$$

$$
E_n^{\max} = \max_{0\le\alpha\le1} E_n(\alpha)\ , \quad E_n^{mean} = \int_0^1 E_n(\alpha)d\alpha\ , \quad n > 1\ .
$$

The aggregator $Y_3(x_1, x_2, x_3; \alpha)$ is appropriate for all practical purposes, because the maximum absolute error of this approximation is sufficiently low:

$$
E_3^{\max} = \max_{0\le\alpha\le1} \left| \alpha - \left[1.5 - 2 \int_0^1\int_0^1\int_0^1 Y_3(x_1,x_2,x_3;\alpha)dx_1dx_2dx_3 \right] \right| = 0.00095.
$$

The presented procedure for 3 variables can be used in the cases of 4 and 5 variables and it yields the following andness-directed aggregators and their approximations:

$$
\begin{aligned}
y_4(x_1,x_2,x_3,x_4;\alpha) &= v_{41}(c(\alpha))x_{(1)} + v_{42}(c(\alpha))x_{(2)} + v_{43}(c(\alpha))x_{(3)} + v_{44}(c(\alpha))x_{(4)} \\
&= V_{41}(\alpha)x_{(1)} + V_{42}(\alpha)x_{(2)} + V_{43}(\alpha)x_{(3)} + V_{44}(\alpha)x_{(4)}
\end{aligned}
$$

$$
D_4^{\max} = \max_{0\le\alpha\le1} \left| \alpha - \left[\frac{4}{3} - \frac{5}{3} \int_0^1\int_0^1\int_0^1\int_0^1 y_4(x_1,x_2,x_3,x_4;\alpha)dx_1dx_2dx_3dx_4 \right] \right| = 0.00458
$$

$$
\begin{aligned}
w_{41}(\alpha) &= -2.4999\alpha^5 + 7.7635\alpha^4 - 8.957\alpha^3 + 5.5225\alpha^2 - 2.83\alpha + 0.9989 \\
w_{42}(\alpha) &= -2.1683\alpha^6 + 10.955\alpha^5 - 19.955\alpha^4 + 17.955\alpha^3 - 9.3454\alpha^2 \\
&\quad + 2.5529\alpha + 0.0033 \\
w_{43}(\alpha) &= -2.0541\alpha^6 + 1.7637\alpha^5 + 2.548\alpha^4 - 4.3959\alpha^3 + 1.8141\alpha^2 \\
&\quad + 0.3297\alpha - 0.0022 \\
w_{44}(\alpha) &= 2.451\alpha^5 - 4.6474\alpha^4 + 2.8613\alpha^3 + 0.2325\alpha^2 + 0.1036\alpha - 0.002 \\
Y_4(x_1,x_2,x_3,x_4;\alpha) &= w_{41}(\alpha)x_{(1)} + w_{42}(\alpha)x_{(2)} + w_{43}(\alpha)x_{(3)} + w_{44}(\alpha)x_{(4)}
\end{aligned}
$$

$$
E_4^{\max} = \max_{0\le\alpha\le1} \left| \alpha - \left[\frac{4}{3} - \frac{5}{3} \int_0^1\int_0^1\int_0^1\int_0^1 Y_4(x_1,x_2,x_3,x_4;\alpha)dx_1dx_2dx_3dx_4 \right] \right| = 0.0053.
$$

$$
\begin{aligned}
y_5(x_1,x_2,x_3,x_4,x_5;\alpha) &= v_{51}(c(\alpha))x_{(1)} + v_{52}(c(\alpha))x_{(2)} + v_{53}(c(\alpha))x_{(3)} \\
&+ v_{54}(c(\alpha))x_{(4)} + v_{55}(c(\alpha))x_{(5)} \\
&= V_{51}(\alpha)x_{(1)} + V_{52}(\alpha)x_{(2)} + V_{53}(\alpha)x_{(3)} + V_{54}(\alpha)x_{(4)} + V_{55}(\alpha)x_{(5)}
\end{aligned}
$$

$$D_5^{\max} = \max_{0\le\alpha\le 1}\left|\alpha - \left[1.25 - 1.5\int_0^1\int_0^1\int_0^1\int_0^1\int_0^1 y_5(x_1,x_2,x_3,x_4,x_5;\alpha)dx_1dx_2dx_3dx_4dx_5\right]\right|$$
$$= 0.0108\,.$$

$$w_{51}(\alpha) = 4.0901\alpha^6 - 17.781\alpha^5 + 29.916\alpha^4 - 24.791\alpha^3 + 11.434\alpha^2 - 3.8697\alpha + 0.9998$$

$$w_{52}(\alpha) = -8.5825\alpha^6 + 31.475\alpha^5 - 45.76\alpha^4 + 34.399\alpha^3 - 14.695\alpha^2 + 3.155\alpha + 0.0065$$

$$w_{53}(\alpha) = 8.6623\alpha^6 - 25.976\alpha^5 + 29.529\alpha^4 - 15.778\alpha^3 + 3.1181\alpha^2 + 0.4444\alpha - 0.0019$$

$$w_{54}(\alpha) = -8.4138\alpha^6 + 19.554\alpha^5 - 16.657\alpha^4 + 5.3454\alpha^3 - 0.0174\alpha^2 + 0.1992\alpha - 0.0027$$

$$w_{55}(\alpha) = 4.2376\alpha^6 - 7.2514\alpha^5 + 2.95\alpha^4 + 0.8358\alpha^3 + 0.1573\alpha^2 + 0.0716\alpha - 0.0017$$

$$Y_5(x_1,x_2,x_3,x_4,x_5;\alpha) = w_{51}(\alpha)x_{(1)} + w_{52}(\alpha)x_{(2)} + w_{53}(\alpha)x_{(3)} + w_{54}(\alpha)x_{(4)} + w_{55}(\alpha)x_{(5)}$$

$$E_5^{\max} = \max_{0\le\alpha\le 1}\left|\alpha - \left[1.25 - 1.5\int_0^1\int_0^1\int_0^1\int_0^1\int_0^1 Y_5(x_1,x_2,x_3,x_4,x_5;\alpha)\,dx_1dx_2dx_3dx_4dx_5\right]\right|$$
$$= 0.00602\,.$$

The andness-directed SItOWA weights for 4 and 5 variables are presented in Fig. 4.

Generally, the andness-directed SItOWA is a version of OWA that has weights that are explicit functions of desired global andness, as follows:

$$Y_n(x_1,...,x_n;\alpha) = w_{n1}(\alpha)x_{(1)} + w_{n2}(\alpha)x_{(2)} + \cdots + w_{nn}(\alpha)x_{(n)}\,,\quad x_{(1)} \ge x_{(2)} \ge \cdots \ge x_{(n)}\,,$$
$$Y_n(x_1,...,x_n;0) = x_{(1)}\,,\quad Y_n(x_1,...,x_n;1) = x_{(n)}\,,\quad Y_n(x_1,...,x_n;1/2) = (x_{(1)} + \cdots + x_{(n)})/n\,.$$

There are no theoretical obstacles to continue the presented process and to create SItOWA aggregators for $n > 5$. However, for practical purposes, taking into account human limitations [12] and the experiences with decision support applications [11], it is sufficient to have aggregators up to $n = 5$.

Additive SItOWA aggregators are soft but they can be used in the whole range of andness, from 0 to 1. As opposed to that, multiplicative ItOWA aggregators (denoted HItOWA) are strictly hard. Consequently, it is meaningful to use them only in the range of hard aggregators, i.e. for $\alpha_\theta \le \alpha \le 1$, where, according to experiments reported in [13], the threshold andness should be $\alpha_\theta \approx 3/4$.

Andness-directed HItOWA aggregators can be realized using the same approximation technique we used for SItOWA. The following results include andness-directed

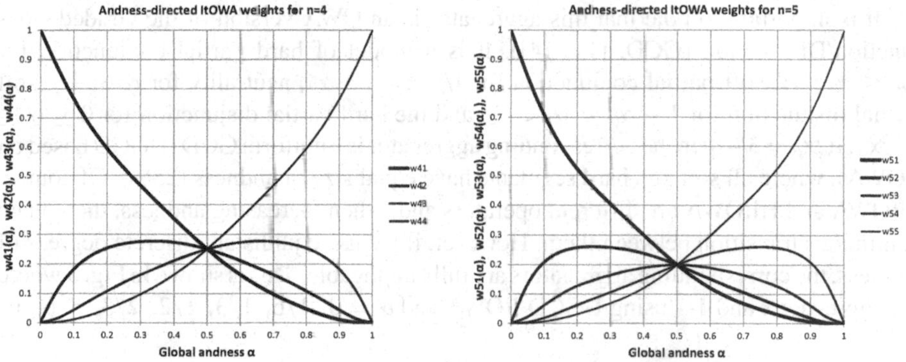

Fig. 4. SItOWA weights in the cases of 4 and 5 variables

HItOWA aggregators $y_n(x_1, \ldots, x_n; \alpha)$ for $n = 2, 3, 4, 5$, and the corresponding mean and maximum absolute differences between the desired and the achieved andness, computed for the whole range $0 \leq \alpha \leq 1$:

$$n = 2: \begin{cases} c(\alpha) = 0.4678\alpha^3 - 0.0409\alpha^2 + 0.5723\alpha - 0.003 \\ y_2(x_1, x_2; \alpha) = x_{(1)}^{1-c(\alpha)} x_{(2)}^{c(\alpha)}; \quad x_{(1)} \geq x_{(2)} \\ D_2^{\text{mean}} = 0.0018, \quad D_2^{\text{max}} = 0.0063 \ . \end{cases}$$

$$n = 3: \begin{cases} c(\alpha) = 5.0604\alpha^5 - 11.131\alpha^4 + 9.4977\alpha^3 - 3.8368\alpha^2 + 1.4027\alpha - 0.0022 \\ y_3(x_1, x_2, x_3; \alpha) = x_{(1)}^{v_{31}(c(\alpha))} x_{(2)}^{v_{32}(c(\alpha))} x_{(3)}^{v_{33}(c(\alpha))}; \quad x_{(1)} \geq x_{(2)} \geq x_{(3)} \\ D_3^{\text{mean}} = 0.0025, \quad D_3^{\text{max}} = 0.0046 \ . \end{cases}$$

$$n = 4: \begin{cases} c(\alpha) = 13.646\alpha^6 - 31.207\alpha^5 + 23.524\alpha^4 - 4.0538\alpha^3 - 2.6779\alpha^2 + 1.7519\alpha + 0.0043 \\ y_4(x_1, x_2, x_3, x_4; \alpha) = x_{(1)}^{v_{41}(c(\alpha))} x_{(2)}^{v_{42}(c(\alpha))} x_{(3)}^{v_{43}(c(\alpha))} x_{(4)}^{v_{44}(c(\alpha))}; \\ x_{(1)} \geq x_{(2)} \geq x_{(3)} \geq x_{(4)}; D_4^{\text{mean}} = 0.0048, \quad D_4^{\text{max}} = 0.0098. \end{cases}$$

$$n = 5: \begin{cases} c(\alpha) = 18.198\alpha^6 - 40.085\alpha^5 + 26.912\alpha^4 - 1.0673\alpha^3 - 5.329\alpha^2 \\ \quad +2.3445\alpha + 0.0074 \\ y_5(x_1, x_2, x_3, x_4, x_5; \alpha) = x_{(1)}^{v_{51}(c(\alpha))} x_{(2)}^{v_{52}(c(\alpha))} x_{(3)}^{v_{53}(c(\alpha))} x_{(4)}^{v_{54}(c(\alpha))} x_{(5)}^{v_{55}(c(\alpha))}; \\ x_{(1)} \geq x_{(2)} \geq x_{(3)} \geq x_{(4)} \geq x_{(5)}; D_5^{\text{mean}} = 0.0093, \quad D_5^{\text{max}} = 0.021 \ . \end{cases}$$

A complete andness-directed ItOWA aggregator can be realized recursively, by combining SItOWA, HItOWA, and De Morgan duality, as shown in the following pseudocode:

ItOWA$(x_1, \ldots, x_n; \alpha)$
{**if** $\alpha \geq \alpha_\theta$ **then return** HItOWA$(x_1, \ldots, x_n; \alpha)$
elsif $\alpha > 1 - \alpha_\theta$ **then return** SItOWA$(x_1, \ldots, x_n; \alpha)$
else return $1 -$ **ItOWA**$(1 - x_1, \ldots, 1 - x_n; 1 - \alpha)$
}

It is important to note that this aggregator is an OWA version of the Graded Conjunction/Disjunction (GCD, [11, 14]). It is a model of hard partial conjunction for $\alpha_\theta \leq \alpha \leq 1$, soft partial conjunction for $1/2 < \alpha < \alpha_\theta$, neutrality for $\alpha = 1/2$, soft partial disjunction for $1 - \alpha_\theta < \alpha < 1/2$, and the hard partial disjunction for $0 \leq \alpha \leq 1 - \alpha_\theta$. If $\alpha_\theta = 3/4$ then the corresponding aggregator is a uniform GCD (UGCD based on ItOWA), where all soft and hard segments have equal size of andness (25%). Of course, SItOWA and HItOWA are different operators and, when increasing andness, there is no continuous transition between them. However, if we use equidistant discrete degrees of andness, the corresponding aggregators are fully applicable. This is shown in Fig. 5 where we aggregate x and 1-x using UGCD/ItOWA and $\alpha = 0,\ 1/6,\ 1/3,\ 1/2,\ 2/3,\ 5/6,\ 1$.

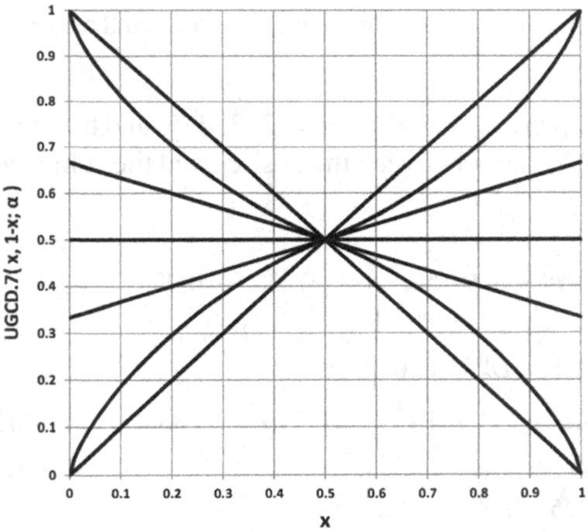

Fig. 5. Uniform UGCD/ItOWA with 7 levels of andness aggregating x and 1-x

4 Weighted ItOWA Aggregators

Let $n = 2$ and let w_{x1}, w_{x2} denote the importance weights of variables x_1 and x_2 respectively. In the case of desired andness α the SItOWA aggregator can be written as follows:

$$z = \begin{cases} \alpha x_1 + (1-\alpha)x_2, & x_1 < x_2 \\ \alpha x_2 + (1-\alpha)x_1, & x_1 \geq x_2 \end{cases}$$

A weighted SItOWA for $n = 2$ can be defined using multiplicative importance weights as follows:

$$z = \begin{cases} \frac{\alpha w_{x1}x_1 + (1-\alpha)w_{x2}x_2}{\alpha w_{x1} + (1-\alpha)w_{x2}}, & x_1 < x_2 \\ \frac{\alpha w_{x2}x_2 + (1-\alpha)w_{x1}x_1}{\alpha w_{x2} + (1-\alpha)w_{x1}}, & x_1 \geq x_2 \end{cases}$$

This multiplicative approach is based on justifiable assumption that both the relative importance of an input, and the highest value of andness/orness ($\alpha \vee \omega$), simultaneously contribute to both the objective ability to produce impact, and the intuitive percept of the overall importance of an input [11]. This concept can be directly applied to the case of SItOWA/HItOWA of n variables, as follows:

$$Y_n(x_1, ..., x_n; \mathbf{w}, \alpha) = \begin{cases} W_{n1}(\alpha)x_{(1)} + W_{n2}(\alpha)x_{(2)} + \cdots + W_{nn}(\alpha)x_{(n)}, & 1 - \alpha_\theta < \alpha < \alpha_\theta \\ x_{(1)}^{W_{n1}(\alpha)} \times x_{(2)}^{W_{n2}(\alpha)} \times \cdots \times x_{(n)}^{W_{nn}(\alpha)}, & \alpha \in [0, 1 - \alpha_\theta] \cup [\alpha_\theta, 1] \end{cases}$$

$$W_{ni}(\alpha) = \frac{w_{x(i)} w_{ni}(\alpha)}{w_{x(1)} w_{n1}(\alpha) + w_{x(2)} w_{n2}(\alpha) + \cdots + w_{x(n)} w_{nn}(\alpha)}, \quad i = 1, ..., n .$$

In the above aggregators we have the following interpretation of weights:

$w_{ni}(\alpha)$ = positional logic weight that corresponds to the ith largest argument.

$w_{x(i)}$ = private importance weight that corresponds the argument that occupies the ith position.

$W_{ni}(\alpha)$ = resulting compound ItOWA weight; $W_{n1}(\alpha) + W_{n2}(\alpha) + \cdots + W_{nn}(\alpha) = 1$.

It is necessary to note that the presented methodology is not the only way to extend simple bivariate means to weighted multivariate means. The problem of extension of bivariate means to weighted means of several arguments can be solved using the method of binary trees [15, 16]. The comparison of these two approaches is one of topics for the future work.

5 Conclusions

The andness-directedness, annihilator selectability, and importance weight adjustability are three indispensable properties of logic aggregators. The andness-directed SItOWA and HItOWA aggregators, and the corresponding ItOWA-based GCD, offer these fundamental properties at the minimum level of mathematical sophistication. The ultimate simplicity of these aggregators makes them attractive for many applications in decision support systems. Future work should compare these aggregators with other members of the OWA family of aggregators, and with aggregators obtained using OWA-based binary trees. It is also necessary to investigate the properties and applicability of compound aggregation structures based on SItOWA and HItOWA. In particular, it is necessary to investigate properties and applicability of ItOWA-based conjunctive and disjunctive partial absorption aggregators.

Our version of HItOWA is based on weighted geometric mean, but it is obvious that the same aggregation effects can be achieved using the weighted harmonic mean, or other similar forms of weighted power mean, as well as any other bivariate mean that supports the annihilator 0 and provides a controlled emphasis of minimum and maximum values. Future work should investigate whether replacing the geometric mean with another similar hard aggregator could yield some new benefits.

ItOWA is an andness-characterized aggregator, and the analysis and comparison of such aggregators presented in [5] shows several competitive aggregators; ItOWA should be compared with all of them. In particular, future theoretical and experimental work should compare ItOWA with segmented andness-directed interpolative GCD, which is the most convenient of general logic aggregators because it covers the whole range from drastic conjunction to drastic disjunction: hyperconjunction, hard and soft partial conjunction, neutrality, soft and hard partial disjunction, and hyperdisjunction.

References

1. Yager, R.R.: On ordered weighted averaging aggregation operators in multi-criteria decision making. IEEE Trans. Syst. Man Cybernet. **18**, 183–190 (1988)
2. Dujmović, J., Preferential neural networks. In: Antognetti, P., Milutinović, V. (eds.) Chapter 7 Neural Networks - Concepts, Applications, and Implementations, vol. II. Prentice-Hall Advanced Reference Series, pp. 155–206. Prentice-Hall, Englewood Cliffs (1991)
3. Dujmović, J.: Weighted conjunctive and disjunctive means and their application in system evaluation, Publikacije Elektrotehničkog Fakulteta, Serija Matematika i Fizika (Journal of the University of Belgrade School of EE, Series Mathematics and Physics) (483), 147–158 (1974). Available through JSTOR
4. Marichal, J.-L.: Aggregation operators for multicriteria decision aid. Ph.D. Dissertation, Institute of Mathematics, University of Liège, Liège, Belgium (1998)
5. Dujmović, J., Torra, V.: Properties and comparison of andness-characterized aggregators. Int. J. Intell. Syst. **36**(3), 1366–1385 (2021)
6. Torra, V.: Andness-directedness for operators of the OWA and WOWA families. Fuzzy Sets Syst. (2020, in press)
7. Dujmović, J.: Continuous preference logic for system evaluation. IEEE Trans. Fuzzy Syst. **15**(6), 1082–1099 (2007)
8. Troiano, L., Díaz, I.: An analytical solution to Dujmovic's iterative OWA. In. J. Uncertain. Fuzziness Knowledge-Based Syst. **24**(Suppl. 2), 165–179 (2016)
9. Ahn, B.S.: An alternative approach to obtaining an analytical solution to Dujmovic's iterative OWA. Internat. J. Uncertain. Fuzziness Knowledge-Based Syst. **26**(2), 261–267 (2018)
10. Dujmović, J., Nine forms of andness/orness. In: Kovalerchuk, B. (ed.) Proceedings of the Second IASTED International Conference on Computational Intelligence, pp. 276–281 (2006). ISBN:Hardcopy:0–88986-602-3/CD: 0-88986-603-1
11. Dujmović, J., Soft Computing Evaluation Logic. John Wiley & Sons, Hoboken (2018)
12. Miller, G.A.: The magical number seven, plus or minus two: some limits on our capacity for processing information. Psychol. Rev. **63**, 81–97 (1956)
13. Dujmović, J.: Weighted compensative logic with adjustable threshold andness and orness. IEEE Transa. Fuzzy Syst. **23**(2), 270–290 (2015). https://doi.org/10.1109/TFUZZ.2014.231 2018
14. Dujmović, J.: Graded logic for decision support systems. Int. J. Intell. Syst. **34**, 2900–2919 (2019). https://doi.org/10.1002/int.22177
15. Beliakov, G., Dujmović, J.: Extension of bivariate means to weighted means of several arguments by using binary trees. Inf. Sci. **331**, 137–147 (2016). https://doi.org/10.1016/j.ins.2015. 10.040
16. Dujmović, J., Beliakov, G.: Idempotent weighted aggregation based on binary aggregation trees. Int. J. Intell. Syst. **32**(1), 31–50 (2017). https://doi.org/10.1002/int.21828
17. Troiano, L., Yager, R.R.: Recursive and Iterative OWA operators. Internat. J. Uncertain. Fuzziness Knowledge-Based Syst. **13**(6), 579–599 (2005)

New Eliahou Semigroups and Verification of the Wilf Conjecture for Genus up to 65

Maria Bras-Amorós(✉) and César Marín Rodríguez

Universitat Rovira i Virgili, Tarragona, Spain
maria.bras@urv.cat

Abstract. We give a graphical reinterpretation of the seeds algorithm to explore the tree of numerical semigroups. We then exploit the seeds algorithm to find all the Eliahou semigroups of genus up to 65. Since all these semigroups satisfy the Wilf conjecture, this shows that the Wilf conjecture holds up to genus 65.

1 Introduction

A *numerical semigroup* is a cofinite submonoid of $\mathbb{N}_0$. See [5] for a general reference on numerical semigroups. The elements in the complement of a numerical semigroup in $\mathbb{N}_0$ are denoted the *gaps* of the semigroup. The *genus* of the semigroup is the number of its gaps.

If a numerical semigroup Λ is $\{\lambda_0 = 0 < \lambda_1 < \dots\}$, define its *multiplicity* as $m(\Lambda) = \lambda_1$. Define its *Frobenius number* $F(\Lambda)$ as its largest gap and its *conductor* $c(\Lambda)$ as its largest gap plus one. If $c(\Lambda) = \lambda_L$, then the elements $\lambda_0, \lambda_1, \dots, \lambda_{L-1}$ are called the *left elements* of Λ.

An element $\lambda_s \geq c(\Lambda)$ is an order-i *seed* of Λ if $\lambda_s + \lambda_i \neq \lambda_j + \lambda_k$ for all $i < j \leq k < s$. The *right primitive elements* of a numerical semigroup are its order-0 seeds. In general, the primitive elements (or minimal generators) of a numerical semigroup are those elements of the semigroup that can not be obtained as a sum of two smaller semigroup elements.

In 1978, Hebert S. Wilf conjectured that for any numerical semigroup with conductor c, with L left elements and with set of primitive elements equal to P, it holds $c \leq L \cdot \#P$ [6]. More than forty years later the conjecture is still open. It has been verified for all semigroups of genus up to 60 by Jean Fromentin and Florent Hivert [4]. An important step to approach the Wilf conjecture is a sufficient condition found by Shalom Eliahou [3]. Semigroups not satisfying it are very unusual. We denote them *Eliahou semigroups.*

If we take away a primitive element from a numerical semigroup we obtain another semigroup with genus increased by one. We can organize all numerical semigroups in an infinite tree rooted at $\mathbb{N}_0$ and such that the children of a node are the semigroups obtained taking away one by one its right primitive elements. In Fig. 1 one can see the lowest genus semigroups organized in the semigroup tree. Each semigroup is represented by its non-gaps which are either colored with dark gray if they are right primitive elements or with light gray if they are not.

V. Torra and Y. Narukawa (Eds.): MDAI 2021, LNAI 12898, pp. 17–27, 2021.
https://doi.org/10.1007/978-3-030-85529-1_2

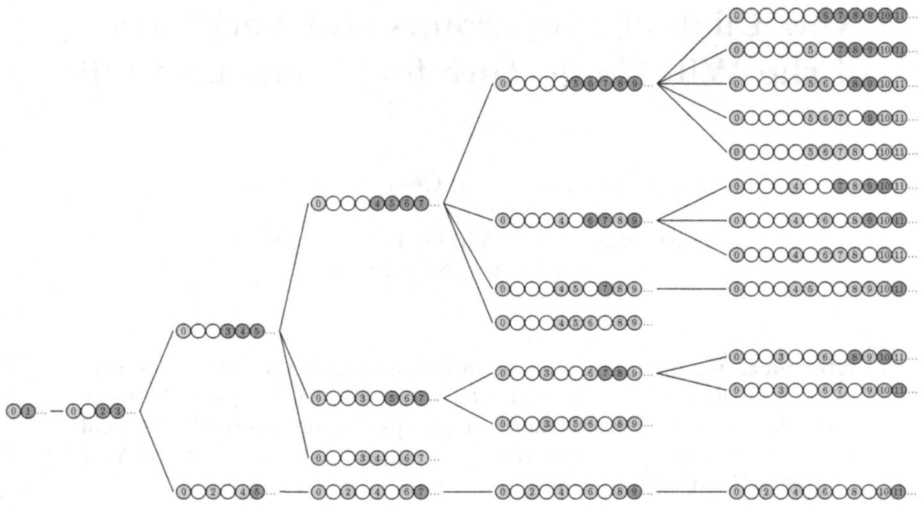

Fig. 1. Lowest depth nodes of the semigroup tree.

In Sect. 2 we recall the *seeds algorithm* [1] to explore the semigroup tree. In Sect. 3 we give a graphical explanation of the algorithm by running it over a particular example. In Sect. 4 we give the list of Eliahou semigroups output by our algorithm. This list allows to verify the Wilf conjecture for genus up to 65.

2 The Bitstream of Gaps and the Bitstream of Seeds of a Numerical Semigroup

A bitstream is a finite sequence $a = a_0 \dots a_\ell$ where a_i is either 0 or 1 for every i. For our purposes, we can indistinctly use a for $a_0 \dots a_\ell$ and for any bitstream of the form $a_0 \dots a_\ell \underbrace{0 \dots 0}_{k}$ for any positive integer k.

Suppose a semigroup Λ has conductor c. We encode its gaps as the bitstream

$$G(\Lambda) = g_0 \cdots g_{c-1}$$

with $g_i = 0$ if $i + 1 \in \Lambda$ is a gap and $g_i = 1$ otherwise. We encode its seeds as the bitstream

$$S(\Lambda) = s_0 \dots s_{c-1}$$

with $s_i = 1$ if $i - \lambda_j$ is an order-j seed of Λ where j is the unique non-negative integer such that $\lambda_j \leq i < \lambda_{j+1}$.

In [1] we presented an algorithm to explore the tree of numerical semigroups by recursively computing the bitstream of gaps and the bitstream of seeds of a numerical semigroup from those of its parent.

In Fig. 2 one can see the sequences G and S for the lowest genus semigroups organized in the semigroup tree. To make it easier to read, we represented each

1 in the sequences with a dark circle with its position written inside, and each 0 in the sequence with a light gray circle with its position also written inside.

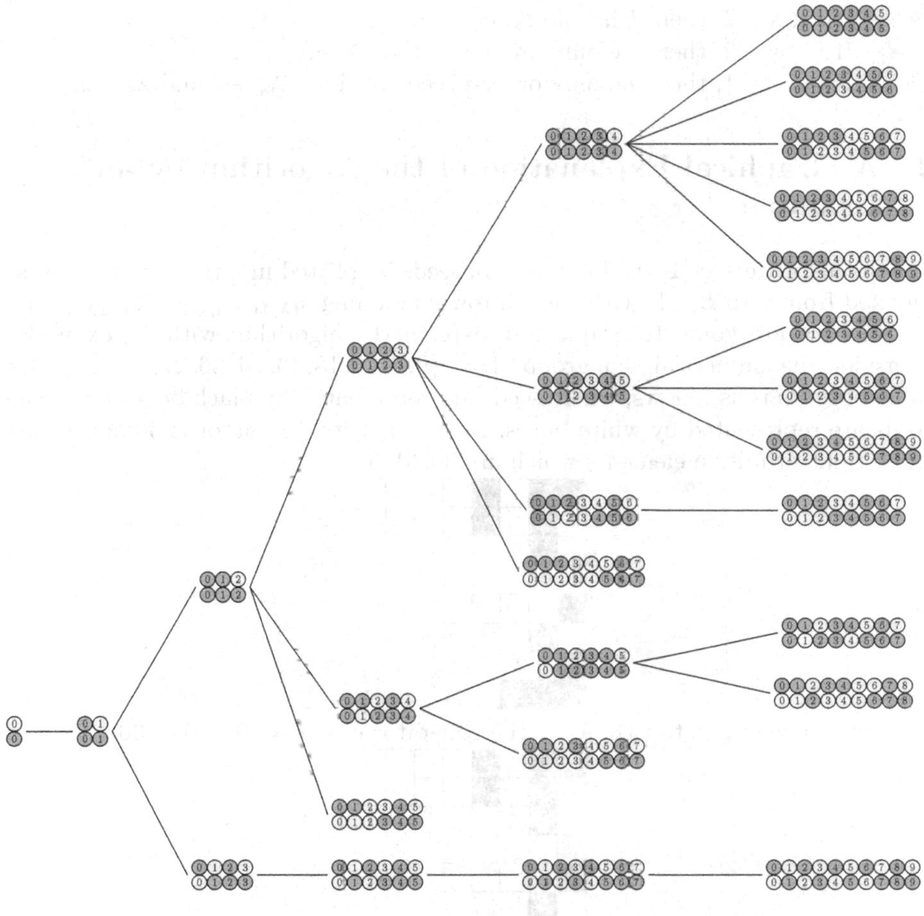

Fig. 2. Bitstream of gaps and bitstream of seeds of the semigroups in the lowest depth nodes of the semigroup tree.

The updating algorithm is based on the next results, which are proved in [1]. Suppose that λ_s is a right primitive element of Λ (hence, $s \geq L$) and let $\tilde{\Lambda} = \Lambda \backslash \{\lambda_s\}$.

1. **Old-order recycled seeds:**
   Suppose $i < L$. Any order-i seed λ_t of Λ with $t > s$ is also an order-i seed of $\tilde{\Lambda}$.
2. **Old-order new seeds:** Suppose $i < L$. Then, $\lambda_t > \lambda_s$, with λ_t not an order-i seed of Λ, is an order-i seed of $\tilde{\Lambda}$ if and only if either
   - $i < L - 1$, $\lambda_t = \lambda_s + \lambda_{i+1} - \lambda_i$ and λ_s is an order-$(i + 1)$ seed of Λ

- $i = L-1$, $\lambda_s = c$, and either $\begin{cases} \lambda_t = \lambda_s + \lambda_L - \lambda_{L-1} \\ \lambda_t = \lambda_s + \lambda_L - \lambda_{L-1} + 1 \end{cases}$
- $i = L-1$, $\lambda_s = c+1$, and $\lambda_t = \lambda_s + \lambda_L - \lambda_{L-1}$.

3. **New-order seeds:** Suppose $i \geq L$. Then,
   - If $i < s-2$, then $\tilde{\Lambda}$ has no order-i seeds.
   - If $i = s-2$, then the only order-i seed of $\tilde{\Lambda}$ is $\lambda_s + 1$.
   - If $i = s-1$, then the only order-i seeds of $\tilde{\Lambda}$ are $\lambda_s + 1$ and $\lambda_s + 2$.

3 A Graphical Explanation of the Algorithm by an Example

For the algorithm in [1] the bitstream of seeds is splitted in a table with L rows, indexed from 0 to $L-1$, with the ith row containing $s_{(\lambda_i)}, s_{(\lambda_i+1)} \dots, s_{(\lambda_{i+1}-1)}$.

Now we are going to graphically explain the algorithm with an example. Consider the numerical semigroup $\Lambda = \{0, 8, 16, 18, 19, 24, 26, 27, 30, \dots\}$. Its table of seeds is as follows, where seeds are represented by black boxes and non-seeds are represented by white boxes. Notice that its conductor is 30 and it has three right primitive elements which are 30, 31, 33.

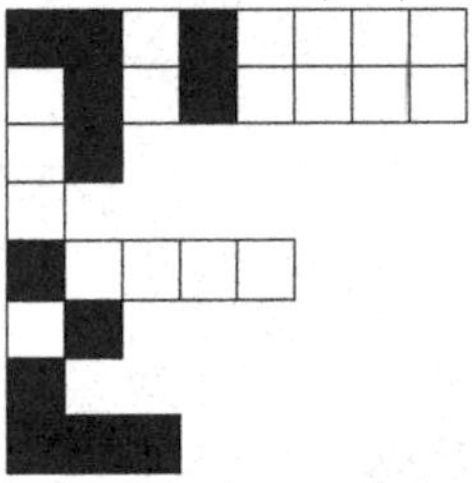

Suppose we want to take away the generator c + 0 = 30 + 0 = 30.

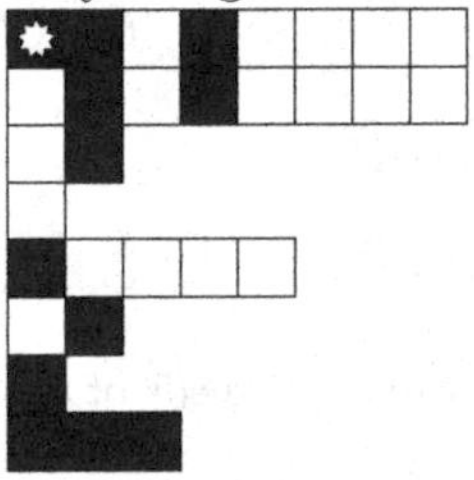

Draw the contour of the new table of seeds.

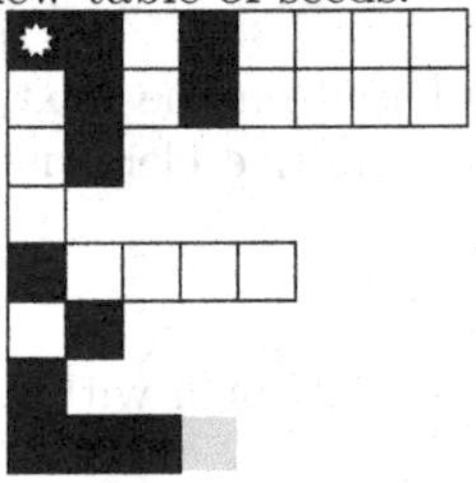

Move the old values to the left of the table and fix the old-order recycled seeds.

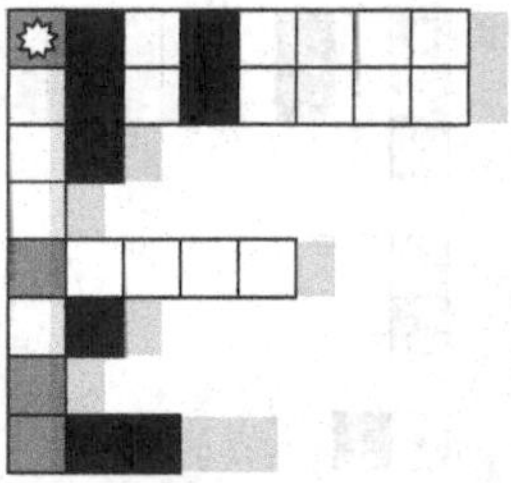

Obtain the old-order new seeds.

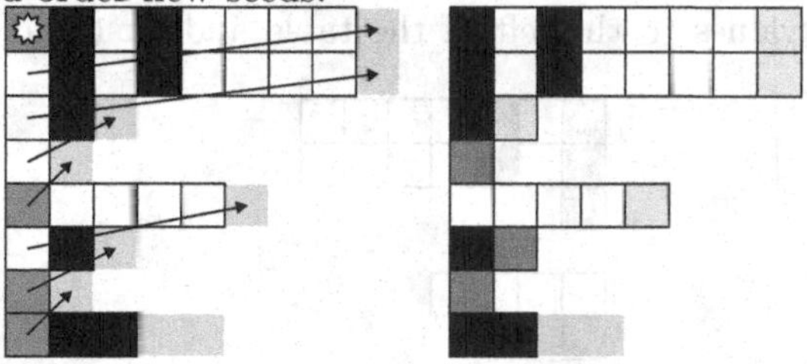

Set the last two elments in the last row of the table as old-order new seeds.

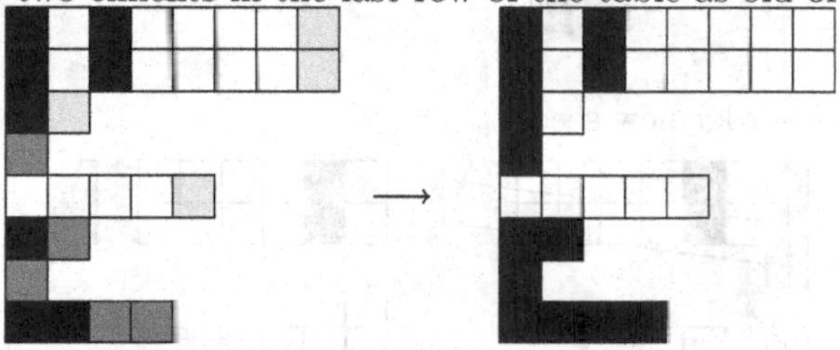

Suppose that now we want to take away the generator $c + 1 = 30 + 1 = 31$.

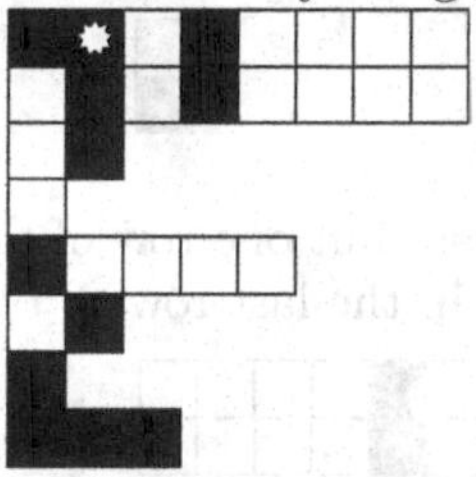

Draw the contour of the new table of seeds.

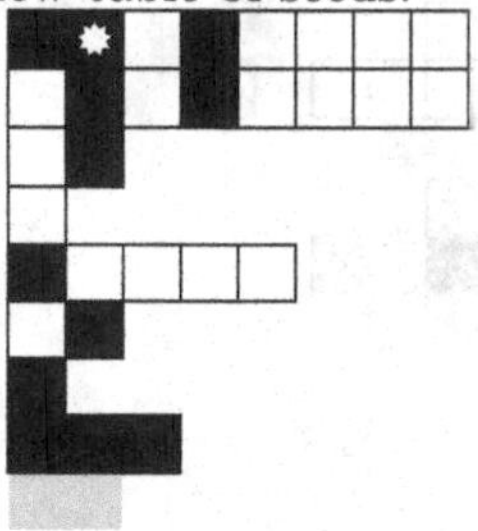

Discard the values corresponding to elements that are smaller than the new Frobenius number, keep shadowed the values corresponding to the new Frobenius number.

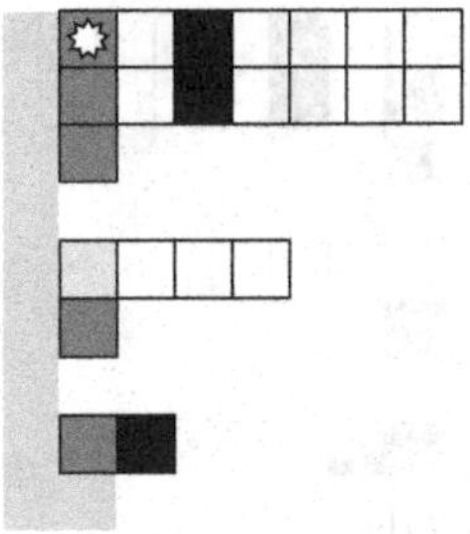

Move the old values to the left of the table and fix the old-order recycled seeds.

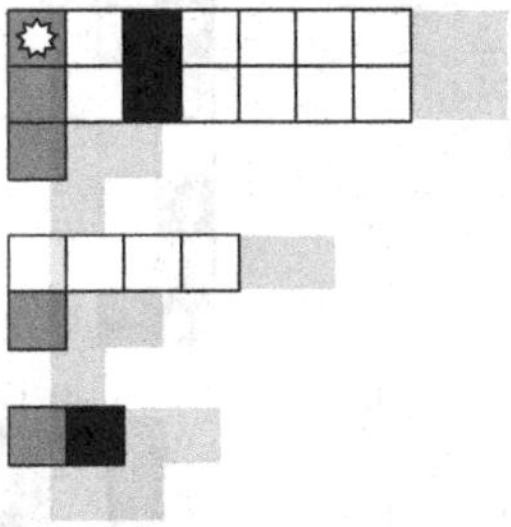

Obtain the old-order new seeds.

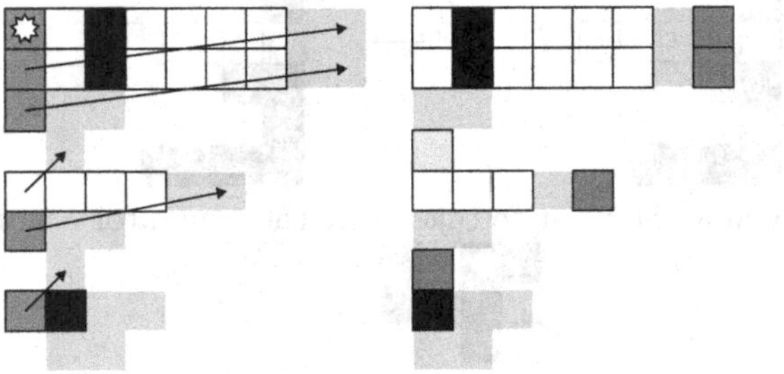

Set the last elment in the last but one row of the table as one old-order new seed and set the two elements in the last row as two new-order seeds.

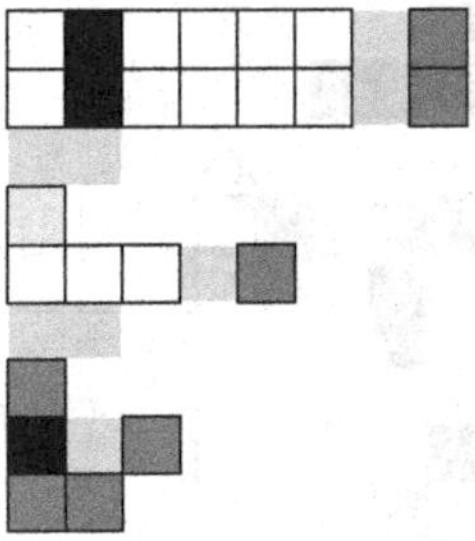

The remaining empty boxes are non-seeds.

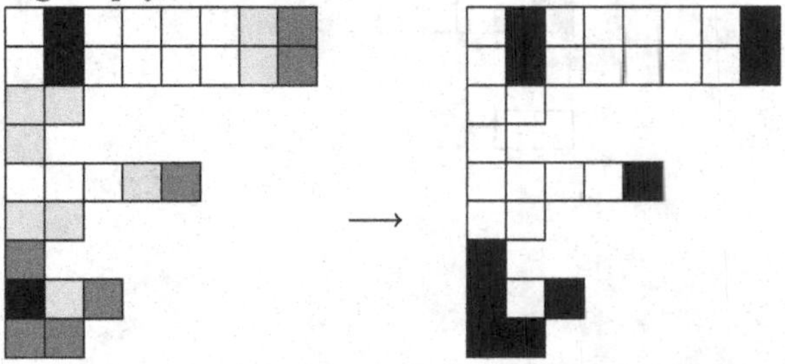

Suppose that now we want to take away the generator c + 3 = 30 + 3 = 33.

Draw the contour of the new table of seeds.

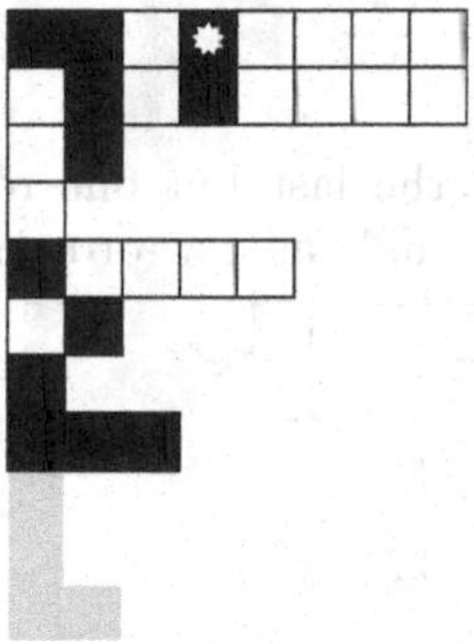

Discard the values corresponding to elements that are smaller than the new Frobenius number, keep shadowed the values corresponding to the new Frobenius number.

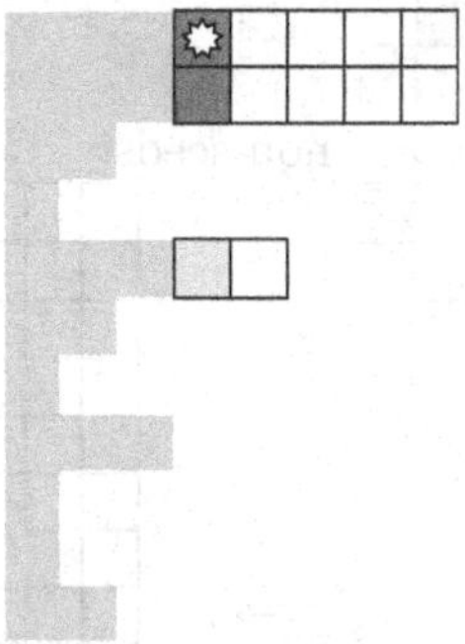

Move the old values to the left of the table and fix the old-order recycled seeds.

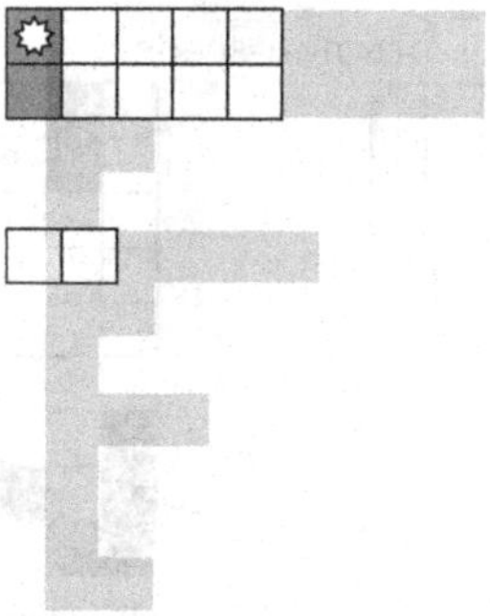

Obtain the old-order new seeds.

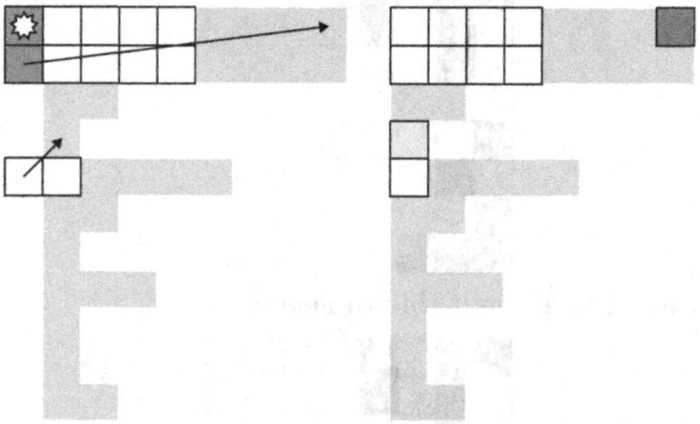

Set the unique element in the last but one row of the table and the two elements in the last row of the table as new-order seeds.

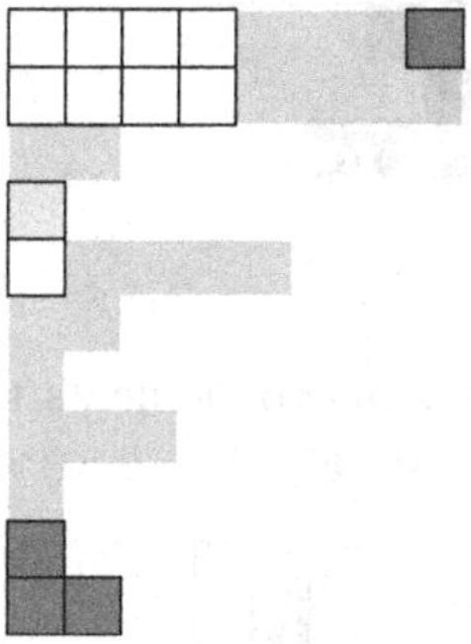

The remaining empty boxes are non-seeds.

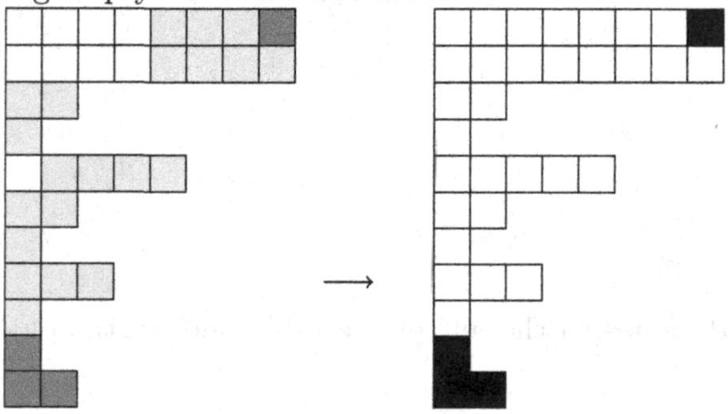

4 Eliahou Semigroups and Wilf Conjecture Verification Extended up to Genus 65

Fix a numerical semigroup Λ with conductor c and multiplicity m. Let $q = \lceil \frac{c}{m} \rceil$, and let $\rho = qm - c$ be the remainder of the division of c by m. Suppose that the left elements of Λ are $\{\lambda_0, \ldots, \lambda_L\}$ and suppose that P is the set of primitive elements of Λ. The Eliahou constant is defined as

$$E(\Lambda) = \#(P \cap \{\lambda_0, \ldots, \lambda_L\})L - q(m - \#(P \backslash \{\lambda_0, \ldots, \lambda_L\})) + \rho.$$

Shalom Eliahou proved that if $E(\Lambda) \geq 0$ then Λ satisfies the Wilf conjecture [3]. Semigroups for which the Eliahou constant is negative are very unusual. We will denote them Eliahou semigroups. According to the same reference, it was computed by Jean Fromentin that the unique Eliahou semigroups of genus $g \leq 60$ are exactly

- $\langle 14, 22, 23 \rangle \mid_{56}$,
- $\langle 16, 25, 26 \rangle \mid_{64}$,
- $\langle 17, 26, 28 \rangle \mid_{68}$,
- $\langle 17, 27, 28 \rangle \mid_{68}$,
- $\langle 18, 28, 29 \rangle \mid_{72}$

for genus 43, 51, 55, 55 and 59, respectively, where $\langle a, b, c \rangle \mid_{\kappa}$ means the minimum semigroup containing a, b, c and all integers larger than or equal to κ.

Using a parallelized version of the seeds algorithm we found that the unique Eliahou semigroups with genus between 61 and 65 are exactly,

$$\{0, 19, 29, 31, 38, 48, 50, 57, 58, 60, 62, 67, 69, 76, \ldots\} = \langle 19, 29, 31 \rangle \mid_{76}$$

and

$$\{0, 19, 30, 31, 38, 49, 50, 57, 60, 61, 62, 68, 69, 76, \ldots\} = \langle 19, 30, 31 \rangle \mid_{76}$$

Using p for the number of primitive elements and r for the number of right primitive elements, the parameters of these semigroups are

- $\{0, 19, 29, 31, 38, 48, 50, 57, 58, 60, 62, 67, 69, 76, \ldots\}$
  - $g = 63$,
  - $c = 76$,
  - $m = 19$,
  - $p = 12$,
  - $r = 9$,
  - $L = 13$,
  - $q = 4$,
  - $\rho = 0$.

– $\{0, 19, 30, 31, 38, 49, 50, 57, 60, 61, 62, 68, 69, 76, \dots\}$
  - $g = 63$,
  - $c = 76$,
  - $m = 19$,
  - $p = 12$,
  - $r = 9$,
  - $L = 13$,
  - $q = 4$,
  - $\rho = 0$.

This allows us to state the next result.

Lemma 1. *The Wilf conjecture holds for all semigroups of genus up to* 65.

We notice that for all Eliahou counterexamples of genus up to 65, $l = 13$, $q = 4$, $\rho = 0$, $g = 3$ modulo 4.

Manuel Delgado constructed in [2], for each integer number, infinite families of numerical semigroups having Eliahou constant equal to that number. In particular, he constructed infinite families of semigroups with negative Eliahou constant. The semigroups in these families are of the form

$$S^{(i,j)}(p,\tau) = \langle m^{(i,j)}, g^{(i,j)}, g^{(i,j)} + 1 \rangle \mid_{c^{(i,j)}},$$

for p an even positive integer and τ, i, j non-negative integers, where

$$\begin{aligned} m^{(i,j)} &= \frac{p^2}{4} + p(\frac{\tau}{2} + 2) + 2 + j\frac{p}{2} \\ g^{(i,j)} &= \frac{p^2}{2} + p(\tau + \frac{7}{2}) - \tau + j(p-1) + im^{(i,j)} \\ c^{(i,j)} &= \frac{p^3}{4} + p^2(\frac{\tau}{2} + 2) + 2p - \tau + j\frac{p^2}{2} + i\left(\frac{p}{2} + 1\right) m^{(i,j)} \end{aligned}$$

One can check that none of the two semigroups listed above is of this kind. In the first case it is easy to see, since the difference between the second and third generator is not one. In the second case, we use that $\rho = \tau$ [2, Lemma 5.4.8] and see, by exhaustive search that there is no combination of a positive even integer p and non-negative integers i, j such that

$$\begin{aligned} 19 &= \frac{p^2}{4} + 2p + 2 + j\frac{p}{2} \\ 30 &= \frac{p^2}{2} + \frac{7}{2}p + j(p-1) + 19i \\ 76 &= \frac{p^3}{4} + p^2(\frac{\tau}{2} + 2) + 2p + j\frac{p^2}{2} + 19i\left(\frac{p}{2} + 1\right) \end{aligned}$$

Acknowledgment. The authors would like to thank Enric Pons Montserrat, Manuel Delgado, and Julio Fernández-González for their contribution in this work. This work was partly supported by the Catalan Government under grant 2017 SGR 00705 and by the Spanish Ministry of Economy and Competitivity under grant TIN2016-80250-R and grant RTI2018-095094-B-C21.

References

1. Bras-Amorós, M., Fernández-González, J.: Computation of numerical semigroups by means of seeds. Math. Comput. **87**(313), 2539–2550 (2018)
2. Delgado, M.: On a question of Eliahou and a conjecture of Wilf. Math. Z. **288**(1–2), 595–627 (2018). https://doi.org/10.1007/s00209-017-1902-3
3. Eliahou, S.: Wilf's conjecture and Macaulay's theorem. J. Eur. Math. Soc. (JEMS) **20**(9), 2105–2129 (2018)
4. Fromentin, J., Hivert, F.: Exploring the tree of numerical semigroups. Math. Comput. **85**(301), 2553–2568 (2016)
5. Rosales, J.C., García-Sánchez, P.A.: Numerical Semigroups. Developments in Mathematics, vol. 20. Springer, New York (2009). https://doi.org/10.1007/978-1-4419-0160-6
6. Wilf, H.S.: A circle-of-lights algorithm for the "money-changing problem". Am. Math. Mon. **85**(7), 562–565 (1978)

Are Sequential Patterns Shareable? Ensuring Individuals' Privacy

Miguel Nunez-del-Prado[1(✉)], Julián Salas[2,4], Hugo Alatrista-Salas[5], Yoshitomi Maehara-Aliaga[1], and David Megías[3,4]

[1] Universidad del Pacífico, Av. Salaverry 2020, Jesús María, Lima, Peru
{m.nunezdelpradoc,ye.maeharaa}@up.edu.pe
[2] Departament d'Enginyeria Informàtica i Matemàtiques, Universitat Rovira i Virgili (URV), Tarragona, Spain
julian.salas@urv.cat
[3] Internet Interdisciplinary Institute (IN3), Universitat Oberta de Catalunya (UOC), Barcelona, Spain
dmegias@uoc.edu
[4] Center for Cybersecurity Research of Catalonia (CYBERCAT), Barcelona, Spain
[5] Pontificia Universidad Católica del Peru, Lima, Peru
halatrista@pucp.pe

Abstract. Individuals' actions like smartphone usage, internet shopping, bank card transaction, watched movies can all be represented in form of sequences. Accordingly, these sequences have meaningful frequent temporal patterns that scientist and companies study to understand different phenomena and business processes. Therefore, we tend to believe that patterns are de-identified from individuals' identity and safe to share for studies. Nevertheless, we show, through unicity tests, that the combination of different patterns could act as a quasi-identifier causing a privacy breach, revealing private patterns. To solve this problem, we propose to use ϵ-differential privacy over the extracted patterns to add uncertainty to the association between the individuals and their true patterns. Our results show that its possible to reduce significantly the privacy risk conserving data utility.

Keywords: Sequential pattern mining · Data privacy · Uniqueness · Edge-differential privacy

1 Introduction

Sequential pattern mining is a technique allowing the extraction of frequent sequences from temporal datasets. To achieve this, events are grouped by time, forming sequences representing the temporal evolution of events. Later, using a threshold σ, the sub-sequences appearing at least σ times are extracted as patterns. This technique was widely used in different domains [1,5,13,27].

Patterns extracted from temporal datasets stand for a part of the whole dataset, which semantically represents frequent behaviors. The utility of these patterns is vast, *e.g.*, for prediction tasks [3,16,24]. However, these patterns

V. Torra and Y. Narukawa (Eds.): MDAI 2021, LNAI 12898, pp. 28–39, 2021.
https://doi.org/10.1007/978-3-030-85529-1_3

can hide sensible information, as each of them represents common behaviors. Indeed, depending on the number of times a pattern appears in the database, it can represent more or fewer objects with the same behavior. For example, if some of the consumption patterns of an individual in terms of purchase categories are known, it is possible to search the original database to find the customers with such patterns. If there is only one such consumer, then these patterns act as an identifier for such a consumer, causing a privacy breach [12]. This privacy breach could outrage the customers if they buy, for instance, anticancer drugs or baldness products. Thus, the question is: how can we publish sequential patterns without compromising data privacy? To answer this question, different works in the literature [6,7,25,26,28] add Laplacian noise to the prefix tree while building the sequential patterns. This approach is complex and computationally time-consuming since depending on the sequence search Depth-first search or Breadth-first search, the first levels of the prefix tree have a strong influence on the pattern sequence construction. This is solved by truncating the sequence pattern length or post-processing the patterns to increase data utility. Nonetheless, the post-processing also adds complexity and computational time to the privacy-aware sequential pattern mining process. To simplify this process, we propose a new approach to apply ϵ-differential privacy over the extracted sequential patterns using a bipartite graph approach. Finally, we measure the performance of our approach in terms of information loss and data utility.

The rest of the paper is organized as follows. Section 2 describes the related work in the literature while Sect. 3 introduces the background for this work. Sections 4 and 5 present how our approach works and the results of the performed experiments. Finally, Sect. 6 concludes the paper and presents new research avenues.

2 Related Work

In the present section, we describe the works on privacy-aware sequential pattern mining. For instance, Chen *et al.* [7] propose a privacy-aware prefix span algorithm for publishing sequential patterns data. The algorithm takes as input a privacy budget ϵ and a tree height h for the Prefixspan algorithm to construct the sanitized prefix tree by adding Laplacian noise to the prefix tree count in each h level. Once the prefix tree is built, the sanitized sequences are extracted. Thus, the authors test their approach using data from individuals count in Canada's metro and bus networks. They evaluate their proposal using count queries and frequent sequential pattern mining to quantify the model's data utility. Cheng *et al.* propose a n-gram algorithms [6] to release sequential patterns providing a privacy guarantee level. The authors use an exploration tree to find the counts of occurrences of grams in the dataset. Then, they applied Laplacian noise to the counts using a privacy budget of ϵ/n, where n is the maximal size of the considered sequences. Authors use the page views of *msnbc.com* (`i.e.,` MSNBC) and the records sequences of stations visited by passengers in time order in the Montreal transportation system (*i.e.,* STM) datasets. Xu *et al.* [25,26] propose the PFS^2 privacy-aware algorithm for extracting frequent patterns. The idea behind the algorithm is to add Laplacian noise to the support of all frequent sequence candidates. Once the noise is applied, the algorithm extracts frequent patterns based on their noisy supports. Authors use MSNBC, BIBLE, and House Power datasets to measure the F-score, and Relative error for the proposed PFS^2 compared to Prefix [7] and n-gram [6] algorithms.

Zhou and Lin [28] propose to truncate frequent patterns to add noise to the pattern support based on a bilateral geometric distribution to satisfy *epsilon*-differential privacy. The authors use two datasets for the experiments, one containing semantic trajectories of 14909 users living in New York and 48564 distinct places; another used dataset is the Internet Information Server logs for *msnbc.com* and news-related portions of *msn.com* for the entire day of September 28, 1999 (Pacific Standard Time). Each sequence in the dataset corresponds to a user's page views during that twenty-four-hour period (MSNBC). The author shows the outperformance of their approach compare to the *PFS2* algorithm [26] in terms of *False Negative Rate* and *Relative Support Error*. Bonomi [4] describes a two-phase algorithm to achieve Laplacian differential privacy in sequential pattern mining. The first phase extracts the prefix tree T' where the noise is added using ϵ budget. The second phase takes as input the k most important patterns from the T' sanitized tree and compares them to patterns extracted from the original dataset to measure the change's sensitivity and calibrate this noise to improve data utility. The drawback is that the author does not present experiments.

In [17] it was proved that the mobility patterns are very unique, and four spatiotemporal points are enough to identify 95% of the individuals in a dataset of fifteen months of a pseudonymized mobile phone dataset of 1.5M of users. The uniqueness of purchase patterns was studied in [18] with similar results, considering the shops and prices of transactions. Considering that unicity quantifies the intrinsic reidentification risk of a data set. However, it has been pointed out in [22] that being unique in a sample does not necessary implies being unique in a population, as it has also been previously observed in [8].

In this regard, Rocher *et al.* [19] propose a generative copula-based method to estimate the likelihood of a given individual to be re-identified, even in an incomplete dataset. The basic idea is to use marginal distributions from sampled datasets to build a copula for re-identifying the complete dataset. Authors use five different census and UCI Machine Learning repository datasets, namely USA, MERNIS, ADULT, MIDUS, and HDV. Their method reaches a uniqueness ranging from 0.84 to 0.97, with a low false-discovery rate.

3 Background

In the present section, we introduce the definitions of Pattern Mining (Subsect. 3.1), Edge-Differential Privacy (Subsect. 3.2), Disclosure Risk (Subsect. 3.3), and Information Loss as well as Utility (Subsect. 3.4).

3.1 Pattern Mining

To illustrate the definitions, we use a sample of sequence database sDB, represented in Table 1. The temporal dimension is represented by parentheses. The domain of the column *ClientID* contains all different clients we have in the database, i.e., $dom(ClientID) = \{Client_1, Client_2, Client_3\}$. The domain of purchases is composed by all different items belonging the database, *i.e.*, $dom(items) = \{I1, I3, I4, I7, I9, I10, I11, I12, I13, I15\}$.

Table 1. Example of sequences database sDB

ClientID	Sequence of itemsets
$Client_1$	[(*I1 I10 I12*) (*I1 I10 I11*) (*I3 I10 I12 I13*)]
$Client_2$	[(*I1 I10 I12*) (*I3 I10 I11*) (*I3 I4 I12*)]
$Client_3$	[(*I3 I10 I9 I15*) (*I1 I7 I11*) (*I3 I7 I9 I13*)]

Definition 1. *Item and Itemset.* *An* item I, *is a literal value for purchase categories. An* itemset, $IS = (I_1\ I_2 \ldots I_u)$, *is a non empty set of* items *such that* $I_i \in dom(items)\ \forall\ i \in [1..u-1]$.

Definition 2. *Inclusion of itemsets.* *An itemset* $IS = (I_1\ I_2 \ldots I_u)$ *is included, denoted* $\subseteq$, *in another itemset* $IS'(I'_1\ I'_2 \ldots I'_v)$, *iff* $\forall I_k \in IS$, $\exists\ i_k$, *such that* $I_k = I'_{i_k}$.

Definition 3. *Sequence.* *A sequence* $\boldsymbol{S}$ *is an ordered list of itemsets, denoted* $S = [IS_1\ IS_2 \ldots IS_v]$ *where* IS_i, IS_{i+1} *satisfy the constraint of temporal sequentiality for all* $i \in [1..v-1]$.

Definition 4. *Inclusion of sequences.* *A sequence* $S = [IS_1\quad IS_2 \ldots IS_u]$ *is included in another sequence* $S' = [IS'_1 IS'_2 \ldots IS'_v]$, *denoted as* $S \subseteq S'$, *iff* $\exists\ i_1 < i_2 < \ldots < i_u$ *such that* $IS_1 \subseteq IS'_{i_1}, IS_2 \subseteq IS'_{i_2}, \ldots, IS_u \subseteq IS'_{i_u}$.

Definition 5. *Support of a sequence.* *We define the support of a sequence* S, *denoted as* $supp(S)$, *as the number of sequences in the database sDB that include* S.

Definition 6. *Problem of sequential pattern mining.* *Given a positive integer* σ *(minimal support) and a sequence in the database sDB, a sequence can be considered frequent if its support* $supp(S)$ *is greater than or equal to* σ, i.e., $supp(S) \geq \sigma$. *All frequent sequences are called sequential patterns and they are stored in a pattern database pDB.*

In Table 1, (*I1 I10 I12*) is an itemset that belongs to $Client_1$'s sequence. The sequence [(*I1 I10 I12*) (*I3 I10 I11*) (*I3 I4 I12*)] represent the temporal evolution of a set of purchases for client C_2 in three different times. The sequence

$[(I1\ I10)\ (I3)]$ is included in the sequence of $Client_2$ because $(I1\ I10) \subseteq (I1\ I10\ I12)$ and $(I3) \subseteq (I3\ I10\ I11)$. Finally, $supp([(I1\ I10)\ (I3)]) = 2$ because it appears in $Client_1$ and $Client_2$. In the same manner $supp([(I1)\ (I13)]) = 3$ because it appears in all sequences. For $\sigma = 3$, sequence $[(I1\ I10)]$ is not frequent because $supp([(I1\ I10)]) = 2$. On the contrary, $[(I1)\ (I13)]$ is frequent because $supp([(I1)\ (I13)]) = 3$.

3.2 Differential Privacy and Noise Graph Mechanism

After extracting the frequent patterns, we represent the pattern database pDB as a graph with n nodes that correspond to the clients in $dom(ClientID)$ and m nodes that correspond to the patterns in pDB. We add an edge between $Client_i \in dom(ClientID)$ and $P_j \in pDB$ if P_j is a pattern of $Client_i$. Using such representation, allows us to apply differential privacy to protect the pattern database pDB.

Intuitively differential privacy [11] tries to reduce the privacy risk when someone has their data in a dataset to the same risk of not giving data at all. For protecting graph structured data, edge-differential privacy may be defined as follows:

Definition 7. *A randomized algorithm A is said to be ϵ-edge-differential private if for two graphs G_1 and G_2 that differ in one edge and all outputs $\mathcal{O} \subseteq Range(A)$:*

$$Pr[A(G_1) \in \mathcal{O}] \leq e^{\epsilon} Pr[A(G_2) \in \mathcal{O}]$$

The larger the value of the ϵ parameter, the weaker the algorithm's privacy guarantee. Therefore, ϵ usually takes a small values [15].

For publishing pDB with differential privacy we will use $\mathcal{A}_{n,m,p}$ the Bipartite Noise-Graph Mechanism [21]. It is based on adding random graphs g sampled from $\mathcal{G}(n, m, p)$, through the Noise-Graph addition technique [23].

Note that for any graph $g \in \mathcal{G}(n, m, p)$, each of the $n \times m$ possible edges in g is present with probability p, this random graph model is known as the Erdös-Rényi (or Gilbert) model for bipartite graphs.

Theorem 1 [21]. *The noise-graph mechanism $\mathcal{A}_{n,m,p}$ is $ln(\frac{1-p}{p})$-edge-differentially private.*

3.3 Disclosure Risk

For the Disclosure Risk assessment of sequential patterns, we consider that an adversary knows a subset $\mathcal{P}$ of a client's patterns $P(Client_x)$ and aims to link this knowledge to the corresponding record in the published data set. Thus, the size of the anonymity set for such adversarial knowledge $\mathcal{P}$ will be useful to measure the risk of disclosure. We define the anonymity set for sequential patterns as follows.

Definition 8. *Anonymity Set.* *The* anonymity set *for* $Client_x$*, given adversarial knowledge* $\mathcal{P} \subset P(Client_x)$*, is the set of* $Client_y$*'s in the database such that* $\mathcal{P} \subset P(Client_y)$.

In this terminology, a unique record is one with an anonymity set of size one, and the term *uniqueness* refers to the amount of unique records in the data set. We remark that a sample unique is not necessarily a population unique; however, if the adversary knows that the client belongs to the published data, finding a unique record will lead to re-identification. Hence, we consider the *identity disclosure risk* as the uniqueness obtained considering that an adversary knows a number of patterns of each user.

Then, we measure the amount of accurate information that an adversary will possibly learn after re-identification. For this evaluation, we consider an adversary that was able to link the patterns that he/she knows about $Client_x$ to the corresponding record in the protected patterns database $\widetilde{pDB}$. We define the Pattern Risk of Disclosure to measure the probability that such an adversary will learn accurate patterns of $Client_x$ by assuming that all of the published patterns $\tilde{P}(Client_x)$ belong to the true patterns $P(Client_x)$.

Definition 9. *Pattern Risk of Disclosure.* *Let* $\widetilde{pDB}$ *the published protected version of the pattern database* pDB*. We define the* Pattern Risk of Disclosure *for a* $Client_x \in \widetilde{pDB}$ *as the proportion of his/her true patterns published, over all published patterns:*

$$dRisk(Client_x) = \frac{|\tilde{P}(Client_x) \cap P(Client_x)|}{|\tilde{P}(Client_x)|}$$

As an example, we suppose that $P(Client_x) = \{P_1, P_2\}$ and the published patterns are $\tilde{P}(Client_x) = \{P_1, P_3\}$, in this case $dRisk(Client_x) = \frac{|\{P_1\}|}{|\{P_1, P_3\}|} = \frac{1}{2}$. This value, shows that $\frac{1}{2}$ of $Client_x$ published patterns are true patterns.

We note that in the case of generalization the published patterns always contain the true patterns, i.e., $P(Client_x) \subset \tilde{P}(Client_x)$ and $dRisk(Client_x) = \frac{|P(Client_x)|}{|\tilde{P}(Client_x)|}$. This is equal to the sensitive attribute risk of disclosure measure defined in [20].

3.4 Information Loss and Utility

In the Statistical Disclosure Control (SDC) literature, simple, generic and intuitive information loss measures can be defined to compare the original and masked datasets, their covariance or correlation matrices. This is done by calculating either their Mean Absolute Error (MAE), Mean Squared Error or their Mean Variation among the corresponding matrices [9,10].

From these possible metrics, we consider that the MAE is the best suited for our evaluation, since it directly measures how many patterns in average where added or removed to each client to obtain the protected pattern database $\widetilde{pDB}$ from pDB. Hence, to carry out this evaluation we represent the client-pattern database pDB as a *sparse matrix* M.

Definition 10. ***Client-Pattern Matrix.*** *We define $M = (x_{ij})$ for pDB, as the matrix with n rows (number of clients in pDB) and m columns (number of patterns in pDB), where row i represents the $Client_i$ and column j represents the pattern P_j, and $x_{ij} = 1$ if and only if $P_j \subset P(Client_i)$.*

Definition 11. ***Information Loss.*** *We define the* Information Loss for $\widetilde{pDB}$ *as:*

$$IL(\widetilde{pDB}) = MAE(M, \widetilde{M}) = 1/n \sum_{i=1,\dots,n} |x_{ij} - \tilde{x}_{ij}|,$$

where x_{ij} and $\tilde{x}_{ij}$ are the entries of the client-pattern matrices M and $\widetilde{M}$ for pDB and $\widetilde{pDB}$, respectively.

Definition 12. ***Data Utility.*** *Let $\widetilde{pDB}$ the published protected version of the pattern database pDB. We use the Normalized Discounted Cumulative Gain (NDCG) [14] to measure the change of relevance, in the top n sequences. Accordingly, We define the* Data Utility for $\widetilde{pDB}$ *as:*

$$DU(\widetilde{pDB}) = \frac{DCG_{\widetilde{pDB}}}{DCG_{pDB}}$$

where DCG is defined as:

$$DCG = \sum_{i=1}^{|p|} \frac{2_i^{rel} - 1}{log_2(i+1)}$$

where rel_i is the relevance of the position of the pattern in position i.

4 Methodology

In the present section, we describe the multi-step process depicted in Fig. 1. Thus, the process has four steps: pattern mining, uniqueness evaluation, privacy protection, and result evaluation.

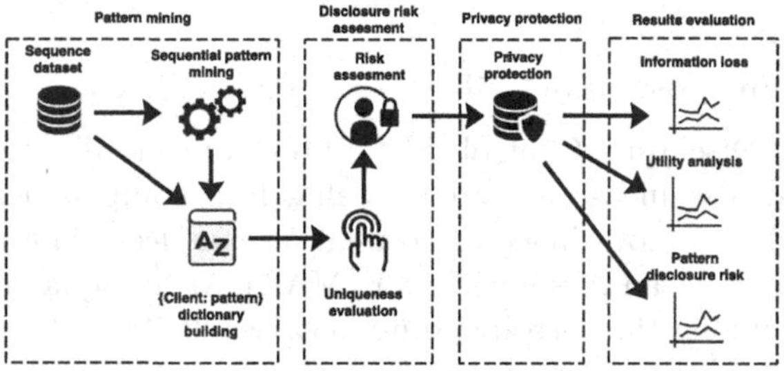

Fig. 1. Process to measure the frequent sequences privacy on data associated to consumer preferences

Pattern Mining - first, we extract sequential patterns from a sequence database sDB. In this step, several minimal supports were used to evaluate this constraint's impact on the privacy guarantees. Later, to identify the individuals following the behavior represented by each pattern, we created the pattern database pDB, which consists of two dictionaries $\{client : pattern\}$ and $\{pattern : client\}$. Indeed, to measure the uniqueness is essential to know which pattern corresponds to a client and vice-versa.

Disclosure Risk Assessment - in this step from the $\{client : pattern\}$ dictionary, clients having patterns greater than or equal to a value s are filtered. Then, a random sample containing t patterns is generated for each customer. After obtaining all the samples, each partition is counted using the $\{pattern : client\}$ dictionary. Once the counts are obtained, they are filtered according to a threshold s, the anonymity set's size. From this filter, it is possible to assess the Disclosure Risk considering an adversary that knows t random patterns of a user for re-identification.

Privacy Protection - to provide differential privacy to the client-pattern database pDB, we generate a bipartite graph G of clients U, patterns P with $|U| = n$, $|P| = m$, from the dictionary $\{client : pattern\}$.

Next, we choose a probability $p = 0.005, 0.05, 0.1, 0.2, 0.3, 0.4$, sample 10 bipartite random graphs $g \in \mathcal{G}(n, m, p)$, and apply the noise-graph mechanism $\mathcal{A}_{n,m,p}$ to G, which by Theorem 1, is ϵ-edge-differentially private for $\epsilon = ln(\frac{1-p}{p})$. This yields the corresponding values of $\epsilon = 5.29, 2.94, 2.19, 1.38, 0.84$ and 0.40. However, we will focus on the values of p, since it is more interpretable, for example, the information loss may be directly calculated as in (1). Then, we get the protected client-pattern database $\widetilde{pDB}$ from $\mathcal{A}_{n,m,p}(G)$.

Results Evaluation - finally, we evaluate the privacy protection step through three metrics: information loss, disclosure risk, and utility analysis of patterns.

5 Experiments and Results

This section describes the dataset we used for experiments and discusses our main results.

In this effort, we use two datasets from a financial institution containing 5 000 and 50 000 individuals' sequences representing the temporal evolution product purchase categories based on Classification of Individual Consumption According to Purpose (COICOP)[1] for July 2017 grouped by days.

The WinCopper algorithm [2] was used to extract sequential patterns under temporal constraints. The algorithm was executed on two datasets through different minimal supports. While the minimal support is high, the extracted patterns are few. When minimal support is low (*e.g.*, 0.1), the number of patterns increases.

[1] https://unstats.un.org/unsd/iiss/Classification-of-Individual-Consumption-According-to-Purpose-COICOP.ashx.

Disclosure Risk Assessment. We evaluate the Disclosure Risk considering the anonymity set sizes depending on the adversarial knowledge prior to any privacy protection. Thus, Fig. 2 shows the cumulative distribution of the number of customers and the size of the anonymity set grouped by the number of patterns that an adversary knows. We notice that the larger the sample of patterns an adversary has, the greater are his opportunities to re-identify unique customers.

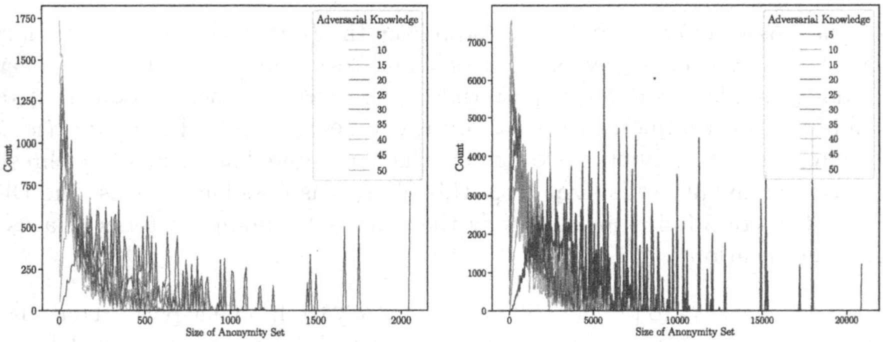

Fig. 2. Size of the Anonymity sets depending on the adversarial knowledge with 5000 and 50000 clients.

We remark that even if an adversary re-identifies a record, the customer will be protected if his/her patterns do not correspond to their actual patterns. Thus we measure the amount of protection provided by the differentially private mechanism applied to the data by measuring the Pattern Disclosure Risk. In this case we are assuming that the adversary has correctly linked his knowledge to a unique record and tries to learn additional patterns of the client.

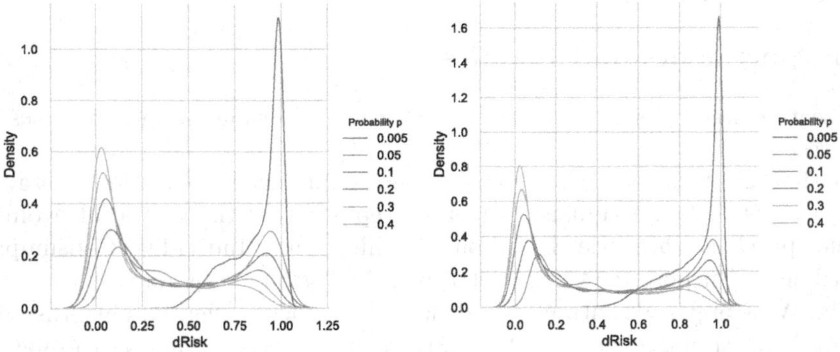

Fig. 3. Pattern Disclosure Risk for datasets with 5000 and 50000 clients.

Figure 3 shows the Pattern Disclosure Risk distributions for the datasets with 5 000 and 50 000 clients. Therefore, increasing the probability of p of randomization moves the density towards the left, decreasing the disclosure risk for the published patterns. We also observe that for the value of $p = 0.005$ ($\epsilon = 5.29$), most of the patterns remain intact. Hence the privacy provided by such ϵ is limited.

Information Loss and Data Utility Evaluation. For this evaluation, we measure the information loss and data utility in function of the $p = \frac{1}{e^{\epsilon}-1}$ for ϵ-differential privacy added to the dataset. The information loss may be measured without the need for any experiments by observing that the $MAE(M, \widetilde{M})$ corresponds to the client nodes' average degree in the bipartite Noise-Graph Mechanism. Since we sampled the noise-graphs g from $\mathcal{G}(n, m, p)$, then, in expectation, the average degree for the client nodes in g is $p \times m$, for $p = 0.005, 0.05, 0.1, 0.2, 0.3, 0.4$ and $m = 190$. That is, the information loss will be:

$$IL(\widetilde{pDB}) = p \times m \tag{1}$$

To quantify the data utility, we rely on the `NDCG` metric. Thus, we measure how the patterns changed in the most frequent rank position based on the noise introduced for the top 10, top 20, top 50, and top 100 most frequent patterns. Figure 4 illustrates the influence of the probability p of randomization in the data utility in both datasets. We note that, the smaller p (*i.e.*, high epsilon ϵ value), the higher the utility. Thus, the patterns' support is less affected by the noise introduced by the differential privacy mechanism. Please note that for every value of p, we measure the average NDCG for the ten repetitions. Finally, we note that the probability p of randomization defines the privacy guarantee to be achieved.

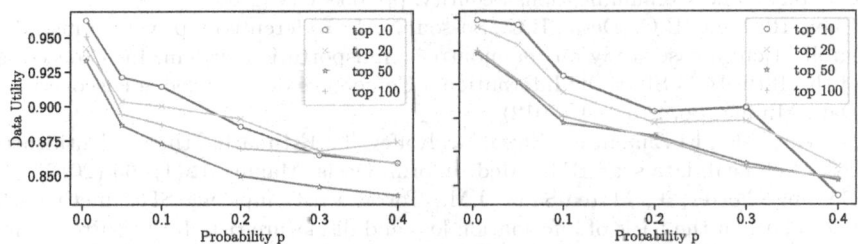

Fig. 4. Pattern Data Utility measure for datasets with 5000 and 50000 clients.

6 Conclusion and Future Work

In the present work, we apply an ϵ-edge-differentially private mechanism to protect a client-pattern database for sequential pattern sharing, built from a dataset containing individual purchases. The mechanism is implemented using bipartite random graphs to add noise over the client-pattern representation. The results show that the privacy mechanism is able to protect users' privacy while keeping low information loss and good data utility. As future work, we plan to compare our method to other privacy mechanisms and use graph embedding to represent the bipartite graph.

Acknowledgements. This research was partly supported by the Spanish Government under projects RTI2018-095094-B-C21 and RTI2018-095094-B-C22 "CONSENT".

References

1. Alatrista-Salas, H., Azé, J., Bringay, S., Cernesson, F., Selmaoui-Folcher, N., Teisseire, M.: A knowledge discovery process for spatiotemporal data: application to river water quality monitoring. Ecol. Inform. **26**, 127–139 (2015)
2. Alatrista-Salas, H., Guevara-Cogorno, A., Maehara, Y., Nunez-del-Prado, M.: Efficiently mining gapped and window constraint frequent sequential patterns. In: Torra, V., Narukawa, Y., Nin, J., Agell, N. (eds.) MDAI 2020. LNCS (LNAI), vol. 12256, pp. 240–251. Springer, Cham (2020). https://doi.org/10.1007/978-3-030-57524-3_20
3. Amiri, M., Mohammad-Khanli, L., Mirandola, R.: A sequential pattern mining model for application workload prediction in cloud environment. J. Netw. Comput. Appl. **105**, 21–62 (2018)
4. Bonomi, L., Xiong, L.: Mining frequent patterns with differential privacy. Proc. VLDB Endow. **6**(12), 1422–1427 (2013)
5. Ceci, M., Lanotte, P.F.: Closed sequential pattern mining for sitemap generation. World Wide Web **24**(1), 175–203 (2020). https://doi.org/10.1007/s11280-020-00839-2
6. Chen, R., Acs, G., Castelluccia, C.: Differentially private sequential data publication via variable-length n-grams. In: Proceedings of the 2012 ACM Conference on Computer and Communications Security, pp. 638–649 (2012)
7. Chen, R., Fung, B.C., Desai, B.C., Sossou, N.M.: Differentially private transit data publication: a case study on the montreal transportation system. In: Proceedings of the 18th ACM SIGKDD International Conference on Knowledge Discovery and Data Mining, pp. 213–221 (2012)
8. Dankar, F.K., El Emam, K., Neisa, A., Roffey, T.: Estimating the re-identification risk of clinical data sets. BMC Med. Inform. Decis. Making **12**(1), 66 (2012)
9. Domingo-Ferrer, J., Mateo-Sanz, J.M., Torra, V.: Comparing SDC methods for microdata on the basis of information loss and disclosure risk. In: Pre-Proceedings of ETK-NTTS, vol. 2, pp. 807–826 (2001)
10. Domingo-Ferrer, J., Torra, V.: A quantitative comparison of disclosure control methods for microdata. In: Confidentiality, Disclosure and Data Access: Theory and Practical Applications for Statistical Agencies, pp. 111–134 (2001)
11. Dwork, C.: Differential privacy. In: Proceedings of the 33rd International Conference on Automata, Languages and Programming - Volume Part II, ICALP 2006, pp. 1–12 (2006)
12. Gambs, S., Killijian, M.O., del Prado Cortez, M.N.: De-anonymization attack on geolocated data. J. Comput. Syst. Sci. **80**(8), 1597–1614 (2014)
13. Guevara-Cogorno, A., Flamand, C., Alatrista-Salas, H.: Copper-constraint optimized prefixspan for epidemiological research. Procedia Comput. Sci. **63**, 433–438 (2015)

14. Järvelin, K., Kekäläinen, J.: Cumulated gain-based evaluation of IR techniques. ACM Trans. Inf. Syst. (TOIS) **20**(4), 422–446 (2002)
15. Lee, J., Clifton, C.: How much is enough? Choosing ε for differential privacy. In: Lai, X., Zhou, J., Li, H. (eds.) ISC 2011. LNCS, vol. 7001, pp. 325–340. Springer, Heidelberg (2011). https://doi.org/10.1007/978-3-642-24861-0_22
16. Lien, Y.C.N., Wu, W.J., Lu, Y.L.: How well do teachers predict students' actions in solving an ill-defined problem in stem education: a solution using sequential pattern mining. IEEE Access **8**, 134976–134986 (2020)
17. de Montjoye, Y.A., Hidalgo, C.A., Verleysen, M., Blondel, V.D.: Unique in the crowd: The privacy bounds of human mobility. Sci. Rep. **3**, 1–5 (2013)
18. de Montjoye, Y.A., Radaelli, L., Singh, V.K., Pentland, A.: Unique in the shopping mall: on the reidentifiability of credit card metadata. Science **347**(6221), 536–539 (2015)
19. Rocher, L., Hendrickx, J.M., De Montjoye, Y.A.: Estimating the success of re-identifications in incomplete datasets using generative models. Nat. Commun. **10**(1), 1–9 (2019)
20. Salas, J.: Sanitizing and measuring privacy of large sparse datasets for recommender systems. J. Ambient Intell. Humaniz. Comput. (2019). https://doi.org/10.1007/s12652-019-01391-2
21. Salas, J., Torra, V.: Differentially private graph publishing and randomized response for collaborative filtering. In: Proceedings of the 17th International Joint Conference on e-Business and Telecommunications, ICETE 2020-V2: SECRYPT, Lieusaint, Paris, France, 8–10 July 2020, pp. 415–422. ScitePress (2020)
22. Sánchez, D., Martínez, S., Domingo-Ferrer, J.: Comment on "unique in the shopping mall: on the reidentifiability of credit card metadata". Science **351**(6279), 1274 (2016)
23. Torra, V., Salas, J.: Graph perturbation as noise graph addition: a new perspective for graph anonymization. In: Pérez-Solà, C., Navarro-Arribas, G., Biryukov, A., Garcia-Alfaro, J. (eds.) DPM/CBT -2019. LNCS, vol. 11737, pp. 121–137. Springer, Cham (2019). https://doi.org/10.1007/978-3-030-31500-9_8
24. Wright, A.P., Wright, A.T., McCoy, A.B., Sittig, D.F.: The use of sequential pattern mining to predict next prescribed medications. J. Biomed. Inform. **53**, 73–80 (2015)
25. Xu, S., Cheng, X., Su, S., Xiao, K., Xiong, L.: Differentially private frequent sequence mining. IEEE Trans. Knowl. Data Eng. **28**(11), 2910–2926 (2016)
26. Xu, S., Su, S., Cheng, X., Li, Z., Xiong, L.: Differentially private frequent sequence mining via sampling-based candidate pruning. In: 2015 IEEE 31st International Conference on Data Engineering, pp. 1035–1046. IEEE (2015)
27. Zheng, Z., Wei, W., Liu, C., Cao, W., Cao, L., Bhatia, M.: An effective contrast sequential pattern mining approach to taxpayer behavior analysis. World Wide Web **19**(4), 633–651 (2016). https://doi.org/10.1007/s11280-015-0350-4
28. Zhou, F., Lin, X.: Frequent sequence pattern mining with differential privacy. In: Huang, D.-S., Bevilacqua, V., Premaratne, P., Gupta, P. (eds.) ICIC 2018. LNCS, vol. 10954, pp. 454–466. Springer, Cham (2018). https://doi.org/10.1007/978-3-319-95930-6_42

Aggregation Operators and Decision Making

On Two Generalizations for k-Additivity

Ryoji Fukuda[1(✉)], Aoi Honda[2], and Yoshiaki Okazaki[3]

[1] Oita University, 700 Dan-noharu, Oita, Oita 870-1192, Japan
rfukuda@oita-u.ac.jp

[2] Kyushu Institute of Technology, 680-4 Kawazu, Iizuka, Fukuoka 820-8502, Japan
aoi@ai.kyutech.ac.jp

[3] Fuzzy Logic Systems Institute, 680-41 Kawazu, Iizuka, Fukuoka 820-0067, Japan
okazaki@flsi.or.jp

Abstract. There are two generalizations for k-additive set functions: constructive k-additivity and formulaic k-additivity. We study some properties around these concepts and their relations. A constructively k-additive set function is always formulaic k-additive. For a distorted measure, these two concepts are equivalent. Under certain conditions of "bounded variation" and "continuity at the $\emptyset$," we prove the constructive k-additivity for a formulaic k-additive set function.

Keywords: Fuzzy measure · Monotone measure · k-order additivity · Möbius transform · Distorted measure

1 Introduction

The concept of k-order additivity was originally introduced for a monotone measure on a finite set using the Möbius transform (see for example [1–3]). This is an important concept to reduce the complexity of non-additive measures. As an example for the identification of a non-additive measure, we need $2^n - 1$ parameters to express a general set function defined on a set with cardinality n; however, only $(n^2 + n)/2$ parameters are required for a two-additive measure. Assuming that one uses linear regression for the identification of a non-additive measure, we use a covariance matrix. For the covariance matrix for a 10 elements' set, the number of elements can be reduced to 3025 from 1048576.

There are two approaches to define the k-additivity for non-discrete set functions. One is a constructive approach and the other uses a formulaic relation among the terms of the Möbius transform. R. Mesiar gave the first generalization for a k-additive non-additive measure [4] through the constructive approach. Fukuda, Honda, and Okazaki [5] proved a monotone decreasing convergence theorem for the Pan Integral with respect to a monotone measure of this type. Such monotone measures can be described by several σ-additive signed measures defined on set spaces, the precise definition of which will be given in Sect. 2, and this is suitable to estimate integral values. Honda, Fukuda, and Okazaki [6] gave the definition of formulaic k-additivity. The Möbius transform was extended to

V. Torra and Y. Narukawa (Eds.): MDAI 2021, LNAI 12898, pp. 43–53, 2021.
https://doi.org/10.1007/978-3-030-85529-1_4

non-discrete set functions for this definition, and the definition and some equivalent conditions were described using this Möbius transform. A distorted measure μ is a set function, which can be represented by $\mu(A) = f(m(A))$ using a probability measure $m(\cdot)$ and a non decreasing function f. The formulaic k-additivity is essentially equivalent to the fact that the distortion function f is a k-order polynomial [6].

Briefly, constructive k-additivity is useful for estimations or calculations, and formulaic k-additivity is a natural extension of k-additivity for the finite element case. The present study attempts to show that these two definitions are essentially the same. This problem is naturally valid on a finite set or on a finite σ-algebra. We analyze the structures of our settings in the finite case and describe the relations between the two generalizations for k-additivity. There are some natural conditions for this extension. A σ-additive measure satisfies continuity from above and below, and this is a key property for the extension of measures. A non-additive set function is not always continuous from above or below, but we assume these continuities.

Under these settings, we show the following properties in this paper.

(a) All constructive k-additive set functions are formulaic k-additive.
(b) A distorted measure is formulaic k-additive if and only if it is constructively k-additive.
(c) Consider a set function μ defined on a countably generated σ-algebra. Then, if a formulaic k-additive set function μ satisfies certain conditions of "bounded variation" and "fine continuity at $\emptyset$," μ is constructively k-additive.

By showing the above problems, we will try to make sure the richness of the concepts of k-additivity.

2 Definitions and Notations

Throughout this paper, $(X, \mathcal{B})$ denotes a measurable space, that is, X is a non-discrete set and $\mathcal{B}$ is a σ-algebra ($\mathcal{B} \subset 2^X$). We assume that any set function μ defined on $\mathcal{B}$ satisfies $\mu(\emptyset) = 0$. We also assume that all one-point sets are measurable. Let n be a positive integer. We define an n-set space $X^{(n)}$ as follows.

$$X^{(n)} = \{\{x_k\}_{k=1}^n \subset X, \ x_j \neq x_k \text{ if } j \neq k\}.$$

$X^{(n)}$ can be also represented by

$$X^{(n)} = \{(x_1, \ldots, x_n) \in X^n : x_j \neq x_k \text{ if } j \neq k\} / \sim,$$

where the equivalence relation $\sim$ is defined by:

$$\begin{aligned} & (x_1, \ldots, x_n) \sim (y_1, \ldots, y_n). \\ \iff & \text{Two sets } \{x_1, \ldots, x_n\} \text{ and } \{y_1, \ldots, y_n\} \text{ are identical.} \end{aligned}$$

Let $\mathcal{B}^n$ denote a standard product σ-algebra on X^n, and let $\mathcal{B}^{(n)}$ be the σ-algebra on $X^{(n)}$ determined by the restriction (each element is different) and the equivalence relation $\sim$. This is one of the easiest way to construct a σ-algebra on the set spaces. These may be too fine to satisfy the uniqueness of σ-additive measures. Then, the structures of the set spaces may not be optimal for a constructively k-additive measure. At this step, we select one possible setting.

For a measurable set $A \in \mathcal{B}$, we define

$$A^{(n)} = \{(x_1, \ldots, x_n) \in X^{(n)} : x_j \in A,\ j \leq n\}.$$

Moreover, for $U \in \mathcal{B}^{(j)}$ and $V \in \mathcal{B}^{(k)}$, we define

$$\begin{aligned} U(\times)V = \{(x_\ell)_{\ell=1}^{j+k} \in X^{(j+k)} : (x_{\phi(1)}, \ldots, x_{\phi(j)}) \in U,\ (x_{\phi(j+1)}, \ldots, x_{\phi(j+k)}) \\ \in V \text{ for some permutation } \phi \text{ of } (1, \ldots, j+k)\}. \end{aligned}$$

We remark that $U(\times)V$ is an element of $\mathcal{B}^{(j+k)}$ if $U \in \mathcal{B}^{(j)}$ and $V \in \mathcal{B}^{(k)}$.

We mainly deal with these set operations for infinite (measurable) sets. For finite sets, if their cardinality is very small, $A^{(n)}$ or $U(\times)V$ can be empty.

Now we are prepared to define constructive k-additivity.

Definition 1 (constructive k-additivity [5]). *A set function μ on $\mathcal{B}$ is constructively k-additive ($k \in \mathbb{N}$), or μ has constructive k-additivity, if there exists a σ-additive signed measure μ_j on $(X^{(j)}, \mathcal{B}^{(j)})$ for each $j = 1, 2, \ldots, k$ such that:*

$$\mu(A) = \sum_{j=1}^{k} \mu_j(A^{(j)})$$

for any $A \in \mathcal{B}$.

Consider the case where $\mathcal{B}$ is a finite family and X essentially is a finite set. Then, the classical Möbius transform and the inverse formula are available in this situation. Our generalization for the Möbius transform is a natural extension of Möbius transform for finite sub σ-algebras.

Definition 2 (generalized Möbius transform [6]). *Let μ be a set function on $\mathcal{B}$ and set $\mathcal{D}_n = \{\{A_1, A_2, \cdots, A_n\} : A_j \in \mathcal{B}, A_j \cap A_k = \emptyset,\quad j,k \leq n, j \neq k\}$ for any $n \in \mathbb{N}$, that is, $\mathcal{D}_n$ is the family of all n disjoint measurable sets' combinations. A generalized Möbius transform $\{\nu_n\}$ of μ, which is a sequence of functions on $\mathcal{D}_n$, is defined as follows.*

(a) $\nu_1(A) = \mu(A), \quad \forall A \in \mathcal{B}$.

(b) $\nu_n(A_1, A_2, \ldots, A_n) = \mu(A_1 \cup A_2 \cup \cdots \cup A_n) - \left\{\sum_{j=1}^{n-1} \sum_{1 \leq i_1 < \cdots < i_j} \nu_j(A_{i_1}, \cdots, A_{i_j})\right\}, \forall \{A_1, A_2, \ldots, A_n\} \in \mathcal{D}_n$.

We call $\nu_j()$ the j-order adjusting function for each $j \in \mathbb{N}$.

Using these concepts, another generalization for k-additivity is given as follows.

Definition 3 (formulaic k-additivity [6]). *Let μ be a set function on $\mathcal{B}$ and $\{\nu_n\}$ be the generalized Möbius transform of μ. μ satisfies formulaic k-additivity (or μ is a formulaic k-additive set function) if $\nu_j(\mathbf{A}) = 0$ for any $\mathbf{A} \in \mathcal{D}_j$ and $j \geq k+1$.*

Formulaic k-additivity satisfies the following equivalent conditions.

Proposition 1 [6] *(Theorem 10)*
Let μ be a set function on $\mathcal{B}$, and let $\{\nu_n\}$ be its generalized Möbius transform. Then, the following are equivalent.

(a) μ is formulaic k-additive.
(b) $\nu_n = 0$ for any $n > k$.
(c) Assume that $(A_1, \cdots, A_{k-1}, B_1), (A_1, \cdots, A_{k-1}, B_2) \in \mathcal{D}_k$, and $B_1 \cap B_2 = \emptyset$. Then

$$\nu_k(A_1, \cdots, A_{k-1}, B_1 \cup B_2) = \nu_k(A_1, \cdots, A_{k-1}, B_1) + \nu_k(A_1, \cdots, A_{k-1}, B_2).$$

Remark. Using condition (b), a formulaic k-additive set function is also formulaic k'-additive for any $k' \geq k$.

3 Formulaic k-Additivity of a Constructively k-Additive Set Function

In this section, we prove the formulaic k-additivity of a constructively k-additive set function.

Proposition 2. *Let μ_n be a σ-additive signed measure on $(X^{(n)}, \mathcal{B}^{(n)})$, and μ be a set function on $\mathcal{B}$ defined by*

$$\mu(A) = \mu_n(A^{(n)}).$$

Then, for each $k \leq n$, ν_k is represented by

$$\nu_k(A_1, \ldots, A_k) = \sum_{\substack{i_1 + \cdots + i_k = n \\ 1 \leq i_1, \ldots, i_k}} \mu_n(A_1^{(i_1)}(\times) \cdots (\times) A_k^{(i_k)}).$$

Proof. We will prove this property by induction on k. For $k = 1$, this property is easily given by the fact that $\nu_1(A_1) = \mu(A_1) = \mu_n(A^{(n)})$. Assume the assertion for $k \leq k_0 - 1$. Then, we have

$$
\begin{aligned}
&\nu_{k_0}(A_1, \cdots, A_{k_0}) \\
&= \mu(A_1 \cup \cdots \cup A_{k_0}) - \sum_{j=1}^{k_0-1} \sum_{1 \le \ell_1 < \cdots < \ell_j \le k_0} \nu_j(A_{\ell_1}, \ldots, A_{\ell_j}) \\
&= \mu_n((A_1 \cup \cdots \cup A_{k_0})^{(n)}) - \sum_{j=1}^{k_0-1} \sum_{1 \le \ell_1 < \cdots < \ell_j \le k_0} \nu_j(A_{\ell_1}, \ldots, A_{\ell_j}) \\
&= \sum_{j_1 + \cdots + j_{k_0} = n} \mu_n(A_1^{(j_1)}(\times) \cdots (\times) A_{k_0}^{(j_{k_0})}) \\
&\quad - \sum_{j=1}^{k_0-1} \sum_{1 \le \ell_1 < \cdots < \ell_j \le k_0,\ i_1 + \cdots + i_j = n} \sum \mu_n(A_{\ell_1}^{(i_1)}(\times) \cdots (\times) A_{\ell_j}^{(i_j)}) \qquad (1) \\
&= \sum_{j_1 + \cdots + j_{k_0} = n} \mu_n(A^{(j_1)}(\times) \cdots (\times) A^{(j_{k_0})}) \\
&\quad - \sum_{i_1 + \cdots + i_{k_0} = n, \exists j, i_j = 0} \mu(A_{\ell_1}^{(i_1)}(\times) \cdots (\times) A_{\ell_{k_0}}^{(i_{k_0})}) \\
&= \sum_{\substack{j_1 + \cdots + j_{k_0} = n \\ 1 \le j_1, \cdots, j_{k_0}}} \mu_n(A_1^{(j_1)}(\times) \cdots (\times) A_{k_0}^{(j_{k_0})})
\end{aligned}
$$

We obtain formula (1) by the induction hypothesis. This implies the assertion for $k = k_0$ and concludes the proof. □

Theorem 1. *For any $n \in \mathbb{N}$, a constructive n-additive set function satisfies formulaic n-additivity.*

Proof. We consider the case of

$$\mu(A) = \mu_n(A^{(n)}).$$

Using Proposition 2, for any disjoint sets $A_1, \ldots, A_{n-1}, B_1, B_2 \in \mathcal{B}$ C we have:

$$
\begin{aligned}
\nu_n(A_1, \ldots, A_{n-1}, B_1 \cup B_2) &= \mu_n(A_1(\times) \cdots (\times) A_{n-1}(\times) B_1 \cup B_2) \\
&= \mu_n(A_1(\times) \cdots (\times) A_{n-1}(\times) B_1) + \mu_n(A_1(\times) \cdots (\times) A_{n-1}(\times) B_2) \\
&= \nu_n(A_1, \ldots, A_{n-1}, B_1) + \nu_n(A_1, \ldots, A_{n-1}, B_2).
\end{aligned}
$$

Then, the above formula follows the formulaic n-additivity of μ using Proposition 1.

Generally, μ can be represented by

$$\mu = \sum_{k=1}^{n} \mu_k(A^{(k)})$$

using signed measures μ_k on $(X^{(k)}, \mathcal{B}^{(k)})$. By the above arguments, $A \mapsto \mu_k(A^{(k)})$ satisfies formulaic k-additivity. Then these are formulaic n-additive since $k \le n$ (Recall the remark after Proposition 1). □

4 *k*-Additivity of Distorted Measure

A set function μ on $\mathcal{B}$ is a distorted measure if there is a probability measure m on $(X, \mathcal{B})$ and non-decreasing continuous function f on $\mathbb{R}$ with $f(0) = 0$ such that

$$\mu(A) = f(m(A))$$

for any $A \in \mathcal{B}$. The non-decreasing function f is called "a distortion function". A distorted measure is monotone measure, that is, $\mu(A) \leq \mu(B)$ if $A \subset B$. If a distorted measure is formulaic k-additive measure, the distortion function must be a polynomial.

Proposition 3 [6] *(Theorem 17)*
Let m be a probability measure on $(X, \mathcal{B})$. Let f be the distortion function of a distorted measure $\mu(A) = f(m(A))$ $(A \in \mathcal{B})$. We assume that, for any $t, s \in \{m(A) : A \in \mathcal{B}\}$ and $A \in \mathcal{B}$ with $m(A) = t$, there exists $B \subset A$ such that $\mu(B) = s$ (this property is called "strong Darboux property"). Then, μ is formulaic k-additive if and only if f is a k-degree polynomial.

In the case where the distortion function of a distorted measure μ is a k-degree polynomial, then μ is constructively k-additive. This property was essentially proven by R. Mesiar [4], and we explain this using our notations.

Proposition 4 *(R. Mesiar [4])*
Let m be a positive finite σ-additive measure on $(X, \mathcal{B})$ and μ be a distorted measure given by $\mu(A) = f(m(A))$ $(A \in \mathcal{B})$ using a distortion function f. If f is a k-th degree polynomial, then μ is constructively k-additive.

Proof. We only need to prove this proposition for $f(x) = x^k$. The product measure m^k (defined on $(X^k, \mathcal{B}^k)$) can be easily reduced to the set space $(X^{(k)}, \mathcal{B}^{(k)})$, which concludes the proof. □

Summing up the propositions in this section, we arrive at the following theorem.

Theorem 2. *Let μ be a distorted measure on $(X, \mathcal{B})$ and $k \in \mathbb{N}$. Assume that the distortion function satisfies the strong Darboux property. Then, μ is constructively k-additive if and only if μ is formulaic k-additive.* □

5 *k*-Additivity in a General Case

We have proved that any constructively k-additive set functions are formulaic k-additive. In this section, we consider whether the reverse statement is true.

First, we consider the case where $\mathcal{B}$ is a finite family. As we mainly deal with infinite measurable spaces, the hypothesis "all one-point sets are measurable" must be removed, and the definition of the n-th power set $A^{(n)}$ should be modified. For an element x of X, let $[x]$ denote the smallest measurable set including x. Then, the definition of $A^{(n)}$ is modified by:

$$A^{(n)} = \{(x_1, x_2, \ldots, x_n) \in A^n : \text{ if } j \neq j',\ x_{j'} \notin [x_j]\}.$$

Remark 1. Let $\mathcal{B}$ be a finite σ-algebra. Then, there exists a family of atoms $\mathbb{D} = \{D_1, D_2, \cdots, D_L\}$, that is, $\mathbb{D} \subset \mathcal{B}$ is a disjoint family satisfying $\mathcal{B} = \sigma(\mathbb{D})$. $\mathcal{B}^{(n)}$ ($n \in \mathbb{N}$) can be expressed as follows:

$$\mathcal{B}^{(n)} = \sigma\left(\{D_{i_1}(\times)\cdots(\times)D_{i_n} \;:\; 1 \leq i_1 < \cdots < i_n \leq n\}\right).$$

$\{D_{i_1}(\times)\cdots(\times)D_{i_n} \;:\; 1 \leq i_1 < \cdots < i_n \leq n\}$ is the family of all atoms in $\mathcal{B}^{(n)}$.

Proposition 5. *Let $(X, \mathcal{B})$ be a measurable space with the finite σ-algebra $\mathcal{B}$. Assume that a set function μ is formulaic k-additive. Then, for each $j \leq k$, we can construct a measure μ_j on each set space $(X^{(j)}, \mathcal{B}^{(j)})$ satisfying*

$$\mu(A) = \sum_{j=1}^{k} \mu_j(A^{(j)}).$$

Proof. Because the σ-algebra $\mathcal{B}$ is a finite set family, there exists a finite partition $\{D_j\}_{j=1}^n$ of X such that $\mathcal{B} = \sigma(\{D_j\}_{j=1}^n)$. Then, any $A \in \mathcal{B}$ can be represented by

$$A = \bigcup_{\ell=1}^{L} D_{i_\ell}, \quad 1 \leq i_1 < \cdots < i_L \leq n.$$

Let $j \leq k$ be a positive integer. Then, $\mathcal{B}^{(j)}$ (the σ-algebra of the set space $X^{(j)}$) can be represented as

$$\mathcal{B}^{(j)} = \sigma\left(\{D_{i_1}(\times)\cdots(\times)D_{i_j} : 1 \leq i_1 < \cdots < i_j \leq n\}\right),$$

and an element in $\mathcal{B}^{(j)}$ can be represented by a finite union of some subset of $\{D_{i_1}(\times)\cdots(\times)D_{i_j} : 1 \leq i_1 < \cdots < i_j \leq n\}$.

Without loss of generality, we assume that $\ell_i = i$ for each $i \leq L$. Then, the j-th power set is given by

$$A^{(j)} = \bigcup_{1 \leq \ell_1 < \ell_2 < \cdots < \ell_j \leq L} D_{\ell_1}(\times)\cdots(\times)D_{\ell_j}.$$

Let $\{\nu_j\}$ be a Möbius transform of the set function μ. Then, $\nu_j = 0$ for any $j \geq k+1$. We define a measure μ_j on $(X^{(j)}, \mathcal{B}^{(j)})$ by

$$\mu_j(D_{i_1}(\times)\cdots(\times)D_{i_j}) = \nu_j(D_{i_1}, \cdots, D_{i_j})$$

for each $(i_1, \ldots, i_j)$ $(1 \le i_1 < \cdots < i_j \le n)$.

$$\begin{aligned}
\mu(A) &= \mu(\bigcup_{\ell=1}^{L} D_\ell) \\
&= \sum_{j=1}^{k} \sum_{1 \le i_1 < \cdots < i_j \le L} \nu(D_{i_1}, \cdots, D_{i_j}) \\
&= \sum_{j=1}^{k} \sum_{1 \le i_1 < \cdots < i_j \le L} \mu_j(D_{i_1}(\times) \cdots (\times) D_{i_j}) \\
&= \sum_{j=1}^{k} \mu_j(A^{(j)}).
\end{aligned}$$

Thus we have proved the proposition. □

For further discussion, we will give some notations. Recall that $\mathcal{D}_j$ $(j \in \mathbb{N})$ denotes the family of j-disjoint measurable sets. Let $\overline{D}$ be an element of $\mathcal{D}_j$ $(\overline{D} = \{D_1, \cdots, D_j\} \in \mathcal{D}_j)$. Set

$$\nu_j(\overline{D}) = \nu(D_1, \cdots, D_j),$$

and

$$(\times\overline{D}) = D_1(\times) \cdots (\times) D_j \subset X^{(j)}.$$

Now, we give the following definitions.

Definition 4. *Let μ be a set function on $\mathcal{B}$ and $\{\nu_j\}$ be its Möbius transform. We define the j-th order total variation of ν_j as follows.*

$$\|\nu_j\| = \sup\{\sum_{\ell=1}^{L} |\nu(\overline{D_\ell})| : L \in \mathbb{N}, \overline{D_\ell} \in \mathcal{D}_j, \ell \le L, (\times\overline{D_\ell}) \cap (\times\overline{D_{\ell'}}) = \emptyset \text{ if } \ell = \ell'\}.$$

Then, μ is said to have k-th order bounded variation if $\|\nu_j\| < \infty$ for any $j \le k$.

Next, we define the fine continuity of μ at $\emptyset$.

Definition 5. *Let μ be a set function on $\mathcal{B}$ and $\{\nu_j\}$ be its Möbius transform. Then, the j-th adjusting function ν_j has fine continuity at $\emptyset$ if, for any sequence $\{\{\overline{D_i^{(\ell)}}\}_{i=1}^{N_\ell}\}_{\ell=1}^{\infty}$ of the disjoint finite set family in $\mathcal{D}_j$ satisfying*

$$\bigcup_{i=1}^{N_\ell}(\times\overline{D_i^{(\ell)}}) \searrow \emptyset \quad \text{as} \quad \ell \to \infty,$$

ν_j satisfies

$$\lim_{\ell\to\infty} \sum_{i=1}^{N_\ell} |\nu_j(\overline{D_i^{(\ell)}})| = 0.$$

Moreover, μ is said to have k-order fine continuity at $\emptyset$ iff ν_j has fine continuity at $\emptyset$ for $j \le k$.

Using these concepts, we will show the constructive k-additivity of a formulaic k-additive set function. To prove the existence of the corresponding σ-additive measure, we use the following extension theorem. This is well known for a non-negative measure (see [7] for example); however, using standard additional arguments, the statement is valid in the following form.

Theorem 3 *[Caratheodory's extension theorem]. Let $\mathcal{A}$ be an algebra on X and μ be a finitely additive signed measure on $(X, \mathcal{A})$. Assume that*

$$\sup\left\{\sum_{i=1}^{n} |\mu(A_i)| : n \in \mathbb{N}, \{A_i\}_{i=1}^{n} \textit{is a disjoint family in } \mathcal{A}\right\} < \infty,$$

and

$$\sum_{i=1}^{N_n} \mu(A_i^{(n)}) \to 0 \quad (n \to \infty),$$

for an arbitrary sequence of the disjoint family $\{\{A_i^{(n)}\}_{i=1}^{N_n}\}_{n=1}^{\infty}$, $N_n \in \mathbb{N}$ for each $n \in \mathbb{N}$, for which the union $\bigcup_{i=1}^{N_n} A_i^{(n)}$ decreases to an empty set. Then, there exists an extension $\widetilde{\mu}$ on $(X, \sigma(\mathcal{A}))$ satisfying $\widetilde{\mu}(A) = \mu(A)$ for any $A \in \mathcal{A}$.

Using these concepts and the above extension theorem, we arrive at the following theorem.

Theorem 4. *Let $\mathcal{A}$ be a countable algebra, and assume that $\mathcal{B} = \sigma(\mathcal{A})$. Let μ be a set function defined on $\mathcal{B}$ and k be a positive integer satisfying the following properties.*

(a) μ is a formulaic k-additive set function on $(X, \mathcal{B})$.
(b) μ has k-oder bounded variation.
(c) μ has k-oder fine continuity at $\emptyset$.
(d) μ is continuous from below and above.

Then, μ is constructively k-additive on $(X, \sigma(\mathcal{A}))$.

Proof. For a countable algebra $\mathcal{A}$, we can construct a sequence $\{\mathcal{A}_n\}_{n\in\mathbb{N}}$ of increasing finite algebras, which satisfies $\mathcal{A} = \bigcup_{n\in\mathbb{N}} \mathcal{A}_n$. Then, $\mathcal{A}^{(j)} = \bigcup_{n=1}^{\infty} \mathcal{A}_n^{(j)}$ for any $j \leq k$. Using Proposition 5, for each $n \in \mathbb{N}$, there exist σ-additive measures $\mu_j^{(n)}$ on $(X^{(j)}, \mathcal{A}_n^{(j)})$ $(j \leq k)$ satisfying

$$\mu(A) = \sum_{j=1}^{k} \mu_j^{(n)}(A^{(j)}), \quad A \in \mathcal{A}_n.$$

Let us define an extension $\widetilde{\mu_j^{(n)}}$ of $\mu_j^{(n)}$ as follows.

$$\widetilde{\mu_j^{(n)}}(U) = \begin{cases} \mu_j^{(n)}(U) & \text{if } U \in \mathcal{A}_n^{(j)} \\ 0 & \text{if } U \notin \mathcal{A}_n^{(j)}, \quad U \in \mathcal{A}^{(j)}. \end{cases}$$

Then, for each $j \leq k$ and $U \in \mathcal{A}$, the sequence $\{\widetilde{\mu_j^{(n)}(U)\}_{n=1}^{\infty}}$ is bounded by the assumption (b). In general, a bounded sequence has a convergent sub-sequence. Thus, by countably selecting sub-sequences many times, there exists a sub-sequence $\mu_j^{(n_\ell)}$ such that $\{\mu_j^{(n_\ell)}(U)\}_\ell$ converges for any $U \in \mathcal{A}$ and $j \leq k$. Thus, we define a set function

$$\mu_j(U) = \lim_{\ell \to \infty} \mu_j^{(n_\ell)}(U).$$

Any element $U \in \mathcal{A}$ belongs to $\mathcal{A}_n$ for a sufficiently large $n \in \mathbb{N}$ and $\mu_j^{(n)}$ is finitely additive on $\mathcal{A}_n$. Then, the limit μ_j is also finitely additive on $\mathcal{A}$. Assumptions (b) and (c) imply that μ_j satisfies the assumptions of Theorem 3 and μ_j can be extended on $(X^{(j)}, \mathcal{B}^{(j)})$ as a σ-additive signed measure for $j \leq k$.

On a finite σ algebra constructive k-additivity is derived from formulaic k-additivity. Thus, constructive k-additivity is valid on $\mathcal{A}$, and using the continuity from above and below (condition (d)), this property can be extend to the minimal monotone class including $\mathcal{A}$. It is well known that this class is same with $\sigma(\mathcal{A})$ (see [8] for example). Then, we obtain constructive k-additivity on $(X, \sigma(\mathcal{A}))$. □

6 Conclusion

In this study, we discussed the relation between constructive and formulaic k-additivity. A constructively k-additive set function is always formulaic k-additive. A distorted measure is constructively k-additive if and only if it is formulaic k-additive, if the corresponding distortion function satisfies the strong Darboux property. We defined "k-order bounded variation" and "fine continuity at $\emptyset$" for a set function, and using these concepts, we gave a sufficient condition for constructive k-additivity for a formulaic k-additive measure.

Constructive k-additivity must be useful for further arguments. The existence of a σ-finite measure is important, for example, to construct an L_p-theory for functional analysis on non additive monotone measure spaces. There remain several problems for the advance of these concepts. To show the uniqueness of the σ-additive measure on the set spaces, to make the structure of σ-algebra of the set spaces clear, and other detailed problems. We have to try to solve these problems.

References

1. Grabisch, M.: k-order additive discrete fuzzy measures and their representation. Fuzzy Sets Syst. **92**, 167–189 (1997)
2. Combarro, E.F., Miranda, P.: On the structure of the k-additive fuzzy measures. Fuzzy Sets Syst. **161**(17), 2314–2327 (2010)
3. Kolesarova, A., Li, J., Mesiar, R.: k-additive aggregation functions and their characterization. Eur. J. Oper. Res. **265**, 985–992 (2018)
4. Mesiar, R.: Generalizations of k-order additive discrete fuzzy measures. Fuzzy Sets Syst. **102**, 423–428 (1999)

5. Fukuda, R., Honda, A., Okazaki, Y.: Constructive k-additive measure and decreasing convergence theorems. In: Torra, V., Narukawa, Y., Nin, J., Agell, N. (eds.) MDAI 2020. LNCS (LNAI), vol. 12256, pp. 104–116. Springer, Cham (2020). https://doi.org/10.1007/978-3-030-57524-3_9
6. Honda, A., Fukuda, R., Okazaki, Y.: Non-discrete k-order additivity of a set function and distorted measure. Fuzzy Sets Syst. (to appear)
7. Tao, T.: An Introduction to Measure Theory. Graduate Studies in Mathematics, American Mathematical Society, Providence (2011)
8. Neveu, J.: Mathematical Foundations of the Calculus of Probability. Holden-Day Series in Probability and Statistics, Holden-Day, San Francisco (1965)

Sequential Decision-Making Under Uncertainty Using Hybrid Probability-Possibility Functions

Didier Dubois[1], Hélène Fargier[1], Romain Guillaume[1(✉)], and Agnès Rico[2]

[1] IRIT - CNRS, Université de Toulouse, Toulouse, France
{dubois,fargier,romain.guillaume}@irit.fr
[2] Eric, Université de Lyon, Lyon, France
agnes.rico@univ-lyon1.fr

Abstract. Probabilistic and possibilistic models of sequential decision problems are known to possess good behavioral and algorithmic properties. In this paper, the range of models of problems of sequential decision under uncertainty that are dynamically consistent, consequentialist and allow for tree reduction is enlarged by considering a representation of uncertainty that is both probabilistic and possibilistic. The corresponding utility functional is expected utility for highly likely states, and an optimistic or pessimistic possibility-based criterion for unlikely states.

Keywords: Decision tree · Possibility theory · Fuzzy measures · Belief functions

1 Introduction

In sequential decision making a strategy is a conditional plan that assigns a (possibly non deterministic) action to each state where a decision has to be made (also called "decision node"). Each strategy leads to a compound lottery, following Von Neuman and Morgenstern's terminology [12] - roughly, it is a tree representing the different possible scenarios, and thus the different possible final states that the plan/strategy may reach. The optimal strategy is then the one that minimizes a criterion whose value depends on utilities of final states and the resulting compound lottery.

Three assumptions are desirable in order to accept an optimal strategy without questioning its meaning. Those assumptions are:

- *Dynamic Consistency*: when reaching a decision node, following an optimal strategy, the best decision at this node is the one that had been considered so when computing this strategy, i.e. prior to applying it.
- *Consequentialism*: the best decision at each step of the problem only depends on potential consequences at this point.
- *Tree Reduction*: a compound lottery is equivalent to a simple one, assigning probabilities to final states.

V. Torra and Y. Narukawa (Eds.): MDAI 2021, LNAI 12898, pp. 54–66, 2021.
https://doi.org/10.1007/978-3-030-85529-1_5

Those three assumptions are instrumental to enable an optimal strategy to be computed using dynamic programming [10].

When the decision maker is able to provide a probability distribution on the possible states and considers that the utilities are additive, a classical approach is the one based on expected utility. Under this strong assumption, the above three assumptions are satisfied.

When the problem is pervaded with possibilistic uncertainty, two families of criteria make sense: a Sugeno integral-based criterion [5] and the Choquet integral [6]. The criterion based on Choquet integral turns out to be incompatible with the above assumptions: it may happen that none of the optimal strategies is dynamically consistent nor consequentialist The possibilistic criteria based on Sugeno integral do not meet such difficulties. Nor does the generalization of both optimistic and pessimistic possibilistic Sugeno proposed by [7]. It aggregates optimistic and pessimistic criteria by means of a uninorm [13], a semi-group operation whose identity plays the role of a degree of optimism. Contrary to other criteria that account for a degree of optimism, like the Hurwicz criterion, the criterion proposed in [7] satisfies the three assumptions governing a good behavior of the decision tree.

In the present paper, we are looking for new decision criteria, beyond expected utility and possibilistic integrals, that can apply to decision trees and respect the three properties recalled above (*Dynamic Consistency, Consequentialism* and *Tree Reduction*).

2 Decision Trees

A convenient language to introduce sequential decision problems is the one of decision trees [10]. This framework proposes an explicit graphical model, representing each possible scenario by a path from the root to the leaves of a tree. Formally, a decision tree $\mathcal{T} = (\mathcal{N}, \mathcal{E})$ is such that $\mathcal{N}$ contains three kinds of nodes (see Fig. 1 for an example):

- $\mathcal{D}$ is the set of decision nodes (depicted by rectangles).
- $\mathcal{LN}$ is the set of leaves, that represent final states in $\mathcal{S}$; such states can be evaluated by a utility function: $\forall s_i \in \mathcal{S}$, $u(s_i)$ is the degree of satisfaction of eventually being in state s_i (of reaching the corresponding node in $\mathcal{LN}$). For the sake of simplicity we assume, without loss of generality, that only leaf nodes are attached utilities.
- $\mathcal{X}$ is the set of chance nodes (depicted by circles).

For any node $n_i \in \mathcal{N}$, $Succ(n_i) \subseteq \mathcal{N}$ denotes the set of its children. In a decision tree, for any decision node d_i, $Succ(d_i) \subseteq \mathcal{X}$: $Succ(d_i)$ is the set of actions that can be chosen when d_i is reached. For any chance node x_i, $Succ(x_i) \subseteq \mathcal{LN} \cup \mathcal{D}$: $Succ(x_i)$ is the set of possible outcomes of action x_i - either a leaf node is observed, or a decision node is reached (and then a new action should be chosen).

Solving a decision tree amounts to building a *strategy*, i.e. a function δ that associates to each decision node d_i an action (i.e. a chance node) in $Succ(d_i)$:

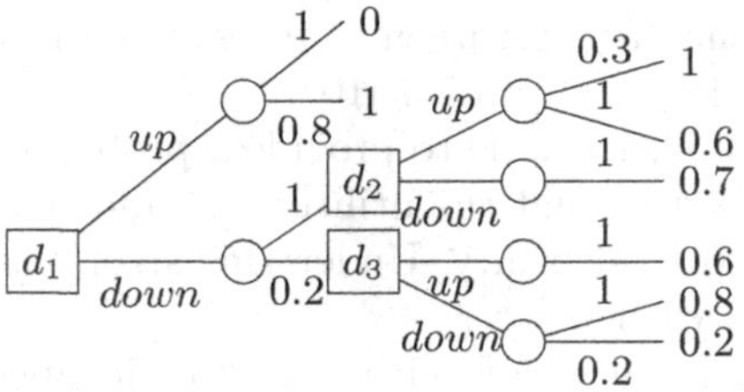

Fig. 1. A possibilistic decision tree.

$\delta(d_i)$ is the action to be executed when decision node d_i is reached. Let Δ be the set of strategies that can be built for $\mathcal{T}$. We shall also consider the subtree $\mathcal{T}_n$ of $\mathcal{T}$ rooted at node $n \in \mathcal{T}$, and denote by Δ_n its strategies: they are subtrategies of the strategies of Δ.

The satisfaction of the decision maker for a consequence (on a leaf of the tree) is captured by a utility degree on a totally ordered scale. The scale $[0,1]$ is generally chosen for these degrees, but any ordered set can be used.

Here, we focus on the case where the information at each chance node is fully captured by a distribution over outcomes of the chance nodes, namely a possibility distribution π and/or a probability distribution p. When bearing on the leaf nodes, such distributions define simple lotteries on the utility degrees. More formally:

- A simple probabilistic lottery L^p [12]: is a probability distribution p on a set of utility degrees, $\Lambda = \{\lambda_1, ..., \lambda_n\}$. The probabilistic lotteries will be written as $L^p = <p_1/\lambda_1, ..., p_n/\lambda_n>$ with $p_i \in [0,1], \sum_{i=1}^{n} p_i = 1$.
- A simple possibilistic lottery L^π [5] is a normalized possibility distribution π on a set of utility degrees, Λ, both being expressed in the same ordered scale. The possibilistic lotteries will be written as $L^p = <\pi_1/\lambda_1, ..., \pi_n/\lambda_n>$ with $\pi_i \in [0,1], \max_{i=1}^{n} \pi_i = 1$.

In a lottery L^π (resp. L^p), the value π_i (resp. p_i) is the possibility (resp. probability) degree of getting utility λ_i according to the decision strategy captured by the lottery. For the sake of brevity, the λ_i's such that $\pi_i = 0$ (resp. $p_i = 0$) are often omitted in the notation of a lottery (e.g., (<1/0.8>) denotes the lottery that provides utility 0.8 for sure, all the other utility degrees being impossible).

The expected utility of probabilistic simple lotteries was proposed by Von Neuman and Morgernstern [12] as a decision criterion under risk: $E(L^p) = \sum_{\lambda_i \in \Lambda} \lambda_i \cdot p_i$. Dubois and Prade [5] proposed to use optimistic and pessimistic possibilistic criteria, denoted by U^{Pes} and U^{Opt} in this paper, to evaluate the global utility of a possibilistic lottery using $U^{Pes}(L^\pi) = \min_{\lambda_i \in \Lambda} \max(1 - \pi_i, \lambda_i)$, and $U^{Opt}(L^\pi) = \max_{\lambda_i \in \Lambda} \min(\pi_i, \lambda_i)$.

Let us now consider full-fledged strategies. A strategy in Δ can be viewed as a connected subtree of $\mathcal{T}$ where there is exactly one edge (and thus one chance node) left at each decision node - skipping the decision nodes, we get a chance tree or, using von Neuwman and Morgernstern's terminology, a compound lottery:

like simple lotteries, which are distributions over utilities, compound lotteries are distributions over (simple or compound) lotteries.

The idea is to define a simple lottery equivalent to the original, compound one and to apply the decision criterion to this simple lottery.

Definition 1 (*Reduction*^{p})**.** *For any probabilistic compound lottery of the form* $L^p = <p_1/L^p_1, \dots, p_m/L^p_m>$, $Red^p(L^p)$ *is the simple lottery that associates to each* λ_i *the probability degree* $\overline{p}_i = \sum_{j=1,\dots,m} p_j \cdot p_i^j$, *where* p_i^j *denotes the probability of getting* λ_i *though lottery* L^p_j *and* p_j *the probability of getting* L^p_j.

Definition 2 (*Reduction*$^\pi$)**.** *For any possibilistic compound lottery of the form* $L^\pi = <\pi_1/L^\pi_1, \dots, \pi_m/L^\pi_m>$, $Red^\pi(L^\pi)$ *is the simple lottery that associates to each* λ_i *the possibility degree* $\overline{\pi}_i = \max_{j=1,\dots,m} \min(\pi_j, \pi_i^j)$, *where* π_i^j *denotes the possibility of getting* λ_i *though lottery* L^π_j *and* π_j *the possibility of getting* L^π_j.

The principle of lottery reduction allows the comparison of compound lotteries: L is preferred to L' iff its reduction is preferred to the one of L' and optimality can then be soundly defined:

- $\delta \in \Delta$ is optimal for a decision tree $\mathcal{T}$ iff $\forall \delta' \in \Delta, C(Red(L_\delta)) \succeq C(Red(L_{\delta'}))$,

for a criterion C. The principle of monotonicity and the one of decomposition [12] are valid for expected utility, and also for U^{Pes} and U^{Opt} in a weak form:

Definition 3 (Weak monotonicity). *A preference criterion* C *over possibilistic/probabilistic lotteries is said to be weakly monotonic iff whatever* L, L' *and* L'', *whatever a normalized possibility/probability distribution* w, v:

$$C(L) \le C(L') \Rightarrow C(<w/L, v/L''>) \le C(<w/L', v/L''>). \tag{1}$$

Importanty, all approaches that satisfy weak monotonicity, and in particular in the approach considered in this paper, also satisfy *Dynamic Consistency*, *Consequentialism* and *Tree Reduction*. This guarantees coherence with the intuition of rationality; this is also important from the algorithmic point of view, since it allows to find an optimal strategy by dynamic programming.

In the following we consider decision trees where uncertainty is captured by set functions that are more general than probability and possibility measures, while taking into account a degree of optimism of the decision maker, although without giving up the monotonicity principle.

3 Hybrid Possibility-Probability Measures

The three previous criteria (expected utility, pessimistic and optimistic possibilistic utility functionals) are particular instances of generalized integrals (Choquet integral for expected utility, Sugeno integral for the other ones) based on fuzzy measures.

Definition 4. *A fuzzy measure is a set function* $\mu : 2^S \to [0,1]$, *satisfying the following axioms:*

- $\mu(\emptyset) = 0$; $\mu(S) = 1$ *(limit conditions)*
- $\mu(A \cup B) \geq \mu(A)$ *(monotony)*

Fuzzy measures include, among others, probability measures, necessity measures and possibility measures.[1]

However, we look for special fuzzy measures representable by means of lotteries (or, equivalently, by distributions on utility values). Hence they must be *decomposable*:

Definition 5 [4]. *A decomposable fuzzy measure is a fuzzy measure* μ *for which there is a t-conorm*[2] S *such that* $\mu(A \cup B) = S(\mu(A), \mu(B))$ *whenever* A *and* B *are disjoint.*

Possibility measures are max-decomposable (for the t-conorm max), while probability measures are additively decomposable using the Łukasiewicz t-conorm $\min(a + b, 1)$. To define lottery reduction, we also need a generalization of the notion of independence between events.

Definition 6. *Let* T *be a triangular norm. Two events* A *and* B *are said to be* T*-separable with respect to a fuzzy measure* μ *if and only if* $\mu(A \cap B) = T(\mu(A), \mu(B))$ *for a t-norm* T.

If A and B are disjoint events T-separable from another event C, then, since $(A \cup B) \cap C = (A \cap C) \cup (B \cap C)$, a decomposable fuzzy measure should satisfy

$$T(S(\mu(A), \mu(B)), \mu(C)) = S(T(\mu(A), \mu(C)), T(\mu(B), \mu(C)),$$

which requires a property called Conditional Distributivity of T over S. This property is essential for enabling the reduction of generalised lotteries.

Definition 7. *A t-norm* T *is conditionally distributive over a t-conorm* S *if for all* $x, y, z \in [0,1]$ *we have*

$$(CD)\, T(x, S(y,z)) = S(T(x,y), T(x,z)), \text{ whenever } S(y,z) < 1.$$

In [3], it has been proved that the only family of pairs (t-conorm, t-norm) that satisfy the condition (CD) are built from the pairs (max, T) and ($\min(a+b, 1)$, $\times$) [3,9], where T is an arbitrary continuous t-norm. It is a parametric family of pairs denoted by (S^α, T^α), with parameter α, where S^α (resp. T^α) is the ordinal sum [9] of max and $\min(x+y, 1)$ (resp. T and product) represented on Fig. 2 for $T = \min$.

---

[1] Given from a possibility distribution π over a set S, the possibility and the necessity of any event $A \subseteq S$ are defined by $\Pi(A) = \max_{s \in A} \pi(s), N(A) = 1 - \Pi(\bar{A}) = 1 - \max_{s \notin A} \pi(s)$.

[2] A t-conorm is a non-decreasing semi-group operation on $[0,1]$ with identity 0 and absorbing element 1. A t-norm is a non-decreasing semi-group operation on $[0,1]$ with identity 1 and absorbing element 0. T-norms and t-conorms are gradual models of conjunction and disjunction. See [9] for more details.

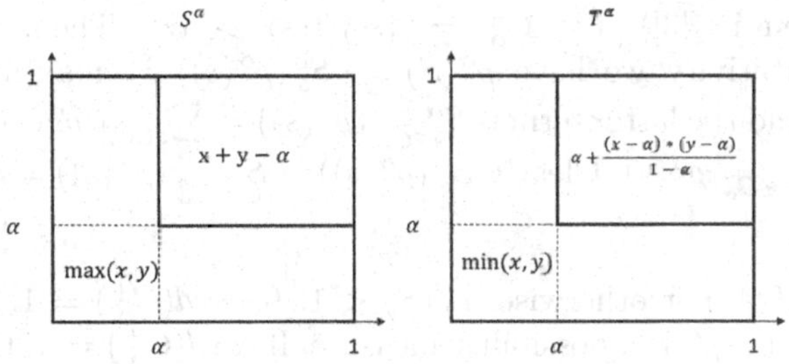

Fig. 2. S^α and T^α

The pairs (S^α, T^α) where T^α is distributive on S^α are thus of the form

$$S^\alpha(x,y) = \begin{cases} x+y-\alpha \text{ if } x>\alpha, y>\alpha \\ \max(x,y) \end{cases} \tag{2}$$

$$T^\alpha(x,y) = \begin{cases} \alpha + \frac{(x-\alpha)(y-\alpha)}{1-\alpha} \text{ if } x>\alpha, y>\alpha \\ \min(x,y) \text{ otherwise.} \end{cases} \tag{3}$$

An S^α-decomposable fuzzy measure, denoted by ρ^α, is a hybrid function between possibility and probability measures defined as follows:

Definition 1 (Hybrid π-p measure [3]**).** *A hybrid possibility-probability measure ρ^α is a fuzzy measure such that, for disjoint sets A and B:*

$$\rho^\alpha(A \cup B) = S^\alpha(\rho^\alpha(A), \rho^\alpha(B)) \tag{4}$$

The fuzzy measure ρ^α is clearly a probability measure (rescaled on $[\alpha, 1]$) for likely events, and a possibility measure if one of the events is unlikely. The hybrid π-p measure is an example of level-dependent capacity [8]. For events that can be considered independent A, B, we have

$$\rho^\alpha(A \cap B) = T^\alpha(\rho^\alpha(A), \rho^\alpha(B)) \tag{5}$$

The limit condition $\rho^\alpha(\Omega) = 1$ comes down to enforcing, for any event A and its complement A^c, the duality condition $\rho^\alpha(A) + \rho^\alpha(A^c) = 1 + \alpha$, when $\min(\rho^\alpha(A), \rho^\alpha(A^c)) > \alpha$ and $\max(\rho^\alpha(A), \rho^\alpha(A^c)) = 1$ otherwise.

Because it is decomposable, ρ^α is completely defined by a distribution of weights ρ^α on the singletons of the referential S. Applying (4) to the union of singletons yields the normalization condition suggested in [3]:

Proposition 1. *A distribution ρ^α on S defines a normalised S^α-decomposable fuzzy measure if and only if*

$$\sum_{s:\rho^\alpha(s)>\alpha} \rho^\alpha(s) = 1 + (card(C_\alpha^+) - 1) \times \alpha \tag{6}$$

where $C_\alpha^+ = \{s, \rho^\alpha(s) > \alpha\}$.

Proof (not given in [3]). Let $C_\alpha^+ = \{s, \rho^\alpha(s) > \alpha\}$. Then, by associativity and commutativity we have $\rho^\alpha(S) = S_s^\alpha(\rho^\alpha(s)) = \max(\max_{s \notin C_\alpha^+} \rho^\alpha(s), S^\alpha_{s \in C_\alpha^+}(\rho^\alpha(s))$, and the latter term is $S^\alpha_{s \in C_\alpha^+}(\rho^\alpha(s)) = \sum_{s \in C_\alpha^+} \rho^\alpha(s) - (card(C_\alpha^+) - 1)\alpha > \alpha \geq \max_{s \notin C_\alpha^+} \rho^\alpha(s)$. Clearly $S_s^\alpha(\rho^\alpha(s)) = S^\alpha_{s \in C_\alpha^+}(\rho^\alpha(s)) = 1$ yields condition (6).

Note that $C_\alpha^+ \neq \emptyset$. For otherwise, $\rho^\alpha(S) < 1$. If $card(C_\alpha^+) = 1$, then $\rho^\alpha(S) = \max_s \rho^\alpha(s) = 1$, i.e. ρ^α is a possibility measure. If $card(C_\alpha^+) = n$, then $\rho^\alpha(S) = 1$ reads $\sum_s \rho^\alpha(s) - (n-1)\alpha = 1$, hence $\sum_s^n \rho^\alpha(s) = 1 + (n-1)\alpha$ and the fuzzy measure ρ^α is additive.

Example 1. Consider the $S^{0.7}$-decomposable fuzzy measure $\rho^{0.7}$ on the set $\{s_a, s_b, s_c, s_d\}$, defined by the distribution $\rho^{0.7}(s_a) = 0.5$ $\rho^{0.7}(s_b) = 0.5$ $\rho^{0.7}(s_c) = 0.8$ $\rho^{0.7}(s_d) = 0.9$. Here, $C_\alpha^+ = \{s_c, s_d\}$. One can check that $\rho^{0.7}(\Omega) = 0.8 + 0.9 - 0.7 = 1$. It is normalized.

More generally it is easy to express the value of ρ^α on general events:

$$\rho^\alpha(A) = \begin{cases} \sum_{s \in A} \rho^\alpha(s) - \alpha(card(A) - 1) \text{ if } A \subseteq C_\alpha^+, \\ \max_{s \in A} \rho^\alpha(s) \text{ if } A \subseteq \overline{C_\alpha^+}, \\ \sum_{s \in A \cap C_\alpha^+} \rho^\alpha(s) - \alpha(card(A \cap C_\alpha^+) - 1) \text{ otherwise.} \end{cases}$$

Note that the third case ($C_\alpha^+ \cap A \neq \emptyset$) covers the first. We can moreover prove that any measure ρ^α is a plausibility measure in the sense of Shafer [11] obtained as a probabilistic mixture between a possibility measure and a probability measure.

Proposition 2. *For any hybrid possibility-probability function ρ^α, there exists a possibility measure Π with possibility distribution π and a probability measure P with distribution p such that $\rho^\alpha(s) = \alpha\pi(s) + (1-\alpha)p(s)$ where $\forall s, \pi(s) < 1$ implies $p(s) = 0$. Moreover $\forall A, \rho^\alpha(A) = \alpha \max_{s \in A} \pi(s) + (1-\alpha)\sum_{s \in A} p(s)$.*

Proof. For events, if $A \cap C_\alpha^+ = \emptyset$, $\rho^\alpha(A) = \alpha \max_{s \in A} \pi(w) = \alpha\Pi(A) + (1-\alpha)P(A)$ since $P(A) = 0$. Otherwise, if $A \cap C_\alpha^+ \neq \emptyset$:

$$\begin{aligned}
\rho^\alpha(A) &= \sum_{s \in A \cap C_\alpha^+} \rho^\alpha(s) - \alpha(card(A \cap C_\alpha^+) - 1) \\
&= \sum_{s \in A \cap C_\alpha^+} (\alpha(1 - p(s)) + p(s)) - \alpha(card(A \cap C_\alpha^+) - 1) \\
&= \alpha card(A \cap C_\alpha^+) - \alpha P(A \cap C_\alpha^+) + P(A \cap C_\alpha^+) - \alpha(card(A \cap C_\alpha^+) - 1) \\
&= \alpha + (1-\alpha)P(A \cap C_\alpha^+) = \alpha\Pi(A) + (1-\alpha)P(A)
\end{aligned}$$

since $\Pi(A) = 1$ and $p(w) = 0$ for $s \notin C_\alpha^+$. □

Note that the dual of ρ^α ($\overline{\rho}^\alpha(A) = 1 - \rho^\alpha(\overline{A})$) is a convex combination of a necessity measure and a probability measure. Indeed, we have $\overline{\rho}^\alpha(A) = 1 -$

$\alpha\Pi(\overline{A})-(1-\alpha)P(\overline{A}) = \alpha(1-\Pi(\overline{A}))+(1-\alpha)(1-P(\overline{A})) = \alpha N(A)+(1-\alpha)P(A)$. When the possibility distribution is vacuous ($C_\alpha^+ = S$), it is a special case of Shafer discounting scheme [11], and is known in the imprecise probability literature as the linear-vacuous model [1]. In particular, $\rho^\alpha = \alpha\pi + (1-\alpha)p$ is a normalized ρ^α distribution.

Hybrid π-p distributions combine models of two extreme behaviors in uncertain contexts, namely when the knowledge can be expressed through a possibility distribution and when it is can be expressed through a probability distribution. Hybrid distributions behave like probabilities only on the states with maximum possibility. As seen above the hybrid model must satisfy the constraint $P(s) = 0$ if $\pi(s) < 1, \forall s$. It is clear that C_α^+ is the core of π. Then from both distributions and a given threshold α, we can build the hybrid one ρ^α using a weighted average $\rho^\alpha(s) = \alpha\pi(s)+(1-\alpha)p(s)$. The idea is that the decision-maker provides a standard probability distribution (e.g., frequencies) on normal states. Then she considers that there is a subjective probability α that the actual state has not been observed, and defines a possibility measure on the states of zero probability, an idea in agreement with the handling of zero probability events in De Finetti approach to subjective probability (see [2]).

4 Decision Making on Hybrid π-p Decision Trees

Back to our problematics of decision making under uncertainty, let us consider hybrid π-p simple lotteries $L = \langle\rho/\lambda_1, ..., \rho_n/\lambda_n\rangle$. We define two utility functionals ES^{Opt} and ES^{Pes} based on such hybrid π-p distributions:

$$ES^{Opt}(L) = S^\alpha_{i=1,\dots,n}(T^\alpha(\rho_i^{\alpha,L}, \lambda_i)) \tag{7}$$

$$ES^{Pes}(L) = 1 - S^\alpha_{i=1,\dots,n}(T^\alpha(\rho_i^{\alpha,L}, 1-\lambda_i)) \tag{8}$$

Considering a lottery $L^\rho = \langle\rho_1/\lambda_1, ..., \rho_n/\lambda_n\rangle$ we define a lottery $(1-L)^\rho$ by $(1-L)^\rho = \langle\rho_1/(1-\lambda_1), ..., \rho_n/(1-\lambda_n)\rangle$. We have the following semi-duality relation

$$ES^{Pes}(L^\rho) = 1 - ES^{Opt}((1-L)^\rho). \tag{9}$$

This property will be useful for some proofs in the following. For the sake of brevity, we will drop α from S^α and T^α in the sequel.

4.1 Hybrid Utility Functional and Decision Maker Behavior

Let us rewrite $ES^{Opt}(L)$ and $ES^{Pes}(L)$ more explicitly so as to lay bare its meaning.

Proposition 3

$$ES^{Opt}(L) = \begin{cases} U^{Opt}(L) \ \textit{if} \ \nexists i \ \textit{s.t.} \ \lambda_i > \alpha \ \textit{with} \ \rho_i^\alpha > \alpha \\ E^{Opt}(L) = \alpha + \frac{\sum_{i|\lambda_i \rho_i^\alpha > \alpha}(\lambda_i - \alpha)(\rho_i^\alpha - \alpha)}{1-\alpha} \ \textit{otherwise} \end{cases} \tag{10}$$

$$ES^{Pes}(L) = \begin{cases} U^{Pes}(L) \text{ if } \nexists i \text{ s.t. } \lambda_i < 1-\alpha \text{ with } \rho_i^\alpha > \alpha \\ E^{Pes}(L) = 1-\alpha - \frac{\sum_{i|1-\lambda_i,\rho_i^\alpha>\alpha}(1-\lambda_i-\alpha)(\rho_i^\alpha-\alpha)}{1-\alpha} \text{ otherwise} \end{cases} \quad (11)$$

Proof. If $\nexists i$ s.t. $\lambda_i > \alpha$ with $\rho_i^\alpha > \alpha$ then $\forall i$, $T = \min$ is used and $S = \max$ too. So we are back to the $U^{Opt}(L)$ criterion. If there is only one i s.t. $\lambda_i > \alpha$ with $\rho_i^\alpha > \alpha$, then $ES^{Opt}(L) = T^\alpha(\rho_i^\alpha, \lambda_i) = \alpha + \frac{(\lambda_i-\alpha)(\rho_i^\alpha-\alpha)}{1-\alpha}$. For the general case where there are several i such that $\min(\rho_i^{\alpha,L}, \lambda_i) > \alpha$, say a set I^+ of indices then $T(\rho_i^{\alpha,L}, \lambda_i) > \alpha, i \in I^+$ only. Then $ES^{Opt}(L) = \sum_{i\in I^+}(\alpha + \frac{(\lambda_i-\alpha)(\rho_i^\alpha-\alpha)}{1-\alpha}) - (card(I^+)-1)\alpha = \alpha + \sum_{i\in I^+}(\frac{(\lambda_i-\alpha)(\rho_i^\alpha-\alpha)}{1-\alpha})$. Using semi-duality between ES^{Pes} and ES^{Opt} we get the expression of the former. □

From Proposition 3 it is easy to check that:

Proposition 4. *$ES^{Opt}(L) \leq \alpha$ iff $ES^{Opt}(L) = U^{Opt}(L)$.*
Likewise, $ES^{Pes}(L) \geq 1-\alpha$ iff $ES^{Pes}(L) = U^{Pes}(L)$.

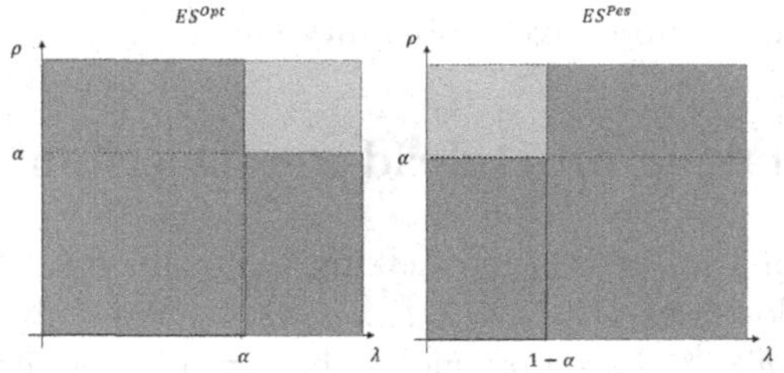

Fig. 3. $ES^{Opt}(L)$ and $ES^{Pes}(L)$

In other words, the criterion $ES^{Opt}(L)$ is possibilistic optimistic ($= U^{Opt}(L)$) so long as entries (utilities or plausibilities) are below the threshold α (distribution included in blue area on Fig. 3). Otherwise, we get an expected value over states with plausibilities and utilities greater than α (see green area in Fig. 3, left). Likewise, with $ES^{Pes}(L)$, we get an expected value over states with utility less than $1-\alpha$ and with high enough plausibility i.e. greater than α. We get the pessimistic possibilistic criterion $U^{Pes}(L)$ otherwise (with either high utilities or low plausibilities); see green area in Fig. 3 right side.

Example 2. let D_1, D_2 and D_3 be decisions with $\rho^{0.70}$ distribution on $\{a, b, c, d\}$ with $\lambda_a = 0.2$, $\lambda_b = 0.6$, $\lambda_c = 0.8$, $\lambda_d = 1$, $D_1 = <0.7/\lambda_a, 0.9/\lambda_b, 0.6/\lambda_c, 0.5/\lambda_d>$, $D_2 = <0.75/\lambda_a, 0.90/\lambda_b, 0.75/\lambda_c, 0.5/\lambda_d>$ and $D_3 = <0.75/\lambda_a, 0.85/\lambda_b, 0.75/\lambda_c, 0.75/\lambda_d>$. If the DM is optimistic, we can see that the D_1 is in the possibility area since $\rho_{D_1}^{0.7}(c)$ and $\rho_{D_1}^{0.7}(c)$ are ≤ 0.7 while D_2 and D_3 are in EU area ES^{Opt}. We have $D_3 \succ D_2 \succ D_1$. Note that D_3 is preferred to D_2 since the expected utility to be in the green square Fig. 3 is greater. If the DM is pessimistic, D_1 is in possibility area while D_2 and D_3 are in EU area since $\rho_{D_2}^{0.7}(a) = \rho_{D_2}^{0.7}(a) < 1-\lambda$ and the preference relation is $D_1 \succ D_2 \sim D_3$.

4.2 Decision Trees with Hybrid π-p Distributions

Consider now decisions tree. We now know how to compare simple lotteries. In order to compare strategies, i.e. compound lotteries, we define a principle of reduction of hybrid π-p compound lotteries:

Definition 8 (*Reduction*$^{\rho}$). *For any compound lottery of the form* $L = <\rho_1^\alpha/L_1^{\rho^\alpha}, \cdots, \rho_m^\alpha/L_m^{\rho^\alpha}>$, $Red^\rho(L)$ *is the simple lottery that associates to each* λ_i *the weight* $\forall\lambda_i, \overline{\rho}_i^\alpha = S_{j=1}^m T(\rho_j^\alpha, \rho_i^{\alpha,j})$, *where* $\rho_i^{\alpha,j}$ *denotes the confidence value of getting* λ_i *though lottery* $L_j^{\rho^\alpha}$ *and* ρ_j^α *the confidence value of getting* $L_j^{\rho^\alpha}$.

It is easy to get the following result:

Proposition 5

$$\overline{\rho}_i^\alpha = \begin{cases} \max_{j=1,\ldots,m} \min(\rho_j^\alpha, \rho_i^{\alpha,j}) \text{ if } \nexists j \text{ s.t. } \rho_j^\alpha > \alpha,\ \rho_i^{\alpha,j} > \alpha \\ \alpha + \frac{\sum_{j|\rho_j^\alpha, \rho_j^{\alpha,i} > \alpha} (\rho_j^\alpha - \alpha)(\rho_j^{\alpha,i} - \alpha)}{1-\alpha} \text{ otherwise} \end{cases} \tag{12}$$

The following proposition states the main result of this paper:

Proposition 6. *Given a decision tree with* ρ^α *measure, the hybrid criteria* ES^{Opt} *and* ES^{Pes} *satisfy the weak monotonicity property.*

Proof. We propose a proof for ES^{Opt}, the proof is similar for ES^{Pes}. For each lottery L, we have two cases: (a) $\nexists i$ such that $\rho_i > \alpha$ and $\lambda_i > \alpha$ and (b) $\exists i$ such that $\rho_i > \alpha$ and $\lambda_i > \alpha$. Let us explore all possible configurations.

1. L, L' and L'' are in (a): $\forall i$ the t-norm min and t-conorm max are used so we are in possibility case and the weak monotonicity property holds.
2. L and L' in (a) and L'' in (b), we need to distinguish two cases:
   i) if $v \leq \alpha$ then we are again in the case with t-norm min and t-conorm max so the weak monotonicity property holds.
   ii) if $v > \alpha$ then $<u/L, v/L''>$ and $<u/L', v/L''>$ are in (b) so $ES^{Opt}(<u/L, v/L''>) = ES^{Opt}(<u/L', v/L''>)$ so the weak monotonicity is satisfied.
3. L, L' and L'' in (b), we need to distinguish three cases:
   i) if $v \leq \alpha$ and $u > \alpha$ then $<u/L, v/L''>$ and $<u/L', v/L''>$ are in (b) so $ES^{Opt}(<u/L, v/L''>) = \alpha + \frac{\sum_{i|\lambda_i, T^P(\rho_i^{\alpha,L}, u) > \alpha} (\lambda_i - \alpha) \times (T^P(\rho_i^{\alpha,L}, u) - \alpha)}{1-\alpha}$ $= \alpha + (u - \alpha)\frac{\sum_{i|\lambda_i, \rho_i^{\alpha,L} > \alpha} (\lambda_i - \alpha) \times (\rho_i^{\alpha,L} - \alpha)}{(1-\alpha)^2} \leq ES^{Opt}(<u/L', v/L''>)$ $= \alpha + (u - \alpha)\frac{\sum_{i|\lambda_i, \rho_i^{\alpha,L'} > \alpha} (\lambda_i - \alpha) \times (\rho_i^{\alpha,L'} - \alpha)}{(1-\alpha)^2}$ so the weak monotonicity is satisfied.
   ii) if $v > \alpha$ and $u \leq \alpha$ we are in a similar situation as in 2) ii)
   iii) if $v > \alpha$ and $u > \alpha$ then we have $ES^{Opt}(<u/L, v/L''>) = (u - \alpha)\frac{ES^{Opt}(L)}{(1-\alpha)} + (v - \alpha)\frac{ES^{Opt}(L'')}{(1-\alpha)} - \alpha \leq$ $ES^{Opt}(<u/L', v/L''>) = (u - \alpha)\frac{ES^{Opt}(L')}{(1-\alpha)} + (v - \alpha)\frac{ES^{Opt}(L'')}{(1-\alpha)} - \alpha$

4. L'' in (a) and L, L' in (b), we need to distinguish two cases: if $u > \alpha$ the proof is similar to the one in 3) i) and if $u \leq \alpha$ similar to the one in 2) i).
5. L, L'' in (a) and L' in (b), we need to distinguish two cases:
   i) if $u > \alpha$, $ES^{Opt}(<u/L, v/L''>) \leq \alpha$ and $ES^{Opt}(<u/L', v/L''>) > \alpha$.
   ii) if $u \leq \alpha$ then $ES^{Opt}(<u/L, v/L''>) \leq ES^{Opt}(<u/L', v/L''>)$ the proof is similar to the one in 2) i) since u/L' is in (a).
6. L in (a) and L', L'' in (b), we need to distinguish three cases:
   i) if $u \leq \alpha$ and $v > \alpha$ then the proof is similar to the one in 2) ii).
   ii) if $u > \alpha$ and $v \leq \alpha$ then $<u/L, v/L''>$ in (a) and $<u/L, v/L''>$ in (b) from Proposition 4 the weak monotonicity property holds.
   iii) if $u > \alpha$ and $v > \alpha$ then $ES^{Opt}(<u/L, v/L''>) = (v - \alpha)\frac{ES^{Opt}(L'')}{(1-\alpha)} \leq ES^{Opt}(<u/L', v/L''>) = (u - \alpha)\frac{ES^{Opt}(L')}{(1-\alpha)} + (v - \alpha)\frac{ES^{Opt}(L'')}{(1-\alpha)} - \alpha$ so the weak monotonicity property holds. □

From Proposition 6, when the decision maker provides a ρ-style decision tree, ES^{Opt} and ES^{Pes} satisfy the three basic properties required in the introduction: consequentialism, dynamic consistency and lottery reduction.

4.3 Composing Possibilistic and Probabilistic Lotteries

According to Proposition 2, a (compound) lottery can be viewed as two (compound) lotteries, a possibilistic one $L^\pi = <\pi_1/L_1^\pi, \ldots, \pi_m^\pi/L_m>$, and a probabilistic one $L^p = <p_1/L_1^p, \ldots, p_m/L_m^p>$, both on the same decision tree. Assume that $\pi_i < 1$ implies $p_i = 0, i = 1, \ldots m$. We are interested in merging them into a hybrid lottery, given the parameter α. Proposition 2 leads to define a fusion operation:

$$F(\pi, p, \alpha) = \begin{cases} \rho_i^\alpha = \alpha\pi_i \text{ if } \pi_i < 1, \\ \rho_i^\alpha = \alpha + (1 - \alpha)p_i \text{ otherwise.} \end{cases} \tag{13}$$

Merging the possibility and probability distributions locally should be equivalent to merging the reduced lotteries at the global level. Moreover, the fusion operation must be distributive over the reduction operator.

Property 1 (Distributivity over reduction). *Let* $L = (L^\pi, L^p)$ *be a pair of possibilistic and probabilistic compound lotteries on the same decision tree. Operator* F *is said to satisfy the distributivity property iff* $F(Red^\pi(L^\pi), Red^p(L^p), \alpha) = Red^\rho(<F(\pi_1, p_1, \alpha)/F(L_1^\pi, L_1^p, \alpha), \ldots, F(\pi_m, p_m, \alpha)/F(L_m^\pi, L_m^p, \alpha)>)$.

Proposition 7. $F(\pi, p, \alpha)$*, defined by Eq. (13), satisfies Property 1.*

Proof. When applying reduction to the probability tree followed by F, we obtain $\rho_i = \alpha + (1 - \alpha)\sum_j p_j \times p_j^i, \forall i$ s.t. $\exists j$ with $p_j^i > 0, p_j > 0$ and $\rho_i = \alpha \max_j \min(\pi_j, \pi_j^i)$ otherwise. When applying F first: if $\exists j$ with $p_j^i >, p_j > 0$ then $\rho_j^{\alpha,i} > \alpha, \rho_j^\alpha > \alpha$. From Proposition 5, the ρ^α-reduction is $\overline{\rho}_i^\alpha = \alpha + (1 - \alpha)\sum_j p_j \times p_j^i$. Otherwise, the ρ^α-reduction is $\max_{j=1,\ldots,n} \min(\rho_j^{\alpha,i}, \rho_j^\alpha)$ with $\min(\rho_j^{\alpha,i}, \rho_j^\alpha) = \alpha\pi_j^i$ or $\alpha\pi_j$ so we obtain $\rho_i = \alpha \max_j \min(\pi_j, \pi_j^i)$. □

This result ensures the dynamic consistency of the hybrid π-p approach to sequential decision-making under uncertainty.

5 Conclusion

In this paper, we try to improve the range of decision trees that can be solved by dynamic programming and respect consequentialism as well as dynamic consistency, beyond standard probabilistic decision trees, and possibilistic ones. We have shown that everything relies on i) defining the uncertainty measure by means of a generalized weight distribution on utility values; 2) the possibility of reducing compound lotteries into simple ones; 3) the definition of a utility functional by means of a generalized integral. This paper proposes a solution to this problem, and shows that it leads to a very restricted family of decomposable measures with respect to a specific family of t-conorms, due to the conditional distributivity property required to ensure lottery reduction. The paper proves that the obtained utility functionals satisfy the weak monotonicity property, which ensures computability of optimal decisions via dynamic programming. The kind of uncertainty function laid bare in this study turns out to be a Shafer plausibility (resp. belief) function obtained as a convex mixture of probability and possibility (resp.necessity) functions, which opens the way to a natural interpretation of these uncertainty measures.

References

1. Augustin, T., Coolen, F., De Cooman, G., Troffaes, M.: Introduction to Imprecise Probabilities. Wiley, Hoboken (2014)
2. Coletti, G., Scozzafava, R.: Probabilistic Logic in a Coherent Setting. Kluwer Academic Publishers, Dordrecht (2002)
3. Dubois, D., Pap, E., Prade, H.: Hybrid probabilistic-possibilistic mixtures and utility functions. In: Fodor, J., De Baets, B., Perny, P. (eds.) Preferences and Decisions Under Incomplete Knowledge. Studies in Fuzziness and Soft Computing, vol. 51, pp. 51–73. Springer, Heidelberg (2000). https://doi.org/10.1007/978-3-7908-1848-2_4
4. Dubois, D., Prade, H.: A class of fuzzy measures based on triangular norms. Int. J. Gen Syst **8**(1), 43–61 (1982)
5. Dubois, D., Prade, H.: Possibility theory as a basis for qualitative decision theory. In: Proceedings of the IJCAI, vol. 95, pp. 1924–1930 (1995)
6. Dubois, D., Rico, A.: New axiomatisations of discrete quantitative and qualitative possibilistic integrals. Fuzzy Sets Syst. **343**, 3–19 (2018)
7. Fargier, H., Guillaume, R.: Sequential decision making under ordinal uncertainty: a qualitative alternative to the Hurwicz criterion. Int. J. of Approx. Reason. **116**, 1–18 (2020)
8. Greco, S., Matarazzo, B., Giove, S.: The Choquet integral with respect to a level dependent capacity. Fuzzy Sets Syst. **175**(1), 1–35 (2011)
9. Klement, E.P., Mesiar, R., Pap, E.: Triangular Norms. Kluwer Academic, Dordrecht (2000)

10. Raiffa, H.: Decision Analysis: Introductory Lectures on Choices Under Uncertainty. Addison-Wesley, Reading (1968)
11. Shafer, G.: A Mathematical Theory of Evidence. Princeton University Press, Princeton (1976)
12. von Neumann, J., Morgenstern, O.: Theory of Games and Economic Behavior. Princeton University Press, Princeton (1944)
13. Yager, R.R., Rybalov, A.: Uninorm aggregation operators. Fuzzy Sets Syst. **80**(1), 111–120 (1996)

Numerical Comparison of Idempotent Andness-Directed Aggregators

Jozo Dujmović(✉)

Department of Computer Science, San Francisco State University, 1600 Holloway Avenue, San Francisco, CA 94132, USA
jozo@sfsu.edu

Abstract. The goal of this paper is to investigate similarities between andness-directed graded logic aggregators. Our analysis is based on idempotent soft disjunctive aggregators because the differences between hard aggregators are provably smaller. We propose three difference indicators and use them to compare three popular andness-directed aggregators: weighted power mean, exponential mean, and OWA. The results of analysis show that differences between aggregators are regularly low, and frequently negligible, taking into account the limited precision of arguments of graded logic aggregators used in decision support systems. This situation has consequences that are also discussed in the paper.

Keywords: Logic aggregators · Comparison of aggregators · Means · Soft partial disjunction · OWA

1 Introduction

Andness-directed aggregators are graded logic functions that are widely used in decision support systems. According to graded logic conjecture [1], in the process of creating logic aggregation structures, it is necessary and sufficient to have ten basic graded logic functions: seven idempotent aggregators (pure conjunction, hard partial conjunction, soft partial conjunction, neutrality, soft partial disjunction, hard partial disjunction, and pure disjunction), two nonidempotent aggregators (hyperconjunction and hyperdisjunction) and the standard negation. All logic aggregators are special cases of the graded conjunction/disjunction (GCD) function. GCD should be andness-directed and weighted: its properties are adjusted by selecting the desired andness (conjunction degree) and the desired weights (importance degrees) of arguments. This process is consistent with observable properties of intuitive human reasoning. So, it is natural to ask the following questions:

(1) What are the basic aggregators that are suitable for building logic aggregation structures and how they compare to each other?
(2) How many basic aggregators are necessary to satisfy practical needs of decision engineering?

V. Torra and Y. Narukawa (Eds.): MDAI 2021, LNAI 12898, pp. 67–77, 2021.
https://doi.org/10.1007/978-3-030-85529-1_6

In this paper our goal is to provide some answers to these questions.

The most frequently used idempotent special case of GCD is the hard (partial or full) conjunction. In the case of n arguments $\boldsymbol{X} = (x_1, \ldots, x_n), n \geq 2, x_i \in I = [0, 1], i = 1, \ldots, n$, this aggregator has the global andness $\alpha \in [\alpha_\theta, 1], \alpha_\theta \in]0.5, 1[$ and normalized weights $\boldsymbol{W} = (w_1, \ldots, w_n),\ w_i \in]0, 1[, i = 1, \ldots, n,\ w_1 + \ldots + w_n = 1$. The hard conjunction satisfies internality $A(x_1, \ldots, x_n; \boldsymbol{W}, \alpha) \in I$, and supports idempotency and annihilator 0, as follows:

$$\forall x \in I,\ \forall \alpha \in I \Rightarrow\ A(x, \ldots, x; \boldsymbol{W}, \alpha) = x;$$

$$\forall \alpha \in [\alpha_\theta, 1], \forall i \in \{1, \ldots, n\}, x_i = 0 \Rightarrow A(x_1, \ldots, x_n; \boldsymbol{W}, \alpha) = 0;$$

$$\alpha = \frac{n}{n-1} - \frac{n+1}{n-1} \int_{I^n} A\big(x_1, \ldots, x_n; \overline{\boldsymbol{W}}, \alpha\big) dx_1 \ldots dx_n,\ \overline{\boldsymbol{W}} = \left(\frac{1}{n}, \ldots, \frac{1}{n}\right).$$

The most frequent (and the most practical) value of the threshold andness α_θ is 0.75; it provides equal space for soft and hard aggregators. The idempotency condition, and the n variables that support annihilators create $n + 1$ restrictive conditions where *all hard conjunctive aggregators are identical.* In the vicinity of $n + 1$ conditions $x_1 = 0, \ldots, x_n = 0$ and $x_1 = \ldots = x_n$ all hard conjunctive aggregators are "almost identical". Consequently, the inevitable question is what happens in the space between these regions of almost complete sameness? Obviously, the differences between various aggregators are limited. Thus, our goal is to investigate the range of these differences and their impact on results of aggregation in decision support systems.

The rest of this paper is organized as follows. In Sect. 2 we briefly survey the reasons for imprecision of arguments of logic aggregators, and estimate the range of corresponding errors. In Sect. 3 we investigate soft idempotent disjunctive aggregators as the class of aggregators that offers most diversity between individual aggregators. Experiments with such aggregators are presented in Sect. 4 and the corresponding conclusions are presented in Sect. 5.

2 Imprecision and Uncertainty of Human Percepts

Decision making is usually the selection of the most suitable among several alternatives. The percept of suitability is not an objectively measurable physical property – it is created in the human mind, and verbally expressed using a rating scale, i.e. a scale of ordered linguistic labels [2–5]. According to the concept of "magical number seven plus or minus two" [6] such scales can be reliably used if they have up to 5 or 7 or 9 labels. The number of labels is regularly odd to have the mean value (labeled *medium* or *average* or *fair*) present as the median of the scale.

In the case of differentiating 5 degrees of suitability, a convenient rating scale would be [*lowest* < *low* < *medium* < *high* < *highest*]. Typical 7 and 9 degrees rating scales are [*unacceptable* < *very poor* < *poor* < *average* < *good* < *very good* < *excellent*] and [*lowest* < *verylow* < *low* < *mid-low* < *medium* < *mid-high* < *high* < *very high*

$<$ *highest*]. Linguistic labels can be interpreted as crisp values, fuzzy sets with interval, triangular or trapezoid membership functions [1], or in the context of type-2 fuzzy sets [5]. The simplest possible interpretation is that the range [0, 100%] is divided in $n \in \{5, 7, 9\}$ equal intervals and the respective ranges of indistinguishable human percepts are 20%, 14%, or 11% of the total range. In other words, the precision of inputs and parameters of aggregation models is regularly rather low and the variations in absolute value of 10% or more can be frequently encountered. This conclusion is consistent with results reported in [7] and [8].

In real life applications, the input values of logic aggregators are imprecise. Consequently, minor differences in aggregator properties cannot discernibly reduce the applicability of results of aggregation of significantly more imprecise inputs. In the case of aggregators that are imprecise and aggregate imprecise inputs, what is the difference of aggregator properties that can be considered insignificant or significant? This question can be answered based on numeric experiments that are presented below.

Imprecision and uncertainty of human percepts also limits the number of inputs of a single aggregator n. The LSP method [1] suggest $2 \leq n \leq 5$ in order to facilitate the comparison and selection of n weights. For larger values of inputs we assume the associative use of multiple aggregators.

3 The Worst Case Analysis: Soft Idempotent Disjunctive Aggregators

In the case of hard aggregators of n arguments (either conjunctive or disjunctive) each aggregator must satisfy $n + 1$ conditions that have geometric interpretation: the surface of idempotent aggregator $z = A(x_1, \ldots, x_n; \mathbf{W}, \alpha)$ must contain $n+1$ straight lines: hard conjunctive aggregators contain n coordinate axes where $x_i = 0 = z$ and hard disjunctive aggregators contain n lines where $x_i = 1 = z$. Then, all such aggregators contain the idempotency line $x_1 = \ldots = x_n = z$. In addition, all idempotent aggregators for $\alpha = 1$ become min $(x_1, \ldots, x_n)$, for $\alpha = 0$ become $\max(x_1, \ldots, x_n)$, and for $\alpha = 1/2$ become the arithmetic mean $(x_1 + \ldots + x_n)/n$. Therefore, the hard (conjunctive or disjunctive) aggregators share $n + 4$ common properties. It is more than obvious that with so many inevitable common properties such functions must be very similar. Thus, it is justifiable to ask whether in practice we need more than one such an aggregator (with adjustable weights and andness) for aggregating imprecise human-generated inputs.

To answer the question of the reason for variety of aggregators, it is useful to investigate the strictly soft idempotent disjunctive aggregators because their variability is higher than the variability of hard aggregators. Indeed, the soft disjunctive aggregators contain only the idempotency line and reduce to disjunction for $\alpha = 0$, and to the arithmetic mean for $\alpha = 1/2$. Therefore, they share only 3 common properties. The presence of significantly smaller number of common properties makes soft disjunctive aggregators convenient for a worst case analysis: if the differences between various strictly soft idempotent disjunctive aggregators are sufficiently small, then it is reasonable to expect that the differences between hard aggregators and hard/soft aggregators are significantly smaller.

The general conjunction/disjunction (GCD, see segmented interpolative examples in [1, 12]) is a continuous graded logic aggregator that usually has the threshold andness $\alpha_\theta = 0.75$ and satisfies the following hard/soft properties:

$$\begin{array}{llll} A(x_1,\ldots,x_n;\boldsymbol{W},\alpha)=0, & \alpha_\theta \le \alpha \le 1, & x_i=0, & i\in\{1,\ldots,n\}\ (\textit{hard con}),\\ A(x_1,\ldots,x_n;\boldsymbol{W},\alpha)>0, & 0.5<\alpha<\alpha_\theta, & x_i>0, & i\in\{1,\ldots,n\}\ (\textit{soft con}),\\ A(x_1,\ldots,x_n;\boldsymbol{W},\alpha)>0, & 1-\alpha_\theta<\alpha<0.5, & x_i>0, & i\in\{1,\ldots,n\}\ (\textit{soft dis}),\\ A(x_1,\ldots,x_n;\boldsymbol{W},\alpha)=1, & 0\le\alpha\le 1-\alpha_\theta, & x_i=1, & i\in\{1,\ldots,n\}\ (\textit{hard dis}). \end{array}$$

By definition, GCD also satisfies De Morgan duality:

$$A(x_1,\ldots,x_n;\boldsymbol{W},\alpha)=1-A(1-x_1,\ldots,1-x_n;\boldsymbol{W},1-\alpha).$$

Some aggregators are naturally strictly soft disjunctive, i.e. they don't have natural hard disjunctive properties. Such aggregators satisfy the following properties:

$$\begin{array}{lllll} A(x_1,\ldots,x_n;\boldsymbol{W},\alpha)>0, & 0<\alpha<0.5, & x_i>0, & i\in\{1,\ldots,n\} & (\textit{strictly soft dis}),\\ A(x_1,\ldots,x_n;\boldsymbol{W},\alpha)<1, & 0<\alpha<0.5, & x_i<1, & i\in\{1,\ldots,n\} & (\textit{strictly soft dis}). \end{array}$$

Examples of such aggregators are the weighted power mean [9, 10], the weighted exponential mean [1, 9], and OWA [11]. Taking into account that the hard aggregators share $n+4$ common properties, and GCD aggregators combine hard and soft aggregation, it follows that the strictly soft disjunctive aggregators, in the whole range $\alpha \in [0, 0.5]$, share only 3 common properties: idempotency, the reduction to the arithmetic mean, and the reduction to the pure disjunction. Thus, the strictly soft disjunctive aggregators permit the highest variability of aggregators and consequently they can be used as the worst case in the numerical analysis of the differences between similar aggregators.

4 Experiments with Strictly Soft Idempotent Disjunctive Aggregators

In decision support applications, the graded logic aggregators are supposed to be andness directed [12], i.e. to have the global andness as explicitly visible and easily adjustable parameter. Consequently, the analysis of differences between aggregators must be performed for all values of global andness, and in the case of disjunctive aggregators that means for $0 \le \alpha \le 0.5$.

In the simplest case, let us compare three single parameter continuous aggregators: the weighted power mean (WPM), the weighted exponential mean (EXM), and OWA. Other candidate means either don't have adjustable parameters, or are strictly bivariate (e.g. Heronian and Centroidal means), have multiple parameters (e.g. Gini and Stolarsky means), or do not support nondecreasing monotonicity (counter-harmonic mean). Consequently, in the case of equal weights and $n = 2$ we can use the following three means:

$$A_{WPM}(x_1,x_2;\alpha)=\left(0.5x_1^{r_2(\alpha)}+0.5x_2^{r_2(\alpha)}\right)^{1/r_2(\alpha)},\ (\text{hard}:\ r_2(\alpha)\le 0,\ \text{soft}:\ r_2(\alpha)>0)$$

$$A_{EXM}(x_1, x_2; \alpha) = \frac{1}{t_2(\alpha)} ln\left[0.5\exp(t_2(\alpha)x_1) + 0.5\exp(t_2(\alpha)x_2)\right],$$

$$A_{OWA}(x_1, x_2; \alpha) = \alpha \min(x_1, x_2) + (1 - \alpha)\max(x_1, x_2).$$

In the case of OWA the global andness is explicitly visible. For WPM and EXM, functions $\alpha \mapsto r_2(\alpha)$ and $\alpha \mapsto t_2(\alpha)$ are polynomial approximations (taken from [1]), that provide the desired value of andness (andness directedness) of WPM and EXM aggregators. The difference between aggregators for $n \geq 2$ can be expressed using the following three indicators (D, M and V):

$$D_{WPM/EXM}(\alpha) = 100 \int_0^1 \ldots \int_0^1 |A_{WPM}(x_1, \ldots, x_n; \alpha) - A_{EXM}(x_1, \ldots, x_n; \alpha)| dx_1 \ldots dx_n$$

$$D_{OWA/EXM}(\alpha) = 100 \int_0^1 \ldots \int_0^1 |A_{OWA}(x_1, \ldots, x_n; \alpha) - A_{EXM}(x_1, \ldots, x_n; \alpha)| dx_1 \ldots dx_n$$

$$D_{WPM/OWA}(\alpha) = 100 \int_0^1 \ldots \int_0^1 |A_{WPM}(x_1, \ldots, x_n; \alpha) - A_{OWA}(x_1, \ldots, x_n; \alpha)| dx_1 \ldots dx_n$$

$$M_{WPM/EXM}(\alpha) = 100 \max_{0 \leq x_1 \leq 1, \ldots, 0 \leq x_n \leq 1} |A_{WPM}(x_1, \ldots, x_n; \alpha) - A_{EXM}(x_1, \ldots, x_n; \alpha)|$$

$$M_{OWA/EXM}(\alpha) = 100 \max_{0 \leq x_1 \leq 1, \ldots, 0 \leq x_n \leq 1} |A_{OWA}(x_1, \ldots, x_n; \alpha) - A_{EXM}(x_1, \ldots, x_n; \alpha)|$$

$$M_{WPM/OWA}(\alpha) = 100 \max_{0 \leq x_1 \leq 1, \ldots, 0 \leq x_n \leq 1} |A_{WPM}(x_1, \ldots, x_n; \alpha) - A_{OWA}(x_1, \ldots, x_n; \alpha)|$$

$$V_{WPM/EXM}(\alpha) = 100 \int_0^1 \ldots \int_0^1 \frac{|A_{WPM}(x_1, \ldots, x_n; \alpha) - A_{EXM}(x_1, \ldots, x_n; \alpha)|}{A_{WPM}(x_1, \ldots, x_n; \alpha) + A_{EXM}(x_1, \ldots, x_n; \alpha)} dx_1 \ldots dx_n$$

$$V_{OWA/EXM}(\alpha) = 100 \int_0^1 \ldots \int_0^1 \frac{|A_{OWA}(x_1, \ldots, x_n; \alpha) - A_{EXM}(x_1, \ldots, x_n; \alpha)|}{A_{OWA}(x_1, \ldots, x_n; \alpha) + A_{EXM}(x_1, \ldots, x_n; \alpha)} dx_1 \ldots dx_n$$

$$V_{WPM/OWA}(\alpha) = 100 \int_0^1 \ldots \int_0^1 \frac{|A_{WPM}(x_1, \ldots, x_n; \alpha) - A_{OWA}(x_1, \ldots, x_n; \alpha)|}{A_{WPM}(x_1, \ldots, x_n; \alpha) + A_{OWA}(x_1, \ldots, x_n; \alpha)} dx_1 \ldots dx_n$$

$$\overline{D}_{WPM/EXM} = 2 \int_0^{0.5} D_{WPM/EXM}(\alpha) d\alpha, \; \overline{D}_{OWA/EXM} = 2 \int_0^{0.5} D_{OWA/EXM}(\alpha) d\alpha,$$

$$\overline{D}_{WPM/OWA} = 2 \int_0^{0.5} D_{WPM/OWA}(\alpha) d\alpha,$$

$$\overline{M}_{WPM/EXM} = 2 \int_0^{0.5} M_{WPM/EXM}(\alpha) d\alpha, \; \overline{M}_{OWA/EXM} = 2 \int_0^{0.5} M_{OWA/EXM}(\alpha) d\alpha,$$

$$\overline{M}_{WPM/OWA} = 2 \int_0^{0.5} M_{WPM/OWA}(\alpha) d\alpha$$

$$\overline{V}_{WPM/EXM} = 2 \int_0^{0.5} V_{WPM/EXM}(\alpha) d\alpha, \; \overline{V}_{OWA/EXM} = 2 \int_0^{0.5} V_{OWA/EXM}(\alpha) d\alpha,$$

$$\overline{V}_{WPM/OWA} = 2 \int_0^{0.5} V_{WPM/OWA}(\alpha) d\alpha,$$

The mean difference indicators $D(\alpha)$ show the average proximity between two aggregators. The maximum difference indicators $M(\alpha)$ show the maximum distance between two aggregators inside the whole aggregation domain $[0, 1]^n$. Similarly to the statistical coefficient of variation, the mean variation $V(\alpha)$ shows the mean ratio of the absolute half-difference of two values and their mean values. E.g., for values p and q, their mean value is $(p+q)/2$ and the deviation from that value is $|p-q|/2$, yielding the percent variation $100|p-q|/(p+q)$. To summarize results for all values of andness, we use the overall mean values $\overline{D}, \overline{M}, \overline{V}$ which express the differences between analyzed aggregators using a single numeric indicator. For convenience, all indicators use the multiplier 100, to be expressed as percentages of the whole range, which is 100%.

The results of aggregator comparison, as functions of andness α, are presented in Fig. 1. The consequences of comparison of the hard partial conjunction (WPM) and the soft partial conjunction (either EXM or OWA) are presented in Figs. 1a and b. Unsurprisingly, for $\alpha > 0.5$, the differences are very large because soft EXM and OWA cannot provide properties expected from hard aggregators. However, for $0 \leq \alpha \leq 0.5$, WPM, EXM, and OWA are soft and the differences between these aggregators are sufficiently small, as shown in Figs. 1c–f. The soft aggregators EXM and OWA are self-dual, and therefore comparable in the whole range $0 \leq \alpha \leq 1$, as shown in Fig. 1c.

Assuming that inputs of aggregators are imprecise at the level of 10% or more, it follows that the analyzed three aggregators of two arguments, in the area of soft partial disjunction, are sufficiently precise and practically equivalent. In other words, there are no reasons to use more than one of them and evaluators can easily adapt to both the properties and the use of the selected aggregator.

In the case of three arguments and $\alpha \in [0, 0.5]$, we use the following andness directed strictly soft aggregators:

$$A_{WPM}(x_1, x_2, x_3; \alpha) = \left(3^{-1}x_1^{r_3(\alpha)} + 3^{-1}x_2^{r_3(\alpha)} + 3^{-1}x_3^{r_3(\alpha)}\right)^{1/r_3(\alpha)},$$

$$A_{EXM}(x_1, x_2, x_3; \alpha) = \frac{1}{t_3(\alpha)} ln\Big[3^{-1}\exp(t_3(\alpha)x_1) + 3^{-1}\exp(t_3(\alpha)x_2) + 3^{-1}\exp(t_3(\alpha)x_3)\Big].$$

$$\begin{aligned} &A_{OWA}(x_1, x_2, x_3; \alpha) \\ &= ((((0.8916\alpha - 1.7754)\alpha + 1.7641)\alpha - 1.8798)\alpha + 0.9994) \max(x_1, x_2, x_3) \\ &+((((-1.7791\alpha + 3.5585)\alpha - 3.5471)\alpha + 1.7677)\alpha + 0.0007) \operatorname{mid}(x_1, x_2, x_3) \\ &+((((0.8876\alpha - 1.7832)\alpha + 1.783)\alpha + 0.1121)\alpha - 0.00009) \min(x_1, x_2, x_3) \end{aligned}$$

We now use the andness directed soft iterative OWA proposed in [13] and derived in the presented form in [14], and polynomial approximations of functions $\alpha \mapsto r_3(\alpha)$ and $\alpha \mapsto t_3(\alpha)$ taken from [1]. The main results of comparison of these aggregators are shown in Fig. 2 and Table 1.

Using the results presented in Fig. 1 and Fig. 2 and their summary shown in Table 1, we can derive the following observations:

1. As expected, all differences between strictly soft idempotent disjunctive aggregators depend on andness and have consistent shapes that attain maximum values in the middle area of the [0, 0.5] domain of andness.

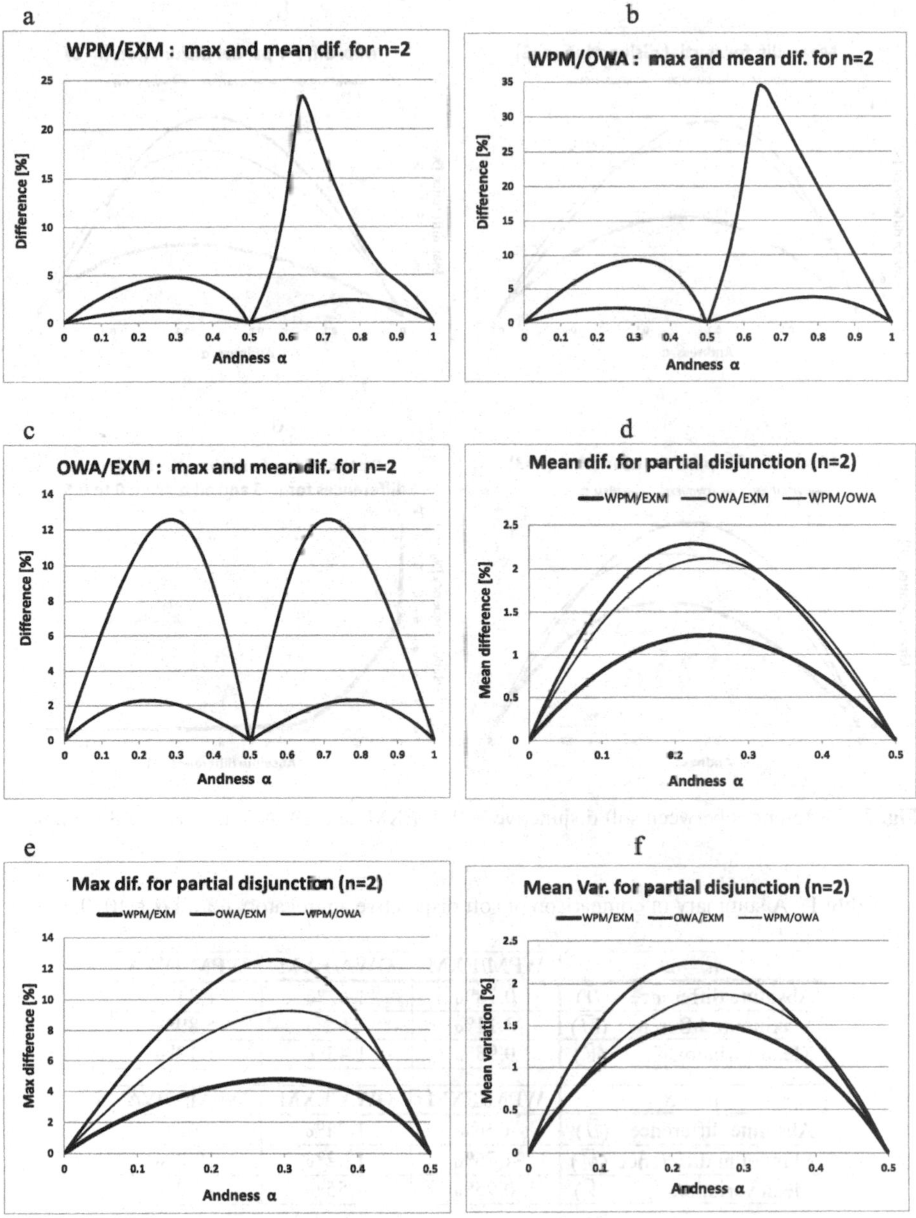

Fig. 1. Differences between WPM, EXM, and OWA in the case of 2 variables

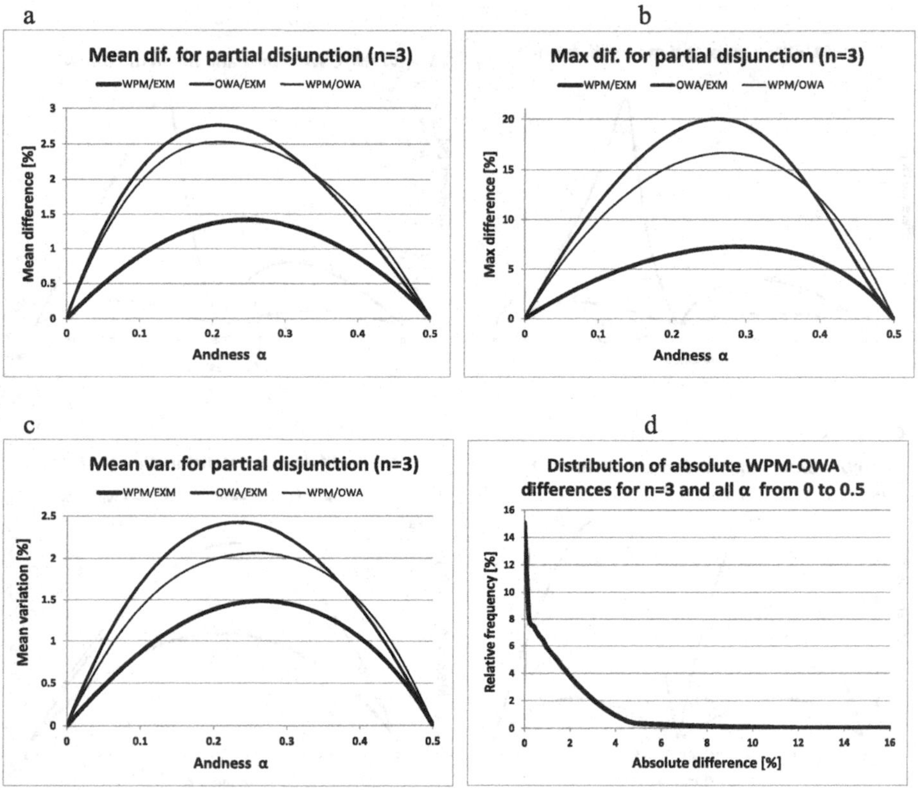

Fig. 2. Differences between soft disjunctive WPM, EXM, and OWA in the case of 3 variables

Table 1. A summary of comparison of soft disjunctive aggregators for all $\alpha \in [0, 0.5]$

$n = 2$	WPM/EXM	OWA/EXM	WPM/OWA
Absolute difference ($\overline{D}$)	0.78%	1.42%	1.34%
Maximum difference ($\overline{M}$)	3.11%	7.66%	5.89%
Mean variation ($\overline{V}$)	0.98%	1.43%	1.19%

$n = 3$	WPM/EXM	OWA/EXM	WPM/OWA
Absolute difference ($\overline{D}$)	0.9%	1.74%	1.67%
Maximum difference ($\overline{M}$)	4.76%	12.2%	10.8%
Mean variation ($\overline{V}$)	0.95%	1.55%	1.39%

2. Mean overall differences between analyzed aggregators ($\overline{D}, \overline{M}, \overline{V}$) for soft partial disjunction are very low (e.g. the overall mean values of absolute difference and variation, computed for all values of andness in Table 1, are below 2%). These values should be compared to imprecision caused by uncertainty of input arguments, which is regularly of the order of 5–15%.

3. Distribution of absolute differences between analyzed aggregators, for all values of andness, has a characteristic shape exemplified in Fig. 2d. Small differences are very frequent and large differences are very rare and can be considered exceptional cases (for $n = 3$ almost all differences are below 4%).
4. Maximum differences between analyzed aggregators (shown in Figs. 1c and e, Fig. 2b and Table 1) can attain values comparable or above the imprecision of input arguments, but according to Fig. 2d they are expected to be unlikely exceptional cases.
5. The most similar disjunctive aggregators are WPM and EXM. Their difference from OWA is slightly larger, but still not significant.
6. The differences between aggregators slightly increase when the number of variables increases, but not to the extent that indicates the possibility of different conclusions of our analysis.
7. Mean difference variation indicators preserve their values when the number of variables increases. Their overall mean values are regularly below 2%.

5 Conclusions

Graded logic aggregators are the class of idempotent aggregators characterized by andness. They can be conjunctive (models of simultaneity) or disjunctive (models of substitutability), and hard (supporters of annihilators) or soft (without support for annihilators). That gives four basic combinations, observable in human intuitive reasoning: hard conjunctive, soft conjunctive, soft disjunctive and hard disjunctive aggregators. Such aggregators can be viewed from two different points of view: theoretical and practical.

Theoretical interest does not need to be justified by high applicability of theoretical results. Thus, the theory of aggregation is a wide area with thousands of research contributions. The practical value of logic aggregators is much more modest, and they are primarily used in decision making, as components of decision support systems. In this modest role, graded logic aggregators must satisfy the minimum set of six fundamental necessary conditions: they must be andness-directed (having andness as an explicitly visible and easily adjustable parameter), and providing four fundamental combinations of conjunctive/disjunctive and soft/hard behavior, as well as noncommutativity based on adjustable importance weights.

The basic logic aggregators are idempotent, and consequently, they are means. There are hundreds of known means and none of them naturally satisfies the minimum set of six necessary conditions of graded logic aggregation. This inconvenient situation is the motivation for building logic aggregators as interpolative combinations of segments that satisfy some of the six necessary conditions [12]. So, we face the problem of selecting the most suitable member functions supporting some of desired properties. In other words, we need to compare similar aggregators and analyze differences between them. That is the motivation for this paper.

The largest differences can be expected from aggregators that have the least restrictive properties. Among logic aggregators the least restrictive are soft aggregators because they are not required to support annihilators. The graded logic aggregators are regularly defined as functions that satisfy De Morgan duality. Consequently, the special case of the soft partial disjunction which is strictly soft in the whole domain of andness from 0 to 0.5, is the most suitable for studying the differences between aggregators because there are several well-known functions in this category. It is convenient to base such analysis on the weighted power mean, exponential mean, and OWA.

The numerical analysis of differences between the selected three popular soft disjunctive aggregators shows a very high similarity between seemingly different functions. Of course, it is necessary to have a criterion for declaring that a difference is high or low. Such criterion is naturally based on imprecision of arguments that are in decision support systems aggregated by logic aggregators. Based on the wide use of rating scales, the imprecision of inputs is estimated to be close to 10% and if the soft aggregators on the average differ for less than 2%, then such aggregators can be considered equivalent and equally applicable from the practical point of view. This conclusion is additionally supported by the fact that real interpolative aggregators include hard segments that yield higher similarity than the strictly soft aggregators. In other words, in practical evaluation projects, the number of candidate graded logic aggregators is very small, and to satisfy necessary logic conditions, they must be combined in segmented interpolative structures. From this standpoint, it is also possible to conclude that individual logic aggregators frequently attract more research attention than they deserve.

The future theoretical work in this area could include more detailed studies of distributions of difference between selected aggregators, and identification of infrequent special cases of larger differences, as well as analyses of differences for $n > 3$, and studying the impact of weights. That might be of interest in some applications where it is provable that it is not necessary to support both soft and hard aggregators. The numerical results of this paper indicate that such attention would be undeserved for most applications in graded logic and decision support systems, where aggregators must be conjunctive and disjunctive, soft and hard, commutative and noncommutative, andness-directed, idempotent and nonidempotent, with adjustable threshold andness and orness. These conditions can only be satisfied by segmented interpolative aggregators, and not by a single aggregation function of n variables. In such cases, all complex aggregation structures are built using superposition of aggregators with small number of inputs that support simplicity, specifiability, and readability.

References

1. Dujmović, J.: Soft Computing Evaluation Logic. Wiley and IEEE Press, Hoboken (2018)
2. Rohrmann, B.: Verbal qualifiers for rating scales: Sociolinguistic and psychometric data. Project report, University of Melbourne, Australia (2007). http://rohrmannresearch.net/pdfs/rohrmann-vqs-report.pdf
3. Friedman, H.H., Amoo, T.: Rating the rating scales. J. Market. Manag. **9**(3), 114–123 (1999)
4. Smith, W.D.: Rating Scale Research relevant to score voting (2013). http://rangevoting.org/RateScaleResearch.html
5. Mendel, J.M., Wu, D.: Perceptual Computing. IEEE Press and J, New York (2010)
6. Miller, G.A.: The magical number seven, plus or minus two: some limits on our capacity for processing information. Psychol. Rev. **63**, 81–97 (1956)
7. Stewart, T.J.: Robustness of additive value function methods in MCDM. J. Multi-Criteria Decis. Anal. **5**, 301–309 (1996)

8. Dujmović, J.J., Fang W.Y.: An empirical analysis of assessment errors for weights and andness in LSP criteria. In: Torra, V., Narukawa, Y. (eds.) Modeling Decisions for Artificial Intelligence. MDAI 2004. Lecture Notes in Computer Science, vol. 3131, pp. 139–150 Springer, Berlin, Heidelberg (2004). https://doi.org/10.1007/978-3-540-27774-3_14
9. Bullen, P.S.: Handbook of Means and Their Inequalities, Kluwer, Dordrecht (2003 and 2010)
10. Beliakov, G., Bustince Sola, H., Calvo Sanchez, T.: A Practical Guide to Averaging Functions. Studies in Fuzziness and Soft Computing, p. 329. Springer, Cham (2016)
11. Yager, R.R.: On ordered weighted averaging aggregation operators in multi-criteria decision making. IEEE Trans. SMC **18**, 183–190 (1988)
12. Dujmović, J., Torra, V.: Properties and comparison of andness-characterized aggregators. Int. J. Intell. Syst. **36**(3), 1366–1385 (2021)
13. Dujmović, J.: Preferential neural networks. In: Antognetti, P., Milutinović, V. (eds.) Chapter 7 in Neural Networks - Concepts, Applications, and Implementations, vol. II. Prentice-Hall, Hoboken, pp. 155–206 (1991)
14. Dujmović, J.: Andness Directed Iterative OWA Aggregators. In: MDAI Proceedings (2021)

Approximate Reasoning

Multiple Testing of Conditional Independence Hypotheses Using Information-Theoretic Approach

Małgorzata Łazęcka[1,2] and Jan Mielniczuk[1,2(✉)]

[1] Institute of Computer Science, Polish Academy of Sciences, Jana Kazimierza 5, 01-248 Warsaw, Poland
{malgorzata.lazecka,jan.mielniczuk}@ipipan.waw.pl
[2] Faculty of Mathematics and Information Science, Warsaw University of Technology, Koszykowa 75, 00-662 Warsaw, Poland

Abstract. In the paper we study the multiple testing problem for which individual hypotheses of interest correspond to conditional independence of the two variables X and Y given each of the several conditioning variables. Approaches to such problems avoiding inflation of probability of spurious rejections are widely studied and applied. Here we introduce a direct approach based on Joint Mutual Information (JMI) statistics which restates the problem as a problem of testing of a single hypothesis. The distribution of the test statistics JMI is established and shown to be well numerically approximated for a single data sample. The corresponding test is studied on artificial data sets and is shown to work promisingly when compared to general purpose multiple testing methods such as Bonferroni or Simes procedures.

Keywords: Conditional independence · Joint mutual information · Multiple testing · Weighted chi square distribution · Dichotomous behaviour · Markov blanket · Dependence analysis

1 Introduction

We focus here on multiple testing problem consisting in testing of conditional independence of two random variables given the third one, the later belonging to a group of variables of interest. The applications in this context are wide ranging. In studying human diseases one might be interested in checking whether occurrences of two diseases are independent given a third disease, where the latter belongs to the group of diseases of interest possibly interacting with the first two. The same question may be asked when conditioning variables are characteristics of a patient such as age, gender or results of medical tests. Formally, the problem can be stated as testing p individual hypotheses

$$H_{0,i}:\ X \text{ and } Y \text{ are conditionally independent given } Z_i, \qquad (1)$$

V. Torra and Y. Narukawa (Eds.): MDAI 2021, LNAI 12898, pp. 81–92, 2021.
https://doi.org/10.1007/978-3-030-85529-1_7

where $X \in \mathcal{X}$, $Y \in \mathcal{Y}$ and $Z_i \in \mathcal{Z}_i$ are some observed discrete random variables for $i = 1, \ldots, p$. We want to construct a test which controls type I error under so called global null $H_0 = \cap_{i=1}^p H_{0,i}$ when all null hypotheses are true; i.e. we stipulate that $P(V \geq 1, H_{0,i} \text{ true}, i = 1, \ldots, p)$ is smaller than the fixed level of significance α when V is the number of rejected hypotheses. Note also that simultaneous testing of (1) may be used as a proxy for testing the hypothesis $\tilde{H}_0 : X \perp Y | Z_1, Z_2, \ldots, Z_p$ i.e. conditional independence of X and Y given all $Z_1, \ldots, Z_p$. This is beneficial in the cases when the number of observations per one cell of $Z_1 = z_1, \ldots, Z_p = z_p$ is small and the conditional independence tests loose power due to the curse of dimensionality. E.g. it is usually advised to have 5 observations per cell while using conditional chi-square test which results in number of observations at least 5×2^p on average when all variables are binary, whereas the use of the proposed test will require much less observations as the conditioning is done given individual variables. As a toy example of such situation consider binary random variables $Z_0, Z_1, \ldots, Z_{p+1}$, where $Z_0 = Y$ and $Z_{p+1} = X$ which form a Markov chain $Z_0 \to Z_1 \ldots \to Z_{p+1}$ such that $P(Z_{i+1} = 1 | Z_i = k)$ is q or $1 - q$ depending on whether $k = 1$ or $k = 0$. Then X and Y are conditionally independent given any individual Z_i, $i = 1, \ldots, p$ but they are dependent.

Note that the problem of testing H_0 is a special case of the multiple testing problem, which due to its importance is analysed intensively in machine learning and statistics [1,2,7,13]. There are several off-the-shelf generic methods of testing multiple hypotheses $H_{0,i}$ such as Bonferroni correction or Simes method described below which are known to perform well when test statistics for individual tests are mutually independent. This in case of testing (1) is hardly realistic and would require having independent samples for testing the individual hypothesis (see [11]). In general such methods may perform rather poorly at detecting violations of H_0 when no strong signal is available for any i resulting in low rejection rate in such situation. Thus true weak associations may be overlooked. In the special case of testing (1) for all i we show that it is possible to design a special purpose test statistic which would control type I error rate and have high true rejection rate when moderate and weak signals occur.

The paper is structured as follows: we introduce some information-theoretic concepts and define Joint Mutual Information (JMI) statistic designed for testing H_0. In Sect. 3 we establish asymptotic distribution of sample JMI which leads to a novel test of H_0 (Sect. 4). In Sect. 5 the behaviour of the test procedure is investigated using synthetic and real data sets. The main contribution is to show that introduced JMI-based test of simultaneous conditional independence usually works better that the generic tests.

2 Preliminaries

2.1 Conditional Mutual Information

We introduce some information theoretic concepts leading to the conditional mutual information definition for discrete random variables.

We denote by $p(x) := P(X = x)$, $x \in \mathcal{X}$ a probability mass function corresponding to X, where $\mathcal{X}$ is a domain of X and $|\mathcal{X}|$ is its cardinality. Joint probability will be denoted by $p(x, y) = P(X = x, Y = y)$ and $p(x, y|z)$ is $P(X = x, Y = y|Z = z)$. The sample estimate of $p(x)$ is denoted by $\hat{p}(x)$.

The mutual information (MI) between X and Y is

$$I(X, Y) = H(X) - H(X|Y) = \sum_{x,y} p(x, y) \log \frac{p(x, y)}{p(x)p(y)}, \tag{2}$$

where $H(X)$ and $H(X|Y)$ are the entropy and the conditional entropy, respectively [5]. This can be interpreted as the amount of uncertainty in X which is removed when Y is known, which is consistent with an intuitive meaning of mutual information as the amount of information that one variable provides about another. MI is non-negative and equals 0 if and only if X and Y are independent. We can extend the definition of $I(X, Y)$ to the conditional mutual information $I(X, Y|Z = z)$ of X and Y given $Z = z$ by replacing unconditional probabilities appearing in (2) by their conditional counterparts given $Z = z$. Then averaging $I(X, Y|Z = z)$ wrt distribution of Z yields the conditional mutual information (CMI)

$$I(X, Y|Z) = E_{Z=z}(I(X, Y|Z = z)) = \sum_{x,y,z} p(x, y, z) \log\left(\frac{p(x, y|z)}{p(x|z)p(y|z)}\right). \tag{3}$$

It follows from $MI = 0$ being equivalent to independence that $I(X, Y|Z) = 0$ if and only if X and Y are conditionally independent given Z which will be denoted by $X \perp Y|Z$ in the following. The construction of the test statistic JMI below relies on this fundamental fact. Moreover, the following chain rule holds:

$$I((X, Z), Y) = I(X, Y) + I(X, Y|Z). \tag{4}$$

For more properties of the basic measures described above we refer to [5].

2.2 Multiple Conditional Independence Testing and JMI Statistic

Intuitively, specially designed statistic should measure the cumulative effect of violating several null hypotheses $H_{0,i}$. In accordance with this heuristics we define

$$JMI = \frac{1}{p}\sum_{i=1}^{p} I(X, Y|Z_i). \tag{5}$$

Note that as the summands in (5) are non-negative, JMI averages violation effects of $H_{0,i}$. Note that for $p = 1$ JMI reduces to CMI. JMI has been introduced in [15] in the context of feature selection when Y is a target variable to be explained by a subset of potentially useful predictors, $(Z_i)_{i=1}^p$ are predictors already chosen and X is a potential candidate. It is also shown to be an approximation of $I(X, Y|Z_1, \ldots, Z_p)$ under certain dependence conditions imposed on $(X, Y, Z_1, \ldots, Z_p)$ [14]. We stress however, that testing H_0

is not equivalent to testing $\bar{H}_0$ of conditional independence of X and Y given $(Z_1,\ldots,Z_p)$ although for many dependence structures the former implies the latter (see e.g. [4], Sect. 13.6). We also note that testing H_0 requires less data than testing $\bar{H}_0$ as the number of elements satisfying $Z_i = z_i$ for fixed z_i is usually larger than satisfying $Z_1 = z_1, Z_2 = z_2, \ldots Z_p = z_p$. The following lemma states some properties of JMI (with the proof confined to the online supplement[1]). Define χ^2 measure of conditional dependence between X and Y given Z_i as

$$\chi_i^2 = \sum_{x,y,z_i} \frac{(p(x,y,z_i) - p(x|z_i)p(y|z_i)p(z_i))^2}{p(x|z_i)p(y|z_i)p(z_i)}.$$

Lemma 1.

$$(i) \qquad JMI = 0 \iff H_0 = \cap_i H_{0i} \text{ holds},$$

$$(ii) \qquad JMI = I(X,Y) + \frac{1}{p}\sum_{i=1}^{p}(I(X,Z_i|Y) - I(X,Z_i)),$$

$$(iii) \qquad \frac{1}{2}\sum_{i=1}^{p}\Big(\sum_{x,y,z_i}|p(x,y,z_i) - p(x|z_i)p(y|z_i)p(z_i)|\Big)^2 \le p \times JMI \le \sum_{i=1}^{p}\log(\chi_i^2 + 1)$$

and both inequalities are tight when H_0 holds.

Observe that statistics defined as $\sum_{i=1}^{p}\chi_i^2$ also enjoys analogous property to (i).

Let us mention that JMI statistic is frequently used in feature selection and Markov blanket discovery (see e.g. [3]) in order to test conditional independence of the response and the candidate predictor given the already chosen predictors. Here our aim is different as we want to test multiple individual conditional independence hypotheses. Given a sample $(X_i, Y_i, Z_i), i = 1,\ldots,n$ of independent observations sampled from distribution $P_{X,Y,Z}$ we denote by $\widehat{JMI}$ plug-in counterpart of JMI defined above obtained by replacing $I(X,Y|Z_i)$ by their empirical versions $\hat{I}(X,Y|Z_i)$. For $p = 1$ $\widehat{JMI}$ reduces to the empirical CMI. In this case, provided conditional independence of X and Y given Z holds, it is asymptotically chi square distributed with $(|\mathcal{X}|-1)(|\mathcal{Y}|-1)(|\mathcal{Z}|)$ degrees of freedom (see e.g. [10]). We will derive the distribution of $\widehat{JMI}$ in the next section: note that it does not follow in straightforward manner from the latter result as the summands $\hat{I}(X,Y|Z_i)$ of $\widehat{JMI}$ are dependent.

3 Main Result: Dichotomous Behaviour of Test Statistic Statistic $\widehat{JMI}$

In the following we explicitly state the asymptotic distribution of $\widehat{JMI}$ when H_0 holds. The general formula for distribution of $\widehat{JMI}$ has been already stated in [9]. We derive below its explicit form which is amenable to computations for

[1] github.com/lazeckam/JMI_GlobalNull.

moderate p and derive some of its properties. Moreover, we indicate that when H_0 fails the behaviour of $\widehat{JMI}$ and its distribution is fundamentally different from that under H_0 suggesting that the resulting test should have a reasonable power.

Let $K = |\mathcal{X}| \times |\mathcal{Y}| \times \prod_{i=1}^{p} |\mathcal{Z}_i|$ be the number of levels of random variable $(X, Y, Z_1, \ldots, Z_p)$ and $z = (z_1, \ldots, z_p)$. Let $A_{x,y,z}^{x',y',z'}$ denote the element of $K \times K$ matrix A with the row index x, y, z and the column index x', y', z'. Finally, $H_{0,i}^c$ is the opposite hypothesis to $H_{0,i}$. $\mathbb{I}(A)$ denotes the indicator function of set A:

Theorem 1. *(i) Assume that the global null H_0 holds. Then*

$$2n\widehat{JMI} \xrightarrow{d} \sum_{i=1}^{K} \lambda_i(M) Z_i^2, \tag{6}$$

where Z_i are independent $N(0,1)$ random variables and $\lambda_i(M), i = 1, \ldots, K$ are eigenvalues of matrix M with the elements

$$M_{x,y,z}^{x',y',z'} = \frac{1}{p} p(x', y', z') \sum_{i=1}^{p} \left[\frac{\mathbb{I}(z_i = z_i')}{p(z_i)} - \frac{\mathbb{I}(x = x', z_i = z_i')}{p(x, z_i)} \right. \tag{7}$$

$$\left. - \frac{\mathbb{I}(y = y', z_i = z_i')}{p(y, z_i)} + \frac{\mathbb{I}(x = x', y = y', z_i = z_i')}{p(x, y, z_i)} \right],$$

where $z = (z_1, \ldots, z_p)$ and $z' = (z_1', \ldots, z_p')$. Moreover, the trace of M equals $p^{-1}(|\mathcal{X}| - 1)(|\mathcal{Y}| - 1) \sum_i |\mathcal{Z}_i|$.

(ii) Assume that the alternative $H_1 = \cup_{i=1}^{p} H_{0,i}^c$ to the global null is valid and Y is binary. Then

$$\sigma_{\widehat{JMI}}^2 = \mathrm{Var}\left(\frac{1}{p} \log \prod_{i=1}^{p} \frac{p(X, Y, Z_i) p(Z_i)}{p(X, Z_i) p(Y, Z_i)} \right) > 0$$

and

$$n^{1/2}(\widehat{JMI} - JMI) \xrightarrow{d} N(0, \sigma_{\widehat{JMI}}^2). \tag{8}$$

The result above states an exact dichotomy of asymptotic behaviour which makes the construction of the test possible: the asymptotic distribution of $\widehat{JMI}$ is either that of quadratic form in normal variables as in (6) or normal (cf. (8)) depending on whether H_0 is satisfied or not.

Proof. (i) Let $f(p) = p^{-1} \sum_{i=1}^{p} p(x, y, z_i) \log(p(x, y, z_i) p(z_i) / p(x, z_i) p(y, z_i))$, where $p = p(x, y, z_1, \ldots, z_p)$. Note that when H_0 holds then $\sigma_{\widehat{JMI}}^2 = 0$ and it follows from the delta method (cf. Corollary 1 in [9]) that the asymptotic distribution of $2n\widehat{JMI}$ is the distribution of $Z^T M Z$ where $Z \in R^p$ has $N(0, I)$ distribution, $M = H\Sigma$, $\Sigma_{xyz}^{x'y'z'} = p(x', y', z')(I(x = x', y = y', z = z') - p(x, y, z))/n$ and $H = D^2 f(p)$ is the Hessian of $f(p)$. By direct calculation we have

$$Df(p)_{xyz} = \frac{1}{p} \sum_{i=1}^{p} \log \left(\frac{p(x, y, z_i) p(z_i)}{p(x, z_i) p(y, z_i)} \right),$$

$$H_{xyz}^{x'y'z'} = D^2 f(p)_{xyz}^{x'y'z'} = \frac{1}{p}\sum_{i=1}^{p}\left[\frac{\mathbb{I}(z_i = z_i')}{p(z_i)} - \frac{\mathbb{I}(x = x', z_i = z_i')}{p(x, z_i)}\right.$$
$$\left. - \frac{\mathbb{I}(y = y', z_i = z_i')}{p(y, z_i)} + \frac{\mathbb{I}(x = x', y = y', z_i = z_i')}{p(x, y, z_i)},\right]$$

where $z = (z_1, \ldots, z_p)$ and M is obtained by the direct multiplication of H and Σ resulting in (7). The trace of M equals $p^{-1}\sum_{x,y,z}\sum_{i=1}^{p}\big(p(x, y|z_i) - p(x|y, z_i) - p(y|x, z_i) + 1\big)$ which yields the result. (ii) is proved in Corollary 1 in [9].

4 Asymptotic Versus Generic Methods

4.1 Asymptotic Method

For a given sample chosen from P_{XYZ} we calculate $\widehat{JMI}$ and plug-in estimator $\widehat{M}$ of matrix M defined in Theorem 1. We use now the fact that the asymptotic distribution W of $\widehat{JMI}$ under H_0 given in (6) is determined by the eigenvalues $\lambda_i(M)$ and we approximate it by $\widehat{W}$ plugging in $\lambda_i(\widehat{M})$ for $\lambda_i(M)$, where $\lambda_i(\widehat{M})$ are numerically calculated. Then the rejection region for a given significance level α is given by $\{\widehat{JMI} \geq q_{\widehat{W},1-\alpha}\}$, where $q_{\widehat{W},1-\alpha}$ is quantile of the order $1-\alpha$ of $\widehat{W}$. A function `eigen` from R package `base` has been used to calculate the eigenvalues and package `CompQuadForm` [6] for quantiles of $\hat{W}$.

4.2 Generic Methods

We use two generic methods to cope with controlling type I error while performing multiple tests, namely Bonferroni correction and Simes method (see e.g. [12] and [7]).

- Bonferroni correction: individual tests are performed with level of significance α/p, where p is the number of tests performed thus bounding probability $P(V \geq 1, \forall_i\, H_{0i}\,\text{true})$ by α. It is known to work well when the test statistics used to test individual hypotheses are independent, but in a general case is conservative leading to the low power when H_0 fails. Individual tests are $\widehat{MI}$-based tests based on chi square benchmark distribution described at the end of Sect. 2.2.
- Simes method: p-values of individual test $p_1, \ldots, p_p$ are calculated and ordered: $p_{(1)} \leq p_{(2)} \leq \cdots \leq p_{(p)}$. H_0 is rejected when for certain $i \leq p$ we have $p_{(i)} \leq i\alpha/p$, or equivalently if $\min_i p_{(i)}/i \leq \alpha/p$. Individual tests considered are the same as for Bonferroni correction method.

5 Simulation Study

5.1 Artificial Data Sets

We discuss first the dependence structures which we use to generate data (see Fig. 1). Below $Z \sim Bern(p)$ stands for Z being distributed as the Bernoulli distribution with probability of success p and Φ is the cumulative distribution function (CDF) of the standard normal distribution.

- **Model A.** Parameters: $\alpha_x \geq 0$, $\alpha_y \geq 0$
  $Z_i \sim Bern(0.5)$ for $i \in \{1, 2, \ldots, p\}$, $\bar{Z} := \frac{1}{p}\sum_{i=1}^{p} Z_i$
  $X|\bar{Z} = z \sim Bern\left(1 - \Phi(\alpha_x(\frac{1}{2} - z))\right)$, $Y|\bar{Z} = z \sim Bern\left(1 - \Phi(\alpha_y(\frac{1}{2} - z))\right)$
  Model A' is a modification of model **Model A** for which $\bar{Z} := \frac{1}{s}\sum_{i=1}^{s} Z_i$ and $s < p$. $Z_i \perp (X, Y)$ for $i \in \{s+1, s+2, ..., p\}$.
- **Model B.** Parameters: $\alpha_x \in [0, 1]$, $\alpha_z \geq 0$
  $X \sim Bern(0.5)$ and $Y \sim Bern(0.5)$,
  $Z_i|(\alpha_x X + (1 - \alpha_x)Y) = w \sim Bern(1 - \Phi(\alpha_z(\frac{1}{2} - w))$ for $i \in \{1, 2, ..., p\}$
  Model B' is a modification of model **Model B** for which the dependence of Z_i on X and Y defined above holds only for $i \in \{1, 2, ..., s\}$, for $i \in \{s+1, s+2, ..., p\}$ $Z_i \perp (X, Y)$ and $Z_i \sim Bern(0.5)$
- **Models C.** Parameters: $q \in [0, 1]$, $q_{XY} \in [0, 1]$
  $\mathbf{C(q)}$: $Y \sim Bern(0.5)$, $Z_1|Y = y \sim Bern(q^y(1-q)^{1-y})$, $Z_{i+1}|Z_i = z \sim Bern(q^z(1-q)^{1-z})$ for $i \in \{1, 2, \ldots, p\}$ and $Z_{p+1} = X$.
  $\mathbf{C(q, q_{XY})}$: while retaining the conditional distribution $P_{X|Z_p}$ as above, the distribution of (X, Y) is modified so that H_1 is satisfied:
  $P(X = z, Y = z|Z_p = z) = q - P(X = z, Y = 1 - z|Z_p = z) = q_{XY}q$,
  $P(X = 1 - z, Y = 1 - z|Z_p = z) = 1 - q - P(X = 1 - z, Y = z|Z_p = z) = q_{XY}(1 - q)$.

Model A corresponds to the situation when variables $Z_1, \ldots, Z_p$ influence X and Y simultaneously. Parameters α_x and α_y control how strong the dependence between the variables Z_i and X or Y is. If at least one of the parameters equals zero then X and Y are independent and conditionally independent given any Z_i, otherwise X and Y are (conditionally) dependent. In Model A$'$ the role of parameter p is taken over by s and the additional variables $Z_i, i = s+1, \ldots, p$ are independent of X and Y. In Model B the dependence structure is reversed and both variables X and Y influence variables Z_i. The parameter α_x measures the strength of influence of X compared to that of Y, whereas the parameter α_z controls the strength of the joint dependence of Y and X on Z_i. In the model X and Y are independent but they are conditionally dependent given Z_i unless $\alpha_x \in \{0, 1\}$ or $\alpha_z = 0$. Model B' is constructed analogously to A'. Model $C(q)$ is a Markov chain for which due to Markov property X and Y are conditionally independent given any in-between variable Z_i. Here, q denotes the probability that the previous variable equals the next one. If $q = 0.5$, then any two adjacent variables are independent and if it increases (decreases) the variables become positively (negatively) dependent. By introduction of an additional parameter q_{XY}, we obtain model $C(q, q_{XY})$ for which H_0 is violated.

Our main aim is to study the actual type I error of the considered procedures (i.e. probability rejection when H_0 is true) and the power (probability of rejection when H_0 is false) for the assumed significance level α using the fractions of rejections for artificial data sampled from the above models. We also studied ROC-type curves for all three considered procedures. ROC-type curves are based on two models: the one for which H_0 holds and the second for which H_1 is true, and the report *the actual* type I error and the power approximated by means of simulations for varying α. In this way y values of three ROC curves for the fixed x value correspond to the power for *the same* actual type I error (see Fig. 5).

We present results in Figs. 2, 3, 4, 5 and Table 1 (the chosen parameters represent various strengths of dependence for the structures considered, see discussion in the online supplement). Figure 2 shows the behaviour of the true asymptotic distribution of $\widehat{JMI}$ (see (6)) and its estimate. The left panel depicts boxplots of sorted eigenvalues $\lambda_i(\widehat{M})$, the right compares averaged CDFs corresponding to $\lambda_i(\widehat{M})$ and 90% confidence bands for the true CDF based on them with the true asymptotic CDF and the empirical CDF based on $\widehat{JMI}$ values. In Figs. 3 and 4 the behaviour of the power of the considered procedures is compared against one of the model's varying parameters when the remaining ones are held fixed and the significance level is set to $\alpha = 0.05$. Table 1 indicates how the power and the type I error for the considered procedures depend on the sample size n.

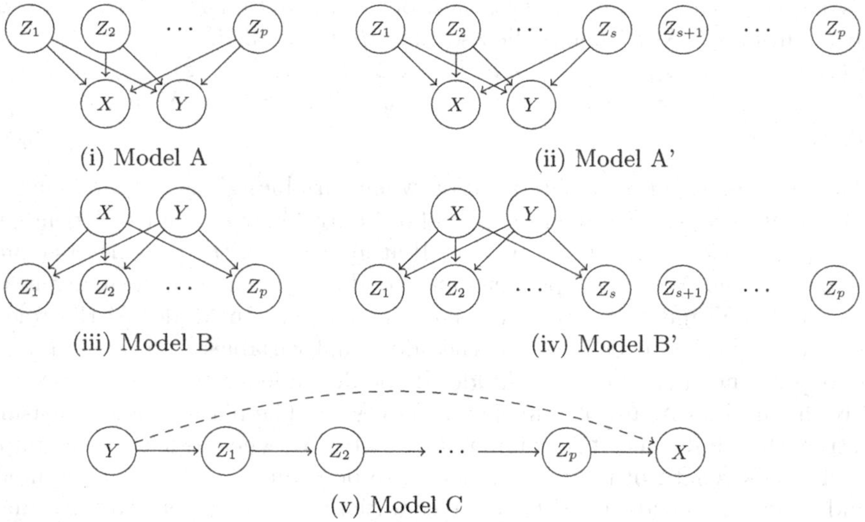

Fig. 1. Graphical representation of the dependence structures

Results. For the sample sizes $n = 500$ and larger the eigenvalues of the estimated matrix $\widehat{M}$ approximate very closely the eigenvalues of the theoretical matrix M and therefore the plots of the averaged CDFs based on eigenvalues $\lambda_i(\widehat{M})$ and CDF using eigenvalues $\lambda_i(M)$ almost overlap (Fig. 2). Such sample sizes are sufficient to ensure the adequate approximation of the distribution of $\widehat{JMI}$ by its asymptotic counterpart. It follows from Table 1 that starting from

Table 1. Estimated powers and type I errors based on $N = 5000$ simulations for varying n and the tests considered. Parameters in Models A, B and C are the same as in Fig. 5.

Mod.	Proc.	Estimated power						Estimated type I error					
		$n = 50$	100	250	500	1000	2000	$n = 50$	100	250	500	1000	2000
A	Bonf.	0.115	0.157	0.368	0.676	0.945	0.999	0.056	0.035	0.037	0.037	0.038	0.033
	Simes	0.129	0.185	0.419	0.732	0.964	1.000	0.063	0.043	0.043	0.043	0.042	0.038
	JMI	0.159	0.249	0.541	0.828	0.983	1.000	0.064	0.052	0.048	0.048	0.052	0.047
B	Bonf.	0.147	0.241	0.487	0.807	0.983	1.000	0.058	0.074	0.053	0.050	0.047	0.046
	Simes	0.162	0.259	0.524	0.837	0.987	1.000	0.062	0.078	0.056	0.052	0.048	0.048
	JMI	0.276	0.333	0.649	0.907	0.997	1.000	0.212	0.089	0.056	0.053	0.049	0.053
C	Bonf.	0.083	0.106	0.180	0.335	0.667	0.952	0.051	0.054	0.040	0.039	0.044	0.039
	Simes	0.096	0.116	0.200	0.363	0.696	0.957	0.060	0.058	0.045	0.042	0.048	0.042
	JMI	0.178	0.139	0.238	0.408	0.747	0.971	0.135	0.072	0.058	0.051	0.050	0.044

the moderate sample sizes ($n \geq 250$) JMI controls well type I error whereas Bonferroni and Simes methods are conservative in some cases (such as Model A for $n = 1000, 2000$). Moreover, it consistently yields the largest power among these three methods. For Fig. 3 H_1 holds and in models A, B (on-line supplement) and C JMI-based test on the whole works better than mutual information-based individual tests with correction applied. As expected, when there is only one strong signal i.e. null hypothesis $X \perp Y|Z_i$ is strongly violated for just one i (model B$'$ with $s = 1$, middle panel of Fig. 4), Bonferroni correction and Simes procedure work well. The novel test does not detect the dependence as frequently as the other two. The situation changes, however, when number of hypotheses that should be rejected increases (see Fig. 4, panels 1 and 3). Comparison of the ROC curves in Fig. 5 indicates that even when *the actual* significance levels of the three tests are matched, JMI-based test remains the most powerful (H_1 hypotheses for the panels correspond to the first column of Fig. 4 for $p = 5$). This is also reflected in the largest values of Area Under Curve (AUC) for JMI.

5.2 Medical Data Set Example

We show an example of the application of the novel test and Bonferroni and Simes procedures to a real medical dataset MIMIC-III [8]. The dataset contains information about patients requiring intensive care and it includes, among others, 10 binary variables representing the presence or absence of the following diseases: `hypertension`, kidney failure (`kidney`), disorders of fluid electrolyte balance (`fluid`), `hypotension`, disorders of lipoid metabolism (`lipoid`), liver disease (`liver`), `diabetes`, thyroid disease (`thyroid`), chronic obstructive pulmonary disease (`copd`) and `thrombosis`. We select two diseases, liver disease and thrombosis for which conditional mutual informations given any of the other eight diseases are approximately the same (see the first panel of Fig. 6) to analyse the situation for which all null hypotheses are rejected with approximately the same strength for

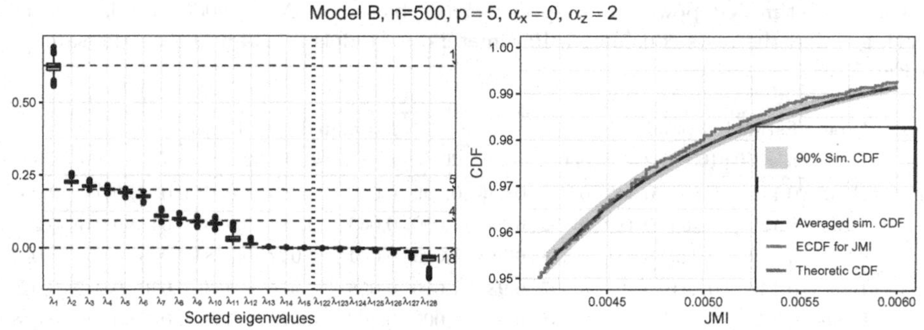

Fig. 2. Left: Box-plots of the empirical values $\lambda_i(\widehat{M}), i = 1, \ldots, 128$ for Model B ($n = 500, p = 5, \alpha_x = 0.5, \alpha_z = 2$). Eigenvalues $\lambda_i(M)$ approximately equal to 0 (multiplicity 118), 0.093 (multiplicity 4), 0.2 (multiplicity 5) and 0.627 (multiplicity 1) are marked by the horizontal lines. Right: values of theoretical CDF, the empirical CDF of $\widehat{JMI}$ and the average of CDFs corresponding to $\lambda_i(\widehat{M})$ for the values of JMI greater than 0.95th quantile of $\widehat{JMI}$.

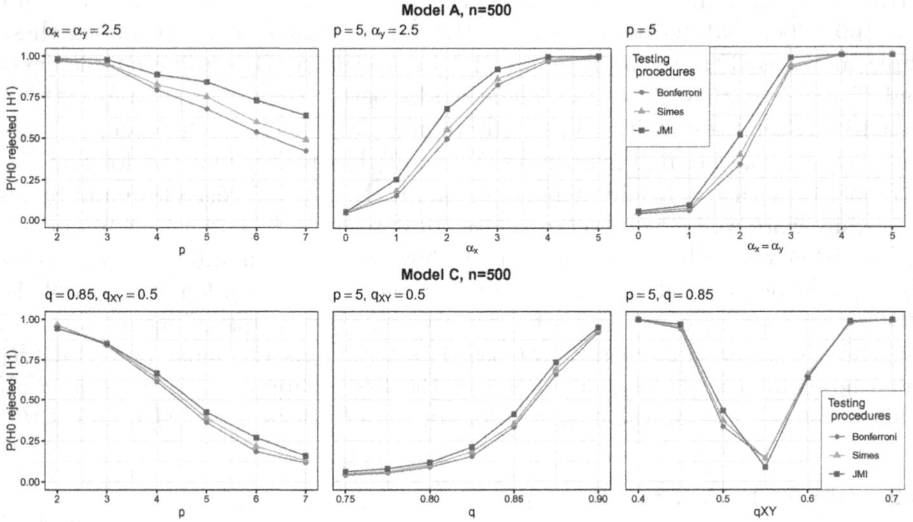

Fig. 3. Power against the changing number of variables and the parameter values for models A and C based on $N = 1000$ simulations.

the whole data set consisting of 10000 observations. In the second panel of Fig. 6 we present how often the null hypothesis that liver disease and thrombosis are conditionally independent is rejected for smaller sample size scenarios for which conditional dependencies are much harder to reject. The estimation is based on $N = 200$ samples randomly sub-sampled from the original data set for each n ranging from 250 to 5000. The asymptotic test works uniformly better than Bonferroni

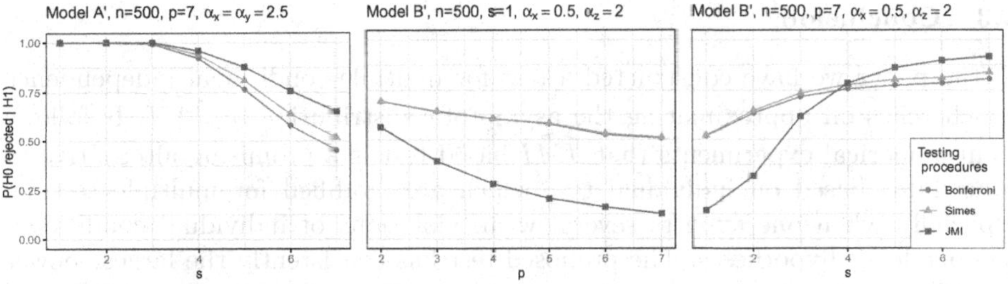

Fig. 4. Power against the number of variables and the number of significant variables in models A′ and B′ (see text) based on $N = 1000$ simulations.

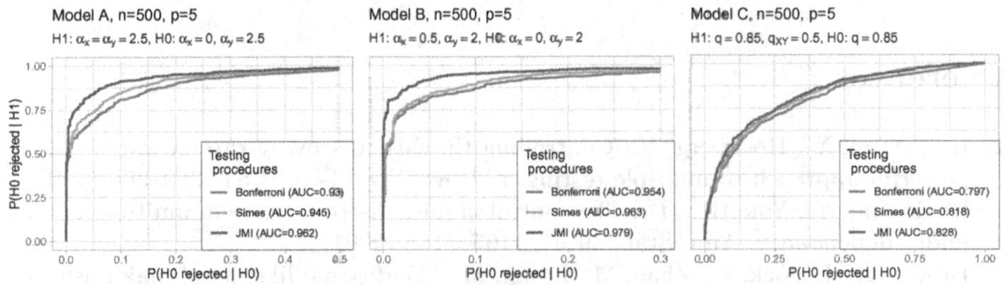

Fig. 5. ROC-type curves and the corresponding values of AUC for models A, B and C and for H_0 and H_1 indicated in the header and $n = 500$.

and Simes procedures. This holds even for small sample sizes for which approximation of the distribution of $\widehat{JMI}$ by its limit is likely to be worse than for larger sample sizes.

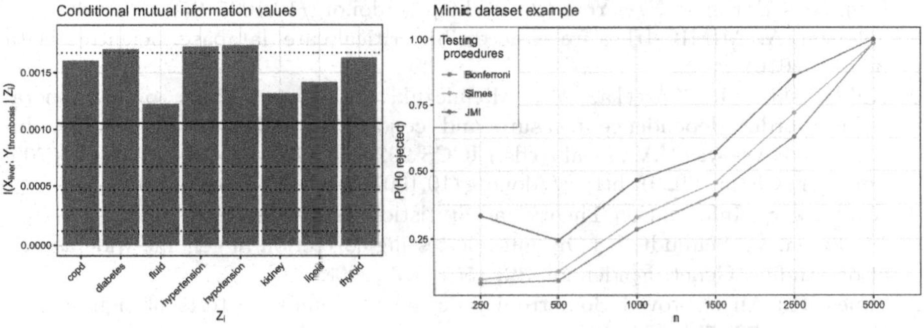

Fig. 6. Left: $\hat{I}(X, Y|Z_i)$, where X and Y denote `liver` and `thrombosis` and Z_i is one of the eight remaining variables. Right: the estimated probability of rejection.

5.3 Conclusion

In the paper we have constructed a test for multiple conditional independence which relies on approximating the asymptotic distribution of $\widehat{JMI}$. It follows from numerical experiments that $\widehat{JMI}$-based test is a promising alternative to procedures based on individual test which are modified for multiple testing, especially when one expects several weak violations of individual conditional independence hypotheses. The proposed test has consistently the largest power in such cases, while controlling for type I error. Its superiority is retained even when Bonferroni and Simes methods are calibrated to have exactly the same value of type I error as JMI-based test. The method is numerically stable and reasonably quick for $p \leq 8$, for larger p `eigen` function has to be modified.

References

1. Benjamini, Y., Hochberg, Y.: Controlling the false discovery rate: a practical and powerful approach to multiple testing. J. Royal Stat. Soc. B **57**, 289–300 (1995)
2. Benjamini, Y., Yakutieli, D.: The control of false discovery rate in multiple testing under dependency. Ann. Stat. **29**(4), 1165–1188 (2001)
3. Brown, G., Pocock, A., Zhao, M., Luján, M.: Conditional likelihood maximisation: a unifying framework for information theoretic feature selection. J. Mach. Learn. Res. **13**(1), 27–66 (2012)
4. Buhlmann, P., de Geer, S.: Statistics for High-Dimensional Data. Springer, New York (2006). https://doi.org/10.1007/978-3-642-20192-9
5. Cover, T.M., Thomas, J.A.: Elements of Information Theory (Wiley Series in Telecommunications and Signal Processing). Wiley-Interscience, New York (2006)
6. Duchesne, P., Lafaye de Micheaux, P.: Computing the distribution of quadratic forms: further comparisons between the Liu-Tang-Zhang approximation and exact methods. Comput. Stat. Data Anal. **54**, 858–862 (2010)
7. Dudoit, S., van der Laan, M.J.: Multiple Testing Procedures with Applications to Genomics. Springer, New York (2009). https://doi.org/10.1007/978-0-387-49317-6
8. Johnson, A.: MIMIC-III, a freely accessible critical care database. Scientific Data **3**, 1–9 (2016)
9. Kubkowski, M., Łazęcka, M., Mielniczuk, J.: Distributions of a general reduced-order dependence measure and conditional independence testing. In: Krzhizhanovskaya, V.V., et al. (eds.) ICCS 2020. LNCS, vol. 12143, pp. 692–706. Springer, Cham (2020). https://doi.org/10.1007/978-3-030-50436-6_51
10. Kullback, S.: Information Theory and Statistics. Smith, P. (1978)
11. Moskvina, V., Schmidt, K.: On multiple-testing correction in genome-wide association studies. Genet. Epidemiol. **32**, 1567–573 (2008)
12. Simes, R.: An improved Bonferroni procedure for multiple tests of significance. Biometrika **73**, 751–754 (1986)
13. Storey, J.: A direct approach to false discovery rates. J. Royal Stat. Soc. B **64**(3), 479–498 (2002)
14. Vergara, J., Estevez, P.: A review of feature selection methods based on mutual information. Neural Comput. Appl. **24**(1), 175–186 (2014)
15. Yang, H., Moody, J.: Data visualization and feature selection: new algorithms for nongaussian data. Adv. Neural. Inf. Process. Syst. **12**, 687–693 (1999)

A Bayesian Interpretation of the Monty Hall Problem with Epistemic Uncertainty

Cristina Manfredotti[1] and Paolo Viappiani[2](✉)

[1] UMR MIA-Paris, AgroParisTech, INRA, University of Paris-Saclay, 75005 Paris, France
cristina.manfredotti@agroparistech.fr
[2] Sorbonne Université, CNRS, LIP6, 75005 Paris, France
paolo.viappiani@lip6.fr

Abstract. The Monty Hall problem is a classic puzzle that, in addition to intriguing the general public, has stimulated research into the foundations of reasoning about uncertainty. A key insight to understanding the Monty Hall problem is to realize that the specification of the behavior of the host (i.e. Monty) of the game is fundamental. Here we go one step further and reason, in Bayesian way, in terms of epistemic uncertainty about the behavior of host, assuming subjective probabilities.

We also consider several generalizations of the classic Monty Hall problem considering different priors for the doors, several doors instead of three, and different ways the host can choose which door to open when several are possible. We show that in these generalized versions, the player faces a sequential decision problem, since the choice of the first door is key. We provide a general solution for the most general case using decision trees and determine the optimal policy.

1 Introduction

The Monty Hall problem [12–14] is a classic puzzle that, in addition to intriguing the general public, has stimulated research [1,2] into the foundations of reasoning about uncertainty. It is stated as follows:

> Suppose you're on a game show, and you're given the choice of three doors: Behind one door is a car; behind the others, goats. You pick a door, say No. 1, and the host, who knows what's behind the doors, opens another door, say No. 3, which has a goat. He then says to you, "Do you want to pick door No. 2?" Is it to your advantage to switch your choice?

The commonly accepted answer is that it is best to switch. Indeed, assuming that the prize is placed behind a door according to a uniform distribution, by choosing to switch the player obtains probability $\frac{2}{3}$ of getting the prize.

This is true however under a particular assumption about the behavior of the host: the host always opens a door; this door is different than the one that the player has chosen and from the one with the prize behind it. Indeed, several

V. Torra and Y. Narukawa (Eds.): MDAI 2021, LNAI 12898, pp. 93–105, 2021.
https://doi.org/10.1007/978-3-030-85529-1_8

authors have argued [1,2,10] that the answer to the puzzle crucially depends on the behavior of the host.

In this paper we go one step further and consider uncertainty over which protocol Monty might be following. We reason about Monty's behavior using subjective probabilities about the possible protocols; therefore *we move from representing uncertainty over the placement of the doors, to representing our epistemic uncertainty over the behavior of the host.* Moreover, we consider some generalizations of the Monty Hall problem supposing that the position of the car might be not distributed uniformly. When considering these generalized settings, we realize that the solution to the problem is a policy dictating which door should we choose at each step of the game. While the Monty Hall problem has been extensively studied before in the computer science and applied mathematics literature [1,5–7,10,11,15], we do not know any works that consider the generalized settings that we address here.

2 Epistemic Uncertainty over Monty's Protocol

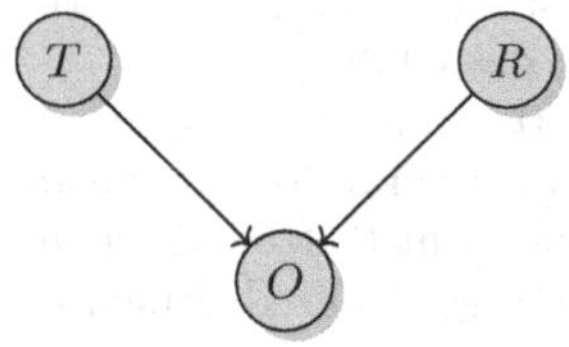

Fig. 1. A Bayesian network formalizing the Monty Hall problem with uncertainty over the host's protocol.

Fig. 2. Simplified Bayesian network for the Monty Hall problem. The uncertainty over Monty's protocol is now integrated in the conditional probability $P(O|T)$.

In this Section we consider the Monty Hall Problem (MHP) with 3 doors and we explicitly reason in terms of epistemic uncertainty about the host's (i.e. Monty's) behavior. We assume for the moment that the car is equally likely to be behind any of the doors. Different assumptions about the host's protocol can be made:

- *AO (always open):* this is the "classic" Monty's behavior. The host always opens a door that has a goat behind it and hasn't been picked by the player (if the player initially picked the door with the car, then he randomly chooses one of the two other doors);
- *RO (open at random):* Monty randomly chooses a door and, if there is no car behind it and it has not been picked by the player, then open it, while no door is opened if the randomly chosen door hides the car;
- *SO (selective open):* the choice of opening a door depends on specific conditions (whether the player picked the door with the prize). In particular we consider "benevolent" Monty (opens a door whenever the player is pointing at a door with a goat, and not when the player picked the door with the car;

this behavior is dubbed SO_+) and "adversarial" Monty (opens a door only when the player is pointing at the car; SO_-)

While under the AO protocol the player has an advantage to switch, under RO, switching gives no advantage, as it has been noticed several times (see, for instance, exercise 3.9 in the book of MacKay [8] on page 57 and its solution on page 61). Obviously under SO_+ it is always beneficial to switch and under SO_- one should never switch (see also Halpern's book [2] on pages 216–217).

We now assume now that Monty's behavior is a situation of epistemic uncertainty: the player does not know exactly which protocol Monty has adopted and this uncertainty is represented by a probability distribution. This means that, from the point of view of the player, Monty is behaving according to a mixture of the protocols above. This mixture is given by the parameters $\boldsymbol{\theta} = (\theta_{AO}, \theta_{RO}, \theta_{SO_+}, \theta_{SO_-})$, where θ_{AO} is the probability of adopting the AO protocol, and so on; in other words $\boldsymbol{\theta}$ is the subjective probability distribution of Monty's behavior. Actually, our model allows for the possibility that Monty itself is behaving according to a mixture of the protocols, but the player has no access to the true mixture parameters and makes use of subjective probabilities instead.[1]

We formalize the Monty Hall problem using the Bayesian network depicted in Fig. 1 with three nodes: T, R, and O. Node T represents the event "the player has pointed to the door with the car behind", R takes value in $\mathcal{R} = \{AO, RO, SO_+, SO_-\}$, that is the set of possible protocols. O is the event "Monty opened a door".

Assuming that the car is uniformly distributed between the three positions, we write the values of the Conditional Probability Tables (CPTs) for the nodes of the Bayesian network. For node T we have:

$$P(T) = \frac{1}{3} \qquad\qquad P(\neg T) = \frac{2}{3}$$

and for R:

$$P(R) = \theta_R \quad \forall R \in \{AO, RO, SO_+, SO_-\}.$$

We now write the probability of the event O (the host opens a door) conditioned on T (the player has pointed at door with the car) and on the protocol. These are the CPT values associated with the node O in Fig. 1.

$$\begin{aligned} P(O|T, AO) &= 1 & P(O|\neg T, AO) &= 1 \\ P(O|T, RO) &= 0.66 & P(O|\neg T, RO) &= 0.33 \\ P(O|T, SO_+) &= 0 & P(O|\neg T, SO_+) &= 1 \\ P(O|T, SO_-) &= 1 & P(O|\neg T, SO_-) &= 0 \end{aligned}$$

[1] A possible extension of this work could investigate the use of Bayesian hierarchical models, adopting prior distributions on the mixture's parameters.

From the belief $\boldsymbol{\theta}$ we can determine the probability of the host opening a door (event O) given the initially chosen door conceals the car (T) or the goat ($\neg T$):

$$P(O|T) = \sum_{r \in \mathcal{R}} \theta_r P(O|T, r) = \theta_{AO} + \frac{2}{3}\theta_{RO} + \theta_{SO_-} \tag{1}$$

$$P(O|\neg T) = \sum_{r \in \mathcal{R}} \theta_r P(O|\neg T, r) = \theta_{AO} + \frac{1}{3}\theta_{RO} + \theta_{SO_+} \tag{2}$$

The above equations allow us to reduce our problem to the simplified Bayesian network given in Fig. 2 (where $\boldsymbol{\theta}$ can be seen as a vector of parameters).

Using basic probability calculus and Bayes theorem, we can derive a condition on $\boldsymbol{\theta}$ for when switching is advantageous. We compute the probability (from the point of view of the player) that the car is behind the initially picked door conditioned to observing that the host has opened another door, using Bayes theorem:

$$P(T|O) = \frac{P(O|T)P(T)}{P(O)} = \frac{P(O|T)P(T)}{P(O|T)P(T) + P(O|\neg T)P(\neg T)} = \frac{P(O|T)}{P(O|T) + 2P(O|\neg T)}$$

If the player sticks to his initial guess, then $P(T|O)$ is the probability of getting the car. If the player switches, the car is found with probability $P(\neg T|O) = 1 - P(T|O)$. Switching is then advantageous when

$$P(\neg T|O) > P(T|O) \iff P(O|\neg T)P(\neg T) > P(O|T)P(T) \tag{3}$$

$$\iff P(O|\neg T) > \frac{1}{2}P(O|T). \tag{4}$$

Since we want to know under what condition with respect to $\boldsymbol{\theta}$ switching is advantageous, we now expand the expression above using Eqs. (1) and (2):

$$\frac{2}{3}(\theta_{AO} + \frac{1}{3}\theta_{RO} + \theta_{SO+}) > \frac{1}{3}(\theta_{AO} + \frac{2}{3}\theta_{RO} + \theta_{SO-}) \tag{5}$$

$$\iff \frac{1}{3}\theta_{AO} + \frac{2}{3}\theta_{SO+} - \frac{1}{3}\theta_{SO-} > 0 \tag{6}$$

We note that in the computation just above, we were only interested in determining when switching is beneficial[2]; that is, we did not considered the situations in which no door is opened, and no choice if offered. Considering a game episode starting with the initial door selection, we are now interested in computing the total expected payoff of the two policies "switch" (switch door when possible) and "keep", where we define payoff as 1 if the player gets the car at the end of the game, and 0 otherwise. Note that the two policies imply the same outcome when Monty does not open a door (and therefore does not offer the possibility to switch choice).

[2] Indeed the original statement of the MHP concerns the specific decision of what to do when offered the possibility of switching.

- The policy "keep" obviously achieves expected payoff $\frac{1}{3}$.
- The policy "switch" achieves expected payoff.

$$P(T, \neg O) + P(\neg T, O) = P(T)P(\neg O|T) + P(\neg T)P(O|\neg T)$$

since the car is won if the player initially picked the right door and the host *does not* open any door (there is no option to switch, and therefore the car is obtained), or if the initial guess is wrong but the host *does* open a door thus offering the chance to switch (the offer is then accepted, since we're following the "switch" policy, and the car is obtained). Therefore, since $P(T) = \frac{1}{3}$, the payoff of the "switch" policy is $\frac{1}{3} - \frac{1}{3}P(O|T) + \frac{2}{3}P(O|\neg T)$ or, equivalently, $\frac{1}{3} + \frac{1}{3}\theta_{AO} - \frac{1}{3}\theta_{SO_-} + \frac{2}{3}\theta_{SO_+}$.

We observe that in the 3-doors setting, with Monty uniformly random when the player chooses the door with the car in the first round, there are really just two parameters: $P(O|T)$ and $P(O|\neg T)$. Given these two values, the distribution over the protocol R is identified according to Eqs. (1) and (2); note however that different $\boldsymbol{\theta}$ may project to the same $P(O|T)$ and $P(O|\neg T)$ values.

The following proposition summarizes our analysis:

Proposition 1. *The payoffs of the two policies "keep" and "switch" are:*

$$V(keep) = \frac{1}{3}$$
$$V(switch) = \frac{1}{3} - \frac{1}{3}P(O|T) + \frac{2}{3}P(O|\neg T) = \frac{1}{3} + \frac{1}{3}\theta_{AO} - \frac{1}{3}\theta_{SO_-} + \frac{2}{3}\theta_{SO_+}$$

Switching is advantageous when $P(O|T) < 2P(O|\neg T)$, or equivalently, when $\frac{1}{3}\theta_{AO} + \frac{2}{3}\theta_{SO+} - \frac{1}{3}\theta_{SO-} > 0$.

Example 1. Assume the player is not given any information about the host's behavior. The player reasons that the host might be following one of the four protocols AO, RO, SO_+ and SO_-. In absence of any prior information, a reasonable way for the player to proceed is to consider a uniform prior on the host's protocol: with $\boldsymbol{\theta} = (0.25, 0.25, 0.25, 0.25)$, thus we have that $\frac{1}{3}\theta_{AO} + \frac{2}{3}\theta_{SO+} - \frac{1}{3}\theta_{SO-} = \frac{1}{6} > 0$, so switching is advantageous according to Proposition 1.

Another reasonable uninformative prior is to suppose $P(O|T) = P(O|\neg T) = 0.5$; this also means that switching is advantageous.

Example 2. Assume now that the player has access to the history of past behaviors of the host in n previous games. The player can use Laplace's rule (equivalent to assuming a Beta prior) to estimate the probability of opening a door. Let n_T the number of episodes where the initially picked door hid the car; $n = n_T + n_{\neg T}$. Let o_T be the number of observations consisting in the host opening a door when the initially picked one is correct. The player estimates the probabilities: $\hat{p}_{O|T} = \frac{o_T+1}{n_T+2}$ and $\hat{p}_{O|\neg T} = \frac{o_{\neg T}+1}{n_{\neg T}+2}$. Equation 4 is then used with these estimations to decide whether to switch or not.

3 Different Priors for Doors

We now consider the situation[3] where each door is associated with a prior probability p_i of concealing the prize (for short we will use the term probability of a door). We still consider that there are three doors and delay the extension to an arbitrary number of doors to Sect. 4.

Unlike the original statement of the puzzle, the choice of the first door is critical (since doors cannot anymore treated as indistinguishable). The behavior of the player is fully specified by a decision policy; with 3 doors a policy is a pair (i, a), where i is the index of a door and a is either "switch" or "keep" (in case Monty offers such possibility). We still assume that the host is not biased, in the sense that, if the player initially picks the door with car, the host, if he decides to open a door, is just as likely to open any of the two remaining doors.

We now compute the expected payoff V of the strategy (i, switch) using the observation that the car is obtained in two cases i) if the door i conceals the car and the host does not open a door (and so he does not offer to switch) and ii) if the door i does not conceal the car and the host does offer to switch; hence:

$$\begin{aligned} V(i, \text{switch}) &= P(\neg O, T) + P(O, \neg T) \\ &= P(\neg O|T)p_i + P(O|\neg T)(1 - p_i) \\ &= (1 - \alpha_T)p_i + \alpha_{\neg T}(1 - p_i) \\ &= (1 - \alpha_T - \alpha_{\neg T})p_i + \alpha_{\neg T} \end{aligned}$$

where we let $\alpha_T := P(O|T)$ and $\alpha_{\neg T} := P(O|\neg T)$. On the other hand, the payoff of strategy (i, keep) is obviously p_i:

$$V(i, \text{keep}) = p_i.$$

The following inequality gives the condition that makes switching beneficial.

$$\begin{aligned} V(i, \text{switch}) > V(i, \text{keep}) &\iff (1 - \alpha_T - \alpha_{\neg T})p_i + \alpha_{\neg T} > p_i & (7) \\ &\iff p_i < \frac{\alpha_{\neg T}}{\alpha_T + \alpha_{\neg T}} & (8) \end{aligned}$$

Equation (8) provides a condition to check to determine whether (i, keep) or (i, switch) is best. However, in order to identify the best policy, we need to account as well the choice of i, i.e. the first door. There are 6 possible policies, but in fact some are dominated: among the "keep" policies, the best one is to pick the door i^+ associated with highest prior $p^+ = \max_{i \in \{1,2,3\}} p_i$. On the other hand, if we switch, it is not so obvious if the initial choice should be a door with high or with low prior probability. We therefore consider several different cases.

- If $\alpha_T + \alpha_{\neg T} > 1$ then the payoff $V(i, \text{switch})$ decreases when p_i increases; hence among all policies that switch door in the second step, the best one is

[3] Rosenhouse [11] also addresses the case where the car is not placed behind the doors with equal probability, but assuming the fixed "always open" protocol for Monty.

to pick, in the first round, the door with lowest prior probability. Therefore to determine the optimal policy we compare p^+ (the payoff of selecting the door with highest p and then keeping this choice) and $(1 - \alpha_T - \alpha_{\neg T})p^- + \alpha_{\neg T}$, the value of the payoff obtained by picking door $i^- = \arg\min_i p_i$ and then switching. The condition to check is

$$(1 - \alpha_T - \alpha_{\neg T})p^- + \alpha_{\neg T} > p^+.$$

In other words: we pick i^- and switch in the case $p^+ + (\alpha_T + \alpha_{\neg T} - 1)p^- - \alpha_T < 0$; otherwise we pick i^+ and keep the same choice.

- If $\alpha_T + \alpha_{\neg T} = 1$ then, assuming that we switch, it does not matter which door we select initially: the payoff $V(i, \text{switch})$ will be always $\alpha_{\neg T}$ for any $i = 1, 2, 3$. Hence, if $p^+ > \alpha_{\neg T}$ we select the door with highest prior and keep this choice, and otherwise choose any door and switch.
- If $\alpha_T + \alpha_{\neg T} < 1$ then the payoff $V(i, \text{switch})$ increases when p_i increases. We therefore initially pick i^+, the door with highest prior, and we compare the payoff of either switching or keeping. Hence, if $p^+ > \frac{\alpha_T}{\alpha_T + \alpha_{\neg T}}$ then the optimal policy is (i^+, keep) otherwise it is (i^+, switch).

Proposition 2. *The payoff of the policies are as follows:*

$$V(i, keep) = p_i$$
$$V(i, switch) = (1 - \alpha_T - \alpha_{\neg T})p_i + \alpha_{\neg T}$$

Obviously this model generalizes that of the previous section. Indeed, if we substitute $p_i = \frac{1}{3}$ in Eq. (8) we determine the condition $\frac{\alpha_T}{\alpha_{\neg T}} < 2$ for switching being advantageous, as shown in the previous section in Eq. (4).

We now consider, as examples, two particular cases.

Example 3. Assume three doors with prior probability p_1, p_2, p_3 and that the host behaves according to the *AO* protocol of Sect. 2 (the player may know this having observed previous games), that means $\alpha_T = \alpha_{\neg T} = 1$. Now, if you pick door i initially, switching gives $1 - p_i$; keeping the same choice gives you p_i. The best policy is to pick the door with least value of the prior probability, wait for the host action and then switch door; the optimal payoff is:

$$V^* = 1 - \min_i p_i.$$

This value is strictly higher than the value of the policy of picking the door with highest p_i and not switching, unless $\max_i p_i = 1$.

Example 4. We now consider, as special case of the scenario studied in this section, that the host does not allow to switch with probability q, regardless of whether the player points at the right door or not; the host opens a door allowing to switch with probability $1 - q$. In other words $\alpha_T = \alpha_{\neg T} = 1 - q$. The payoff of (i, switch), the policy "pick door i and switch when offered", is then:

$$V(i, \text{switch}) = qp_i + (1 - q)(1 - p_i) = (2q - 1)p_i + 1 - q.$$

We analyze the different cases:

- If $q < 0.5$ and $p^+ \geq (2q-1)p^- + 1 - q$ then (i^+, keep) is an optimal policy
- If $q < 0.5$ and $p^+ \leq (2q-1)p^- + 1 - q$ then (i^-, switch) is an optimal policy
- If $q = 0.5$ and $p^+ \leq 0.5$ then (i, switch), for all $i \in \{1,2,3\}$, are optimal policies
- If $q = 0.5$ and $p^+ \geq 0.5$ then (i^+, keep) is an optimal policy
- If $q > 0.5$ and $p^+ \geq 0.5$ then (i^+, keep) is an optimal policy
- If $q > 0.5$ and $p^+ \leq 0.5$ then (i^+, switch) is an optimal policy.

For the last two cases, notice that when $q > 0.5$, the condition $V(i^+, \text{keep}) \geq V(i^+, \text{switch})$ simplifies to $p^+ \geq 0.5$.

4 General Setting: *n* Doors and General Response Model

In this section we analyze the general formulation of the MHP and develop a model based on sequential decision making. We consider the general situations with n doors and arbitrary prior probabilities p_i. Monty may decide not to open any door. Moreover, in this section we allow for Monty to be biased with respect to which door to open when he can choose among several unopened doors. Note that the models discussed in the previous Sections can be seen as special cases of this general model.

The problem is solved with a decision tree; an excerpt of the general tree is shown in Fig. 3. Note that we use a different notation from previous Sections, since the generalized problem does not enjoy the symmetries that simplified the treatment of the former models. Each node of the tree is labeled with the variable (either a decision or a random variable) that it represents.

The decision node S represents the initial door choice, with possible choices in $\{S_1, \ldots, S_n\}$. For each S_j, there is a chance node O, with outcomes in $\{O_\emptyset\} \cup \{O_i\}_{i \neq j}$ representing whether and which door the host opens; in our notation $O_\emptyset$ is the event no door is opened, and O_i means that the door i is opened; then:

- In case no door is opened, $O = O_\emptyset$, the position of the car is revealed to be at a position k in a chance node T, with outcomes in $\{T_1, \ldots, T_n\}$. In the leaf nodes, utility is 1 if $j = k$, and 0 otherwise.
- If, instead, the event O_i happens, we face the decision node F, with possible choices in $\{F_l\}_{l \neq i}$, representing the final door choice[4], with the choice that must be different from i. Then, the chance node T, with outcomes in $\{T\}_{k \neq i}$, reveals the car's position; utility is 1 if the choice for node F is the same as the outcome of T.

Let $\alpha_{i,j,k} := P(O_i | S_j, T_k)$ to be the probability of Monty opening door i given player's selection of door j and car in door k. The vector

$$(\alpha_{i,j,k})_{i,j,k \in \{1,\ldots,n\}}$$

[4] In this generalized model, switching occurs when the choice at node F is a different door from the one chosen at the root S.

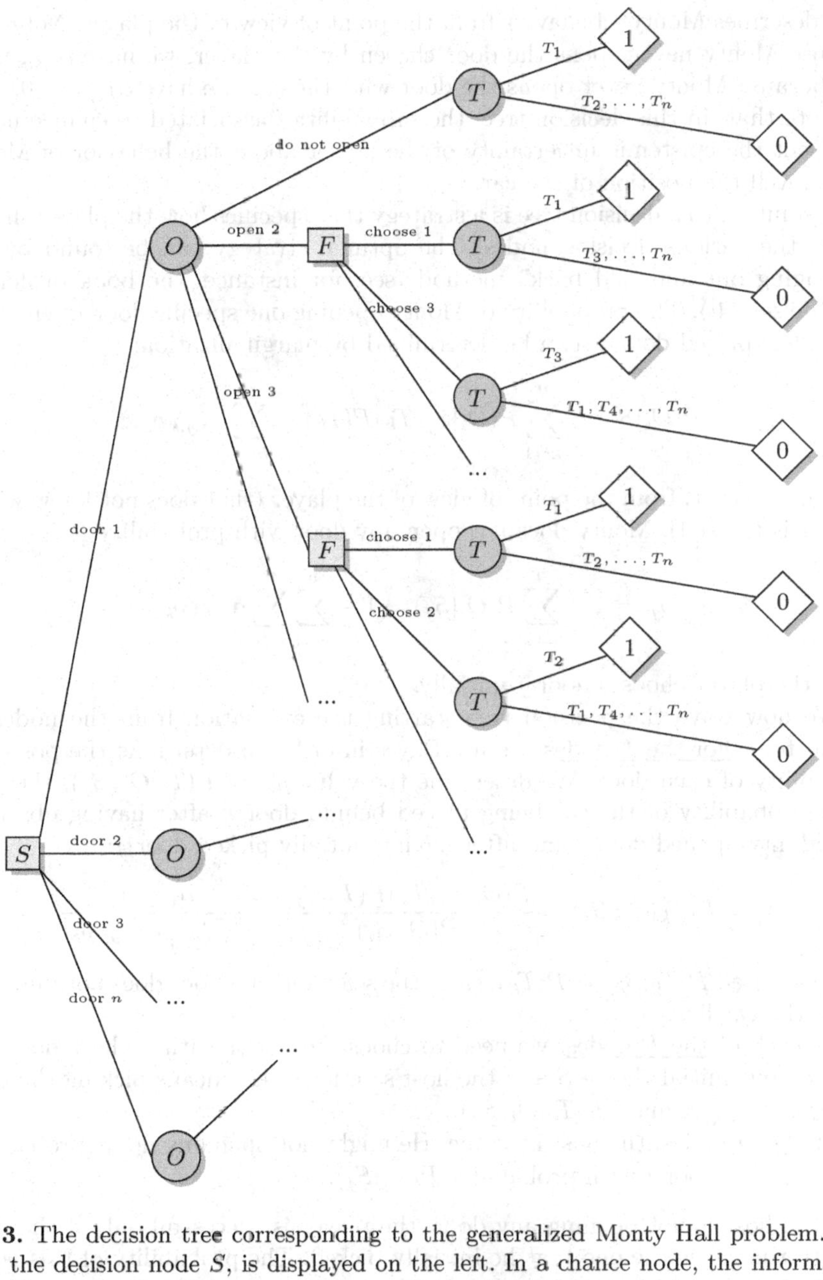

Fig. 3. The decision tree corresponding to the generalized Monty Hall problem. The root, the decision node S, is displayed on the left. In a chance node, the information available up to that point is used to condition the distribution; for example, if the player selected door 1 initially, the variable O is distributed according to $P(O|S_1)$, that can be computed using Eq. 9. Similarly, if the player chooses door i and the host does not open any door, the probability of T_i is given by $P(T_i|S_i, O_\emptyset)$.

fully describes Monty's behavior from the point of view of the player. Note that because Monty never opens the door chosen by the player, we have $\alpha_{i,i,k} = 0$, and because Monty never opens the door with the car, we have $\alpha_{k,j,k} = 0$.

Note that, in this decision tree, the probabilities associated to chance nodes represent the epistemic uncertainty of the player about the behavior of Monty and as well the position of the car.

A solution to a decision tree is a strategy that specifies how the player should act at the various decision nodes. The optimal strategy can be found by the "averaging out and fold back" method (see, for instance, the book of Jensen and Nielsen [4]). The probability of Monty opening one specific door i, when the player has picked door j, can be determined by marginalization:

$$P(O_i|S_j) = \sum_{k=1}^{n} P(O_i|S_j, T_k)P(T_k) = \sum_{k=1}^{n} \alpha_{i,j,k} p_k. \tag{9}$$

This means that, from the point of view of the player (that does not know where the car is located), Monty does not open any door with probability

$$\beta_j := 1 - \sum_{i=1}^{n} P(O_i|S_j) = 1 - \sum_{k=1}^{n} \sum_{i=1}^{n} \alpha_{i,j,k} p_k$$

when the player chooses door j initially.

We now solve the decision tree starting the evaluation from the nodes at the bottom. For the T nodes, we use Bayes in order to determine the posterior probability of each door. We determine the value $p'_k := P(T_k|O_i, S_j)$, the posterior probability of the car being placed behind door k after having observed that Monty opened door i and after having initially picked door j:

$$p'_k = P(T_k|O_i, S_j) = \frac{P(O_i|S_j, T_k)P(T_k|S_j)}{P(O_i|S_j)} = \frac{\alpha_{i,j,k} p_k}{\sum_{k'=1}^{n} \alpha_{i,j,k'} p_{k'}}$$

where we used $P(T_k|S_k) = P(T_k)$, since the selection of a door does not influence where the car lies.

At each of the F nodes, we need to choose the door with highest posterior p'_k given our initial choice S and the host's action. This means picking the door giving $\max_k p'_k = \max_k P(T_k|O_i, S_j)$.

At the O nodes, the host is acting. He might not open any door (probability β_j) or open a door i with probability $P(O_i|S_j)$.

- If the host is not opening any door, the player is successful only if the door with the car is the one that he initially picked. The probability of this is

$$P(T_j|O_\emptyset, S_j) = \frac{P(O_\emptyset|S_j, T_j)P(T_j)}{P(O_\emptyset|S_j)} = \frac{(1 - \sum_i \alpha_{i,j,j})p_j}{\beta_j}$$

  and the contribution to the O node is $P(T_j|O_\emptyset, S_j)$ times $P(O_\emptyset|S_j)$.
- If, instead, the host opens door i, the contribution to the value of the node is $P(O_i|S_j) \max_k P(T_k|O_i S_j)$.

This gives the following value for a node of type O:

$$P(O_\emptyset|S_j)P(T_j|O_\emptyset,S_j)+\sum_{i=1}^{n}P(O_i|S_j)\max_k P(T_k|O_i,S_j) \tag{10}$$

$$=\beta_j\frac{(1-\sum_i\alpha_{i,j,j})p_j}{\beta_j}+\sum_{i=1}^{n}P(O_i|S_j)\max_k\frac{\alpha_{i,j,k}p_k}{P(O_i|S_j)} \tag{11}$$

$$=\Big(1-\sum_{i=1}^{n}\alpha_{i,j,j}\Big)p_j+\sum_{i=1}^{n}\max_k\alpha_{i,j,k}p_k \tag{12}$$

At the root, we have the decision node S where we take the door j that maximizes the value of Eq. (12).

Proposition 3. *The optimal policy achieves expected payoff:*

$$V^*=\max_{j=1,\ldots,n}\Big[\Big(1-\sum_{i=1}^{n}\alpha_{i,j,j}\Big)p_j+\sum_{i=1}^{n}\max_k\alpha_{i,j,k}p_k\Big].$$

Example 5. We now consider *classic Monty with response bias*, that is the scenario with 3 doors and uniform priors, $P(T_i)=\frac{1}{3}$, the host always open one door (AO protocol), but when the player chooses the door with the car behind in the first step, then the host may not be following an uniform distribution in deciding which door to open (see also [2], pages 216–217).

In the following description let i, j and k to be distinct; i.e. (i,j,k) is a permutation of $(1,2,3)$. We have $\alpha_{i,j,k}=1$ and $\alpha_{i,j,j}+\alpha_{k,j,j}=1$. Now, assume that the player selects door j and the host opens door i. Observe that the total probability of opening door i is $P(O_i|S_j)=\frac{1}{3}(1+\alpha_{i,j,j})$. We then determine the posterior probabilities for positions j and k (the car cannot be behind door i since this door was opened): $P(T_k|S_j,O_i)=\frac{\alpha_{i,j,k}p_k}{\alpha_{i,j,j}p_j+\alpha_{i,j,k}p_k}=\frac{1}{\alpha_{i,j,j}+1}$ and $P(T_j|S_j,O_i)=\frac{\alpha_{i,j,j}p_j}{\alpha_{i,j,j}p_j+\alpha_{i,j,k}p_k}=\frac{\alpha_{i,j,j}}{\alpha_{i,j,j}+1}$. The best decision in the second stage of the game consist in picking the door j or k associated with the higher posterior. Now, consider the decision at the root. The payoff $V(S_j)$ of selecting door j, assuming that then choosing optimally in the second step, is:

$$V(S_j)=\sum_{i\neq j}\frac{\alpha_{i,j,j}+1}{3}\max\{\frac{1}{\alpha_{i,j,j}+1},\frac{\alpha_{i,j,j}}{\alpha_{i,j,j}+1}\}=\frac{1}{3}\sum_{i\neq j}\max\{1,\alpha_{i,j,k}\}=\frac{2}{3}$$

Since this value does not depend on j, the first door can be chosen in an arbitrary way. It turns out that the best policy in this case is "pick any door randomly and then, after that the host opens a door, switch choice to other unopened door". The optimal value of the optimal policy is $V^*=\frac{2}{3}$.

5 Discussion and Conclusions

The Monty Hall Problem (MHP) is a puzzle that has raised a lot of attention and is frequently used as a didactic tool for explaining how to reason with subjective probabilities. Some interesting variations of the Monty Hall problem have been analyzed by Lucas et al. [6,7]; we refer the reader to the book of Rosenhouse that provide an excellent review of materials on the Monty Hall problem [11].

In computer science, the Monty Hall problem has stimulated a variety of research activities, including works on epistemic logic [5] and reasoning about uncertainty [2]; we also mention the interpretation given by Viappiani and Boutilier (in the appendix of [16]) in terms of preferences and choice. On the other hand, psycologhists have used the Monty Hall problem to study how human people reason with probabilities [15].

In this paper we provided an analysis of the Monty Hall problem and some of its extensions emphasizing the role of dealing with epistemic uncertainty. We have considered policies that determine which door to select in the first round, and whether to keep the same choice or to switch in the second. We provided the characterization of the optimal policy in several generalizations of the MHP: considering different prior subjective probabilities for the position of the prize behind the doors, considering uncertainty over the possible host's behaviors and considering n doors. We mention some interesting further extensions of the MHP worth studying: considering the generalization m rounds, and the case where the number of rounds is uncertain.

We now provide some brief comments on how the Monty Hall problem is related to several areas of artificial intelligence. First of all, notice that the tools we have used (Bayesian reasoning, Bayesian networks, and decision trees) are typically used in AI. Moreover, some of the ideas behind our work are relevant to research in *multi-agent systems* since agents often have to reason about other agents' behaviour. In some sense, the MHP can be seen as an emblematic case of an agent reasoning about another agent's behavior, a key aspect of multi agent system research; we advocate that it often worth to consider a wide variety of possible behaviors and not just a single one, and to consider mixture of such possible behaviors (as we did in our treatment of the MHP). This could be of relevance for opponent modeling in games, for instance.

The Monty Hall problem has connections with the statistical areas of selectively reported data and missing data; in particular the missing at random hypothesis in machine learning [3]. In the case of *recommender systems* based on collaborative filtering where users rate items such as movies, the missing at random hypothesis imputes missing ratings as the result of a random process that selects the items that are rated or not. This assumption might not be valid [9], causing the system to underperform. Indeed it is possible that an item, let's say a movie, is watched and then rated for a variety of reasons:

- the movie is popular (and the user often watches popular movies; although he might not necessarily like them),
- the movie is perceived by the user as similar to others seen in the past,
- the user thinks (based on his knowledge) that he might like the movie and therefore decide to watch it,
- the movie was recommended to the user (perhaps by a competitor), etc.

Therefore, instead of a simple probabilistic model, one could consider a richer model accounting for a mixture of all such different "user protocols" and the associated uncertainties in terms of subjective probabilities (allowing to model the interplay between the user habits, the popularity of movies, the beliefs of the user about which movies he might like, etc.). Of course, learning such a probabilistic model would be challenging. We believe that the design of recommender systems dealing with such "protocol uncertainty" is an important research direction.

References

1. Grünwald, P., Halpern, J.Y.: Updating probabilities. J. Artif. Intell. Res. **19**, 243–278 (2003). https://doi.org/10.1613/jair.1164
2. Halpern, J.Y.: Reasoning About Uncertainty. MIT Press, Cambridge (2005)
3. Jaeger, M.: On testing the missing at random assumption. In: Machine Learning: ECML 2006, 17th European Conference on Machine Learning, Berlin, Germany, 18–22 September 2006, Proceedings, pp. 671–678 (2006)
4. Jensen, F.V., Nielsen, T.D.: Bayesian Networks and Decision Graphs, 2nd edn. Springer, New York (2007). https://doi.org/10.1007/978-0-387-68282-2
5. Kooi, B.P.: Probabilistic dynamic epistemic logic. J. Logic Lang. Inform. **12**(4), 381–408 (2003)
6. Lucas, S.K., Rosenhouse, J.: Optimal strategies for the progressive Monty Hall problem. Math. Gaz. **93**(528), 410–419 (2009). https://doi.org/10.1017/S0025557200185158
7. Lucas, S.K., Rosenhouse, J., Schepler, A.: The Monty Hall problem, reconsidered. Math. Mag. **82**(5), 332–342 (2009). http://www.jstor.org/stable/27765931
8. MacKay, D.J.C.: Information Theory, Inference, and Learning Algorithms. Cambridge University Press, New York (2003)
9. Marlin, B.M., Zemel, R.S.: Collaborative prediction and ranking with non-random missing data. In: Proceedings of the 2009 ACM Conference on Recommender Systems, RecSys 2009, New York, NY, USA, 23–25 October 2009, pp. 5–12. ACM (2009)
10. Mueser, P., Granberg, D., Mueser, K., Nickerson, R.: The Monty Hall dilemma revisited: Understanding the interaction of problem definition and decision making. University of Missouri Working Paper, February 1999
11. Rosenhouse, J.: The Monty Hall Problem: The Remarkable Story of Math's Most Contentious Brainteaser. Oxford University Press, Oxford (2009)
12. vos Savant, M.: Ask Marilyn. Parade, p. 15, September 1990
13. vos Savant, M.: Ask Marilyn. Parade, p. 25, December 1990
14. Selvin, S., et al.: Letters to the editor. Am. Stat. **29**(1), 67–71 (1975). http://www.jstor.org/stable/2683689
15. Stibel, J., Dror, I., Ben-Zeev, A.: The collapsing choice theory: Dissociating choice and judgment in decision making. Theory Decision **66**, 149–179 (2009). https://doi.org/10.1007/s11238-007-9094-7
16. Viappiani, P., Boutilier, C.: On the equivalence of optimal recommendation sets and myopically optimal query sets. Artif. Intell. J. **286**, 103328 (2020)

How the F-Transform Can Be Defined for Hesitant, Soft or Intuitionistic Fuzzy Sets?

Jiří Močkoř[1,2](✉)

[1] Institute for Research and Applications of Fuzzy Modeling, University of Ostrava, Ostrava, Czech Republic
Jiri.Mockor@osu.cz
[2] Centre of Excellence IT4Innovations, 30. Dubna 22, 701 03 Ostrava 1, Czech Republic
http://irafm.osu.cz/

Abstract. Classical F-transform for lattice-valued fuzzy sets can be defined using the Zadeh's power set monad and a special monadic relation. In this paper we use the fact that power set structures of hesitant fuzzy sets, fuzzy soft sets or intuitionistic fuzzy sets also define power set monads and we show how by selecting appropriate monadic relations the F-transforms for these fuzzy type structures can be defined.

1 Introduction

Fuzzy transform represents a method in fuzzy set theory, which is used in many applications in signal and image processing [4,5], signal compressions [22,26], numerical solutions of ordinary and partial differential equations [9,27], data analysis [6,23] and many other applications. This concept was introduced for the first time in [21] both for classically defined $[0,1]$-valued fuzzy sets and L-valued fuzzy sets, where L is a complete residuated lattice. The F-transform method represents a special transformation map based on a system of fuzzy sets defined on a given universe, which is called a *fuzzy partition*. In general, any variant of a fuzzy partition then represents a pair $(X, \mathcal{A})$, where X is a set and $\mathcal{A} = \{A_i : i \in I\}$ is a set of fuzzy sets in X. Based on a fuzzy partition $(X, \mathcal{A})$, the F-transform is then a special map transforming fuzzy sets from a set X to fuzzy sets in the index set I of a fuzzy partition $\mathcal{A}$. This procedure makes it possible to significantly simplify the work with the original fuzzy sets, especially in those areas, such as methods for image processing, where the original set X is huge, while the index set I from the fuzzy partition can be significantly smaller.

Fuzzy sets, both classical and with values in lattices, are not the only tool that allows us to work with uncertainty, both theoretically and with a number of practical applications. Currently, there is a whole range of theories and

This work was partly supported from ERDF/ESF project CZ.02.1.01/0.0/0.0/17-049/0008414.

V. Torra and Y. Narukawa (Eds.): MDAI 2021, LNAI 12898, pp. 106–117, 2021.
https://doi.org/10.1007/978-3-030-85529-1_9

theoretical structures which are based on the principles of fuzzy set theory but create their own tools and methods for solving theoretical and practical problems. These theories undoubtedly include the *theory of intuitionistic fuzzy sets, the theory of fuzzy soft sets and the theory of hesitant fuzzy sets.* The common feature of these three theories is, among other things, a large number of current publications dealing with theoretical properties and application possibilities of these theories. For a basic overview of these theories and their applications, see, e.g., [1,2,30,32] for intuitionistic fuzzy sets, [7,11,12,19] for fuzzy soft sets and [25,28,29,31] for hesitant fuzzy sets.

In our previous paper [17] we tried to unify some of methods used in all these theories. For this purpose, we used a special tool from the theory of categories, namely the theory of power set monads (see, e.g., [14,16]), which allows to unify not only various types of relations but also transformation operators defined by these relations. In this paper we use this result about the existence of power set monads defined by fuzzy soft sets, hesitant fuzzy sets or intuitionistic fuzzy sets and using suitable monadic relations in these monads we define F-transform operators for these fuzzy type structures. We will also show how these F-transform operators can be defined axiomatically. All these results then represent generalizations and extensions of results from the theory of F-transform for classical lattice-valued fuzzy sets. Transformation operations based on the F-transform principle in these fuzzy type structures make it possible to simplify the use of these structures in real applications, because with the help of these transformations the size of the sets on which these structures are defined can be significantly reduced. For example, F-transform for fuzzy soft sets can be effectively used in the method of the color segmentation of images, published in [18]. For hesitant fuzzy sets theory, F-transform can be used in clustering [33], where the reduction of data structures for clustering could be of importance, and for intuitionistic fuzzy sets the importance of F-transform is practically the same as for classical fuzzy sets.

In the paper we present only theoretical background of this theory, other details and proofs will be published elsewhere.

2 Preliminaries and Categorical Tools

A basic membership structure of fuzzy sets in the paper is a *complete residuated lattice* (see e.g. [20]), i.e. a structure $\mathcal{L} = (L, \wedge, \vee, \otimes, \rightarrow, 0_L, 1_L)$ such that $(L, \wedge, \vee)$ is a complete lattice, $(L, \otimes, 1_L)$ is a commutative monoid with operation $\otimes$ isotone in both arguments and $\rightarrow$ is a binary operation which is residuated with respect to $\otimes$. Recall that a negation of an element a in $\mathcal{L}$ is defined by $\neg a = a \rightarrow 0_L$.

In the case of intuitionistic fuzzy sets we use a special example of a residuated lattice $\mathcal{L}$, namely, an MV-algebra [3], i.e., a structure $\mathcal{L} = (L, \oplus, \otimes, \neg, 0_L, 1_L)$ satisfying the following axioms:

(i) $(L, \otimes, 1_L)$ is a commutative monoid,
(ii) $(L, \oplus, 0_L)$ is a commutative monoid,

(iii) $\neg\neg x = x$, $\neg 0_L = 1_L$,
(iv) $x \oplus 1_L = 1_L$, $x \oplus 0_L = x$, $x \otimes 0_L = 0_L$,
(v) $x \oplus \neg x = 1_L, x \otimes \neg x = 0_L$,
(vi) $\neg(x \oplus y) = \neg x \otimes \neg y$, $\neg(x \otimes y) = \neg x \oplus \neg y$,
(vii) $\neg(\neg x \oplus y) \oplus y = \neg(\neg y \oplus x) \oplus x$,

for all $x, y \in X$.

If we put

$$x \vee y = (x \oplus \neg y) \otimes y, \quad x \wedge y = (x \otimes \neg y) \oplus y, \quad x \to y = \neg x \oplus y,$$

then $(L, \wedge, \vee, \otimes, \to, 0_L, 1_L)$ is a residuated lattice. MV-algebra is called a complete, if that lattice is a complete lattice.

If $\mathcal{L}$ is a complete residuated lattice, an $\mathcal{L}$-fuzzy set in a crisp set X is a map $f : X \to L$.

We recall the notions of hesitant, soft and intuitionistic $\mathcal{L}$-fuzzy sets.

Definition 1 *[2, 11, 29]. Let X be a set.*

1. *Let K be a set of all possible criteria. A pair (E, s) is called an $\mathcal{L}$-fuzzy soft set in a space (X, K), if $\emptyset \neq E \subseteq K$ and $s : E \to L^X$. By $T(X, K)$ we denote the power set of all $\mathcal{L}$-fuzzy soft sets in (X, K).*
2. *A hesitant $\mathcal{L}$-fuzzy set in X is a mapping $h : X \to 2^L$, i.e., for $x \in X$, $h(x) \subseteq L$. By $H(X)$ we denote the power set of all hesitant $\mathcal{L}$-fuzzy sets in X.*
3. *An intuitionistic $\mathcal{L}$-fuzzy set in a set X is a pair (u, v) of $\mathcal{L}$-fuzzy sets on X, such that $\neg u \geq v$. By $J(X)$ we denote the power set of all intuitionistic $\mathcal{L}$-fuzzy sets in X.*

For basic information about the category theory see [8, 10]. In what follows, categories will be denoted by bold letters and morphisms in a category **K** will be called **K**-morphisms.

As we mentioned in the introduction, the main tool from the category theory we will use is the *monad*. We use the following modified version called the power set monad, which is a generalization of standard Zadeh's power set structure.

Definition 2 *[17]. A structure $\mathbf{T} = (T, \Diamond, \eta, W)$ is called a power set monad in a category **K**, if*

1. *$T : obj(\mathbf{K}) \to obj(\mathbf{K})$ is mapping between objects of **K**,*
2. *$W : \mathbf{K} \to \mathbf{Set}$ is a functor,*
3. *For arbitrary object X in **K**, a structure of a complete $\bigvee$-semilattice is defined on a set $W(T(X))$,*
4. *For **K**-morphisms $f : X \to T(Y)$ and $g : Y \to T(Z)$ there exists their composition $g \Diamond f : X \to T(Z)$, (called the Kleisli composition) which is associative,*
5. *For arbitrary **K**-morphisms $f, f' : X \to T(Y), g, g' : Y \to T(Z)$, the following implications hold*

$$W(g) \leq_Y W(g') \Rightarrow W(g \Diamond f) \leq_Z W(g' \Diamond f),$$
$$W(f) \leq_Y W(f') \Rightarrow W(g \Diamond f) \leq_Z W(g \Diamond f'),$$

where $\leq_Y, \leq_Z$ are point-wise pre-order relations defined by ordering on $W(T(Y))$ or $W(T(Z))$, respectively.

6. *η is a system of* $\mathbf{K}$*-morphisms* $\eta_X : X \to T(X)$*, for any object* X *of* $\mathbf{K}$*,*
7. *For any* $\mathbf{K}$*-morphism* $f : X \to Y$*, the* $\mathbf{K}$*-morphism*

$$f_{\mathbf{T}}^{\rightarrow} := \eta_Y.f \Diamond 1_{T(X)} : T(X) \to T(Y)$$

is such that $W(f_{\mathbf{T}}^{\rightarrow})$ *is also* $\bigvee$*-preserving map with respect to ordering defined in 3, where* $1_{T(X)}$ *is the identity* $\mathbf{K}$*-morphism* $T(X) \to T(X)$ *in* $\mathbf{K}$*.*
8. *For any* $\mathbf{K}$*-morphism* $f : X \to T(Y)$*,* $\eta_Y \Diamond f = f$ *holds,*
9. $\Diamond$ *is compatible with composition of* $\mathbf{K}$*-morphisms, i.e., for* $\mathbf{K}$*-morphisms* $f : X \to Y$*,* $g : Y \to T(Z)$*, we have* $g \Diamond (\eta_Y.f) = g.f$*.*

Let us consider the following classical example of a monadic power set theory.

Example 1 [24]. Let $\mathcal{L}$ be a complete residuated lattice. The power set monad $\mathbf{Z} = (Z, \boxplus, \chi, 1_{\mathbf{Set}})$ is defined by

1. $Z : \mathbf{Set} \to \mathbf{Set}$ is an object function defined by $Z(X) = L^X$ and $1_{\mathbf{Set}} : \mathbf{Set} \to \mathbf{Set}$ is the identity functor,
2. On L^X the order relation is defined point-wise,
3. For each $X \in \mathbf{Set}$, $\chi^X : X \to Z(X)$ is the characteristic map of elements from X, i.e., for $x, y \in X$, $\chi^X(x)(y) = \chi^X_{\{x\}}(y) = \begin{cases} 1_L, & x = y, \\ 0_L, & x \neq y \end{cases}$,
4. For each $f : X \to Z(Y)$ and $g : Y \to Z(V)$ in $\mathbf{Set}$, $g \boxplus f : X \to Z(V)$ is defined by

$$(g \boxplus f)(x)(z) = \bigvee_{y \in Y} f(x)(y) \otimes g(y)(z).$$

In the paper [17] we proved the existence of power set monads based on power sets $T(X, K), H(X)$ and $J(X)$, respectively. Due to the limited scope of the paper, we will present here only the basic structure of these power set monads, without any technical details (for more information see [17]).

Let $\mathbf{Set}_*$ be the subcategory of the product category $\mathbf{Set} \times \mathbf{Set}$, where objects of $\mathbf{Set}_*$ are all pairs (X, K) such that K contains a special object $\star$ (called a *trivial criterium*) and morphisms are pairs $(f, \alpha) : (X, K) \to (Y, M)$ such that $f : X \to Y$ and $\alpha : K \twoheadrightarrow M$ is a surjective map with $\alpha(\star) = \star$.

Theorem 1 *[17]. Let* $\widetilde{\mathbf{T}} = (\widetilde{T}, \square, \xi, W)$ *be defined by*

1. $\widetilde{T} : obj(\mathbf{Set}_*) \to obj(\mathbf{Set}_*)$ *is a mapping defined by* $\widetilde{T}(X, K) = (T(X, K), K)$*, where*

$$T(X, K) = \{(E, s) : \star \in E \subseteq K, s : E \to L^X\}.$$

2. $W : \mathbf{Set}_* \to \mathbf{Set}$ *is the functor such that* $W(X, K) = X$*,* $W(f, \alpha) = f$*.*
3. *An order relation* $\sqsubseteq$ *is defined on* $T(X, K) = W(\widetilde{T}(X, K))$ *by* $(E, s), (F, t) \in T(X, K)$*,* $(E, s) \sqsubseteq (F, t)$ *iff* $E \subseteq F$*,* $(\forall e \in E) s(e) \leq t(e)$ *in* L^X*.*

4. If $(f,\alpha) : (X,K) \to \widetilde{T}(Y,M)$ and $(g,\beta) : (Y,M) \to \widetilde{T}(Z,N)$ are morphisms in $\mathbf{Set}_*$, the Kleisli composition $\Box$ is defined by

$$(g,\beta)\Box(f,\alpha) = (g \vartriangle f, \beta.\alpha) : (X,K) \to \widetilde{T}(Z,N),$$

where $g \vartriangle f : X \to T(Z,N)$ is defined in [17]; Theorem 1.
5. For $(X,K) \in \mathbf{Set}_*$, the $\mathbf{Set}_*$-morphism $\xi_{(X,K)} : (X,K) \to \widetilde{T}(X,K)$ is defined by $\xi_{(X,K)} = (\xi_X, 1_K)$, $\xi_X : X \to T(X,K)$, $\xi_X(x) = (K, \eta_x^X)$, where $\eta_x^X : K \to L^X$ is defined by $\eta_x^X(k)(z) = \chi_{\{x\}}^X(z)$, $k \in K, z \in X$.

Then $\widetilde{\mathbf{T}}$ is a power set monad in the category $\mathbf{Set}_*$.

Theorem 2 *[17]. The structure $\mathbf{H} = (H, \diamond, \sigma, 1_{\mathbf{Set}})$ is defined by*

1. The object function $H : obj(\mathbf{Set}) \to obj(\mathbf{Set})$ is defined by $H(X) = \{h | h : X \to 2^L\}$.
2. The set $H(X)$ is ordered by the relation

$$h, g \in H(X), h \preceq g \Leftrightarrow (\forall x \in X) h(x) \subseteq g(x).$$

3. If $f : X \to H(Y)$ and $g : Y \to H(Z)$ are $\mathbf{Set}$-morphisms, for arbitrary $x \in X, z \in Z$ we set

$$g \diamond f : X \to H(Z),$$
$$g \diamond f(x)(z) = \bigcup_{y \in Y} f(x)(y) \otimes g(y)(z) \subseteq L,$$

where for $A, B \subseteq L$, $A \otimes B = \{\alpha \otimes \beta | \alpha \in A, \beta \in B\}$ and $A \otimes \emptyset = \emptyset$.
4. For $X \in \mathbf{Set}$, $\sigma_X : X \to H(X)$ is defined by

$$x, z \in X, \quad \sigma_X(x)(z) = \begin{cases} \{1_L\}, & x = z \\ \emptyset, & x \neq z \end{cases}.$$

Then $\mathbf{H}$ is a power set monad in the category $\mathbf{Set}$.

Theorem 3 *[17]. Let $\mathcal{L}$ be an MV-algebra and let the structure $\mathbf{J} = (J, \boxtimes, \eta, 1_{\mathbf{Set}})$ be defined in the category $\mathbf{Set}$ by*

1. $J : obj(\mathbf{Set}) \to obj(\mathbf{Set})$ is a mapping defined by

$$J(X) = \{(u,v) | u, v \in L^X, \neg u \geq v\}.$$

2. The set $J(X)$ is ordered by the relation $\sqsubseteq$ such that

$$(u,v), (s,t) \in J(X), \quad (u,v) \sqsubseteq (s,t) \Leftrightarrow u \leq s, v \geq t,$$

where $\leq$ is a point-wise order relation on L^X.

3. *If $f : X \to J(Y)$ and $g : Y \to J(Z)$ are* **Set**-*morphisms, $g \boxtimes f : X \to J(Z)$ is defined by*

$$x \in X, \quad g \boxtimes f(x) = ((g \boxtimes f)^x, (g \boxtimes f)_x) \in J(Z),$$

where for $z \in Z$,

$$(g \boxtimes f)^x(z) = \bigvee_{y \in Y} f^x(y) \otimes g^y(z),$$

$$(g \boxtimes f)_x(z) = \bigwedge_{y \in Y} f_x(y) \oplus g_y(z).$$

4. *For $X \in$* **Set**, *$\eta_X : X \to J(X)$ is defined by*

$$x \in X, \quad \eta_X(x) = (\chi^X_{\{x\}}, \neg\chi^X_{\{x\}}).$$

Then **J** *is a power set monad in the category* **Set**.

With the help of monadic power set theory in a category, we can now define the concept of a *monadic relation*. This construction was first explicitly mentioned in the paper of Manes [15] and has recently proven to be a universal construction of relations for many fuzzy type structures (e.g., see [16]). We use the following form of a monadic relation in a category.

Definition 3 *[15]. Let* **K** *be a category and let $\mathbf{T} = (T, \Diamond, \eta, W)$ be a power set monad in* **K**.

1. *A* **T**-*relation R from an object X to an object Y in* **K**, *in symbol $R : X \rightsquigarrow Y$, is a* **K**-*morphism $R : X \to T(Y)$ in the category* **K**.
2. *If $R : X \rightsquigarrow Y$ and $S : Y \rightsquigarrow Z$ are* **T**-*relations, their composition is a* **T**-*relation $S \Diamond R : X \rightsquigarrow Z$.*

In fuzzy mathematics and its applications, various types of approximation and transformation operators are very often used, which convert fuzzy objects defined over the basic structure X to fuzzy objects over the other structure Y. Many of these transformation operators are special examples of a general transformation operator defined by **T**-relations as it is defined in the following definition.

Definition 4 *[16]. Let $\mathbf{T} = (T, \Diamond, \xi, W)$ be a power set monad in a category* **K** *and let $R : X \rightsquigarrow Y$ be a* **T**-*relation from X to Y. Then a R-transformation of objects from $T(X)$ is a* **K**-*morphism*

$$R^{\uparrow} = R \Diamond 1_{T(X)} : T(X) \to T(Y).$$

We recall a basic definition of a F-transform for L-fuzzy sets.

Definition 5 *[21]. Let X be a set and let $\mathcal{A} = \{A_i : i \in I\} \subseteq L^X$. Then*

1. *$\mathcal{A}$ is called a fuzzy partition, if $\{core(A_i) : i \in I\}$ is a partition of X, where $core(A) = \{x \in X : A(x) = 1_L\}$, i.e., $\bigcup_{i \in I} core(A_i) = X$, $core(A_i) \cap core(A_j) = \emptyset$, if $i \neq j$.*
2. *A transformation $F_{X,\mathcal{A}} : L^X \to L^I$ is called a F-transform, if for $s \in L^X, i \in I$, $F_{X,\mathcal{A}}(s)(i) = \bigvee_{x \in X} s(x) \otimes A_i(x)$.*

3 F-Transform for Hesitant, Soft and Intuitionistic $\mathcal{L}$-Fuzzy Sets

In order to be able to define the F-transform for the mentioned three fuzzy type structures, it is important to realize that the F-transform for $\mathcal{L}$-fuzzy sets can be equivalently defined by a power set monad $\mathbf{Z} = (Z, \boxplus, \chi, 1_{\mathbf{Set}})$ in the category **Set** from Example 1. In fact, for a fuzzy partition $(X, \mathcal{A})$, $\mathcal{A} = \{A_i : i \in |\mathcal{A}|\}$, and the monad $\mathbf{Z}$ we can consider a $\mathbf{Z}$-relation $Z_{\mathcal{A}} : X \rightsquigarrow |\mathcal{A}|$, such that for $x \in X, i \in |\mathcal{A}|$, $Z_{\mathcal{A}}(x)(i) = A_i(x)$. Then, it can be proven that for the F-transform $F_{X,\mathcal{A}} : L^X \to L^{|\mathcal{A}|}$, it holds

$$F_{X,\mathcal{A}} = Z_{\mathcal{A}} \boxplus 1_{Z(X)}.$$

In order to be able to use this construction to define a monadic version of an F-transform in a category $\mathbf{K}$ with a power set monad $\mathbf{T} = (T, \eta, \Diamond, W)$, we should emphasise that this monadic F-transform will be defined by the composition functor $W.T : \mathbf{K} \to \mathbf{Set}$. To define a F-transform, we need

1. For an arbitrary object $X \in \mathbf{K}$ to define a suitable fuzzy type partition $\mathcal{A} \subseteq W.T(X)$,
2. For $(X, \mathcal{A})$ to define an appropriate $\mathbf{T}$-relation $T_{\mathcal{A}} : X \rightsquigarrow |\mathcal{A}|$.

Then a F-transform based on $(X, \mathcal{A})$ can be defined by using a $\mathbf{K}$-morphism

$$F_{X,\mathcal{A}} = T_{\mathcal{A}} \Diamond 1_{T(X)} : T(X) \to T(|\mathcal{A}|).$$

In the rest of this section we want to show how these two constructions can be realized for the above mentioned fuzzy structures and how the resulting F-transform is defined. Details of these constructions and proofs will be published elsewhere.

3.1 Hesitant F-Transform

We will now apply this procedure to the power set monad $\mathbf{H} = (H, \diamond, \sigma, 1_{\mathbf{Set}})$ from Theorem 2. Let $X \in \mathbf{Set}$ and $h \in H(X)$. We define the core of h by

$$core(h) = \{x \in X : 1_L \in h(x)\}.$$

Then a subset $\mathcal{A} = \{h_i : i \in |\mathcal{A}|\} \subseteq H(X)$ is called a *hesitant* $\mathcal{L}$ *-fuzzy partition* of X, if $\{core(h_i) : i \in |\mathcal{A}|\}$ is a partition of a set X. It follows that there exists a surjective mapping

$$i_{\mathcal{A}} : X \twoheadrightarrow |\mathcal{A}|,$$

such that $x \in core(h_{i_{\mathcal{A}}(x)})$. Then according to [17]; Lemma 2.1, we can define a $\mathbf{H}$-relation $H_{\mathcal{A}} : X \rightsquigarrow |\mathcal{A}|$ such that

$$x \in X, i \in |\mathcal{A}|, \quad H_{\mathcal{A}}(x)(i) = h_i(x) \subseteq L.$$

Definition 6. *If $\mathcal{A} = \{h_i : i \in |\mathcal{A}|\}$ is a hesitant $\mathcal{L}$-fuzzy partition of X, then a hesitant F-transform $HF_{X,\mathcal{A}} : H(X) \to H(|\mathcal{A}|)$ defined by $(X, \mathcal{A})$ is such that for $h \in H(X), i \in |\mathcal{A}|$,*

$$HF_{\mathcal{A}} = H_{\mathcal{A}}^{\uparrow} = H_{\mathcal{A}} \diamond 1_{H(X)},$$

$$HF_{X,\mathcal{A}}(h)(i) = (H_{\mathcal{A}} \diamond 1_{H(X)})(h)(i) = \bigcup_{x \in X} h(x) \otimes h_i(x) \subseteq L.$$

Hesitant F-transform can be equivalently defined axiomatically as a general hesitant transformation system:

Definition 7. *A triple (Y, u, G) is called a hesitant transformation system of $X \in \mathbf{Set}$, if*

1. *$Y \in \mathbf{Set}$ and $u : X \twoheadrightarrow Y$ is a surjective mapping,*
2. *$G : H(X) \to H(Y)$ is a mapping, such that*
   (a) G is $\bigvee$-preserving with respect to ordering in $H(X)$,
   (b) For all $\alpha \in L, g \in H(X)$, $\quad G(\alpha \otimes g) = \alpha \otimes G(g)$,
   (c) For all $x \in X, y \in Y$, $1_L \in G(\sigma_X(x)) \Leftrightarrow u(x) = y$.

Then the following theorem describes an axiomatic definition of a hesitant F-transform.

Theorem 4. *Let $X, Y \in \mathbf{Set}$, $u : X \twoheadrightarrow Y$ and $G : H(X) \to H(Y)$. Then the following statements are equivalent.*

1. *(Y, u, G) is a hesitant transformation system of X.*
2. *There exists a hesitant $\mathcal{L}$-fuzzy partition $\mathcal{A} = \{h_y : y \in Y\}$ of X such that*
   (a) For all $x \in X, y \in Y$, $u(x) = y \Leftrightarrow x \in core(h_y)$,
   (b) $G = HF_{X,\mathcal{A}}$.

3.2 Intuitionistic F-Transform

The same procedure we can apply to the power set monad $\mathbf{J} = (J, \boxtimes, \eta, 1_{\mathbf{Set}})$ from Theorem 3. Let $\mathcal{L}$ be a MV-algebra and let $X \in \mathbf{Set}$ and $(u, v) \in J(X)$. We define the core of (u, v) by

$$core(u, v) = \{x \in X : u(x) = 1_L\}.$$

If $x \in core(u, v)$, then we have $v(x) = 0_L$. Then a subset $\mathcal{A} = \{(s_i, t_i) : i \in |\mathcal{A}|\} \subseteq J(X)$ is called an *intuitionistic $\mathcal{L}$ -fuzzy partition* of X, if $\{core(s_i, t_i) : i \in |\mathcal{A}|\}$ is a partition of X. It follows that there exists a surjective map

$$i_{\mathcal{A}} : X \twoheadrightarrow |\mathcal{A}|$$

such that $x \in core(s_{i_{\mathcal{A}}(x)}, t_{i_{\mathcal{A}}(x)})$. We use the following notation:

1. For $(u,v) \in J(X), x \in X$, $(u,v)(x) = (u(x), v(x)) \in L^2$,
2. For $\alpha \in L, (u,v) \in J(X)$, $\alpha \otimes (u,v) = (\alpha \otimes u, \neg\alpha \oplus v) \in J(X)$,
3. For $(\alpha,\beta) \in L^2, \neg\alpha \geq \beta$, $(\alpha,\beta) \otimes (u,v) = (\alpha \otimes u, \beta \oplus v) \in J(X)$.

Then, according to [17]; Lemma 3, we can define an **J**-relation $J_{\mathcal{A}} : X \rightsquigarrow |\mathcal{A}|$ by

$$x \in X, i \in |\mathcal{A}|, \quad J_{\mathcal{A}}(x)(i) = (s_i(x), t_i(x)).$$

Using the monad **J** and the **J**-relation $J_{\mathcal{A}}$, according to [17]; Proposition 5, we can define an intuitionistic F-transform $JF_{X,\mathcal{A}}$.

Definition 8. *If $\mathcal{A} = \{(s_i,t_i) : i \in |\mathcal{A}|\}$ is an intuitionistic $\mathcal{L}$-fuzzy partition of X, then an intuitionistic F-transform $JF_{X,\mathcal{A}} : J(X) \to J(|\mathcal{A}|)$ defined by $(X,\mathcal{A})$ is such that for $(u,v) \in J(X), i \in |\mathcal{A}|$,*

$$JF_{X,\mathcal{A}} = J_{\mathcal{A}}^{\uparrow} = J_{\mathcal{A}} \boxtimes 1_{J(X)},$$

$$JF_{X,\mathcal{A}}(u,v)(i) = (J_{\mathcal{A}} \boxtimes 1_{J(X)})(u,v)(i) = (\bigvee_{x\in X} u(x) \otimes s_i(x), \bigwedge_{x \in X} v(x) \oplus t_i(x)).$$

Intuitionistic F-transform can be equivalently defined axiomatically as a general intuitionistic transformation system:

Definition 9. *A triple (Y,u,G) is called an intuitionistic transformation system of $X \in$ **Set**, if*

1. *$Y \in$ **Set** and $u : X \twoheadrightarrow Y$ is a surjective mapping,*
2. *$G : J(X) \to J(Y)$ is a mapping, such that*
   - *(a) G is $\bigvee$-preserving with respect to ordering in $J(X)$,*
   - *(b) For all $(\alpha,\beta) \in L^2, \neg\alpha \geq \beta, (u,v) \in J(X)$, $G((\alpha,\beta) \otimes (u,v)) = (\alpha,\beta) \otimes G(u,v)$,*
   - *(c) For all $x \in X, y \in Y$, $G(\eta_X(x))(y) = (1_L, 0_L) \Leftrightarrow u(x) = y$.*

Then the following theorem describes an axiomatic definition of an intuitionistic F-transform.

Theorem 5. *Let $\mathcal{L}$ be a MV-algebra and let $X, Y \in$ **Set**, $u : X \twoheadrightarrow Y$ be a surjective map and $G : J(X) \to J(Y)$. Then the following statements are equivalent.*

1. *(Y,u,G) is an intuitionistic transformation system of X.*
2. *There exists an intuitionistic $\mathcal{L}$-fuzzy partition $\mathcal{A} = \{(s_y,t_y) : y \in Y\}$ of X such that*
   - *(a) For all $x \in X, y \in Y$, $u(x) = y \Leftrightarrow x \in core(s_y,t_y)$,*
   - *(b) $G = JF_{X,\mathcal{A}}$*

3.3 Soft F-Transform

To define a soft version of a F-transform we use the power set monad $\widetilde{\mathbf{T}}$ in the category $\mathbf{Set}_*$ from Theorem 1. In that case the soft F-transform will be defined by the functor $W.\widetilde{T} = T$ from Theorem 1. We use the following notations for $(E,s) \in T(X,K)$, $\alpha \in L$ and $\varphi \in L^E$ and $x \in X, e \in E$:

$$(E,s)(e)(x) = s(e)(x), \quad \alpha \otimes (E,s) = (E, \alpha \otimes s),$$
$$\varphi \otimes (E,s) = (E, \varphi \otimes s), \quad (\varphi \otimes s)(e)(x) = \varphi(e) \otimes s(e)(x),$$
$$core(E,s) = \{x \in X : \forall e \in E, s(e)(x) = 1_L\},$$
$$\chi^E_{\{-\}}(e) : E \to L, \quad \chi^E_{\{-\}}(e)(n) = \begin{cases} 1_L, & n = e, \\ 0_L, & n \neq e. \end{cases}$$

By a *$\mathcal{L}$-fuzzy soft partition* of (X,K) we understand a set $\mathcal{A} = \{(E_i, p_i) : i \in |\mathcal{A}|\} \subseteq T(X,K)$, such that

1. $\{core(E_i, p_i) : i \in |\mathcal{A}|\}$ is a partition of X and
2. $\bigcup_{i \in |\mathcal{A}|} E_i = K$.

It follows that there exists a surjective map

$$i_{\mathcal{A}} : X \twoheadrightarrow |\mathcal{A}|$$

such that $x \in core(E_{i_{\mathcal{A}}(x)}, p_{i_{\mathcal{A}}(x)})$ For an $\mathcal{L}$-fuzzy soft partition $\mathcal{A}$ of (X,K) we can defined a $\widetilde{\mathbf{T}}$-relation $\widetilde{T_{\mathcal{A}}} = (T_{\mathcal{A}}, 1_K) : (X,K) \rightsquigarrow (|\mathcal{A}|, K)$ by

$$T_{\mathcal{A}} : X \to T(|\mathcal{A}|, K),$$
$$x \in X, \quad T_{\mathcal{A}}(x) = (E_{i_{\mathcal{A}}(x)}, i_{\mathcal{A}}^{\rightarrow} . p_{i_{\mathcal{A}}(x)}),$$

where $i_{\mathcal{A}}^{\rightarrow}$ is the Zadeh's extension of a mapping $i_{\mathcal{A}} : X \to |\mathcal{A}|$ to the mapping $L^X \to L^{|\mathcal{A}|}$. Now, using a fuzzy soft partition $\mathcal{A}$ and a $\widetilde{\mathbf{T}}$ relation $\widetilde{T_{\mathcal{A}}}$ we can defined the notion of a soft F-transform $SF_{(X,K),\mathcal{A}}$ of (X,K) defined by $\mathcal{A}$ as follows:

Definition 10. *If $\mathcal{A} = \{(E_i, p_i) : i \in |\mathcal{A}|\}$ is an $\mathcal{L}$-fuzzy soft partition of (X,K), then a soft F-transform of (X,K) is a mapping $SF_{(X,K),\mathcal{A}} : T(X,K) \to T(|\mathcal{A}|, K)$ such that*

$$SF_{(X,K),\mathcal{A}} = T^{\uparrow}_{\mathcal{A}} = T_{\mathcal{A}} \triangle 1_{T(X,K)},$$
$$(E,s) \in T(X,K), \quad SF_{(X,K),\mathcal{A}}(E,s) = (E, \overline{s}) \in T(|\mathcal{A}|, K),$$
$$\overline{s}(n)(i) = (E, \overline{s})(n)(i) = \bigvee_{z \in X, n \in E_{i_{\mathcal{A}}(z)}} s(n)(z) \otimes i_{\mathcal{A}}^{\rightarrow}(p_{i_{\mathcal{A}}(z)}(n))(i)$$

for $n \in E, i \in |\mathcal{A}|$.

The soft F-transform can be equivalently defined axiomatically as a general soft transformation system:

Definition 11. *A triple* (Y, u, G) *is called a soft transformation system of* $(X, K) \in \mathbf{Set}_*$, *if*

1. $Y \in \mathbf{Set}$ *and* $u : X \twoheadrightarrow Y$ *is a surjective mapping,*
2. $G : T(X, K) \to T(Y, K)$ *is a mapping, such that*
   (a) G *is* $\bigvee$*-preserving with respect to ordering in* $T(X, K)$,
   (b) *For all* $\alpha \in L$, $G(\alpha \otimes (E, s)) = \alpha \otimes G(E, s)$,
   (c) *For all* $e \in K$, $G((E, s) \otimes \chi^E_{\{-\}}(e)) = G(E, s) \otimes \chi^E_{\{-\}}(e)$,
   (d) $G(K, \eta^X_x)(e)(y) = 1_L \Leftrightarrow w(x) = y$,
   (e) G *preserves the criteria set of objects from* $T(X, K)$.

In the following theorem we present an equivalent axiomatic definition of a soft F-transform.

Theorem 6. *Let* $X, Y \in \mathbf{Set}$, $u : X \twoheadrightarrow Y$ *be a surjective map and* $G : T(X, K) \to T(Y, K)$ *be a map. Then the following statements are equivalent.*

1. (Y, u, G) *is a soft transformation system of* (X, K).
2. *There exists a* $\mathcal{L}$*-fuzzy soft partition* $\mathcal{A} = \{(E_y, p_y) : y \in Y\}$ *of* (X, K) *such that*
   (a) *For all* $x \in X, y \in Y$, $u(x) = y \Leftrightarrow x \in core(E_y, p_y)$,
   (b) $G = SF_{(X,K),\mathcal{A}}$

References

1. Aggarwal, H., Arora, H.D., Kumar, V.: A decision-making problem as an applications of intuitionistic fuzzy set. Int. Eng. Adv. Technol. **9**(2), 5259–5261 (2019)
2. Atanassov, K.T.: Intuitionistic fuzzy sets. Fuzzy Sets Syst. **20**(1), 87–96 (1986)
3. Cignoli, R.L., d'Ottaviano, I.M., Mundici, D.: Algebraic Foundations of Many-Valued Reasoning. TREN, Springer, Heidelberg (2000). https://doi.org/10.1007/978-94-015-9480-6
4. Di Martino, F., et al.: An image coding/decoding method based on direct and inverse fuzzy tranforms. Int. J. Approx. Reason. **48**, 110–131 (2008)
5. Di Martino, F., et al.: A segmentation method for images compressed by fuzzy transforms. Fuzzy Sets Syst. **161**(1), 56–74 (2010)
6. Di Martino, F., et al.: Fuzzy transforms method and attribute dependency in data analysis. Inf. Sci. **180**(4), 493–505 (2010)
7. Feng, F., Jun, Y.B., Zhao, X.Z.: Soft semirings. Comput. Math. Appl. **56**, 2621–2628 (2008)
8. Herrlich, H., Strecker, G.E.: Category Theory, 3rd edn. Heldermann Verlag, Berlin (2007)
9. Khastan, A., Perfilieva, I., Alijani, Z.: A new fuzzy approximation method to Cauchy problem by fuzzy transform. Fuzzy Sets Syst. **288**, 75–95 (2016)
10. MacLane, S.: Categories for the Working Mathematician. GTM, vol. 5. Springer Verlag, Berlin (1998). https://doi.org/10.1007/978-1-4612-9839-7
11. Maji, P.K., et al.: Fuzzy soft-sets. J. Fuzzy Math. **9**(3), 589–602 (2001)
12. Maji, P.K., et al.: An application of soft sets in a decision making problem. Comput. Math. Appl. **44**, 1077–083 (2002)

13. Majumdar, P., Samanta, S.K.: Similarity measure of soft sets. New Math. Nat. Comput. **4**(1), 1–12 (2008)
14. Manes, E.G.: Algebraic Theories. GTM, Springer Verlag, Berlin (1976). https://doi.org/10.1007/978-1-4612-9860-1
15. Manes, E.G.: Book review fuzzy sets and systems, theory and applications. Bull. (New Ser.) Am. Math. Soc. **7**(3), 603–612 (1982)
16. Močkoř, J.: Fuzzy type relations and transformation operators defined by monads. Int. J. Comput. Intell. Syst. **13**(1), 1530–1538 (2020)
17. Močkoř, J., Hýnar, D.: On unification of methods in theories of fuzzy sets, hesitant fuzzy set, fuzzy soft sets and intuitionistic fuzzy sets. Mathematics **9**(4), 447, 1–26 (2021)
18. Močkoř, J., Hurtík, P.: Approximations of fuzzy soft sets by fuzzy soft relations with image processing application. Soft Comput. **25**(10), 6915–6925 (2021)
19. Mushrif, M.M., Sengupta, S., Ray, A.K.: Texture classification using a novel, soft-set theory based classification algorithm. In: Narayanan, P.J., Nayar, S.K., Shum, H.-Y. (eds.) ACCV 2006. LNCS, vol. 3851, pp. 246–254. Springer, Heidelberg (2006). https://doi.org/10.1007/11612032_26
20. Novák, V., Perfilijeva, I., Močkoř, J.: Mathematical Principles of Fuzzy Logic. Kluwer Academic Publishers, Boston, Dordrecht, London (1991)
21. Perfilieva, I.: Fuzzy transforms: theory and applications. Fuzzy Sets Syst. **157**, 993–1023 (2006)
22. Perfilieva, I.: Fuzzy transforms and their applications to image compression. In: Bloch, I., Petrosino, A., Tettamanzi, A.G.B. (eds.) WILF 2005. LNCS (LNAI), vol. 3849, pp. 19–31. Springer, Heidelberg (2006). https://doi.org/10.1007/11676935_3
23. Perfilieva, I., Novak, V., Dvořak, A.: Fuzzy transforms in the analysis of data. Int. J. Approx. Reason. **48**, 36–46 (2008)
24. Rodabaugh, S.E.: Power set operator foundation for poslat fuzzy set theories and topologies. In: Höhle, U., Rodabaugh, S.E. (eds.) Mathematics of Fuzzy Sets: Logic, Topology and Measure Theory. The Handbook of Fuzzy Sets Series, vol. 3, pp. 91–116. Kluwer Academic Publishers, Boston, Dordrecht (1999)
25. Rodríguez, R.M., et al.: Hesitant fuzzy sets: state of the art and future directions. Int. J. Intell. Syst. **29**(6), 495–524 (2014)
26. Stefanini, L.: F-transform with parametric generalized fuzzy partitions. Fuzzy Sets Syst. **180**, 98–120 (2011)
27. Tomasiello, S.: An alternative use of fuzzy transform with application to a class of delay differential equations. Int. J. Comput. Math. **94**(9), 1719–1726 (2017)
28. Torra, V., Narukawa, Y.: On hesitant fuzzy sets and decision. In: Proceedings of the 2009 IEEE International Conference on Fuzzy Systems (FUZZ-IEEE), Jeju Island, Korea, pp. 1378–1382 (2009)
29. Torra, V.: Hesitant fuzzy sets. Int. J. Intell. Syst. **25**(6), 529–539 (2010)
30. Yahya, M., Begum, E.N.: A study on intuitionistic L-fuzzy metric spaces. Ann. Pure Appl. Math. **15**(1), 67–75 (2017)
31. Zeshui, X.: Hesistant Fuzzy Sets Theory. STUDFUZZ, Springer, Cham (2014). https://doi.org/10.1007/978-3-319-04711-9
32. Zhang, H.: Linguistic intuitionistic fuzzy sets and application in MAGDM. J. Appl. Math. (2014). Article ID 432092
33. Zhang, X.L., Xu, Y.S.: A MST clustering analysis method under hesitant fuzzy environment. Control Cybern. **41**(3), 645–666 (2012)

Enhancing Social Recommenders with Implicit Preferences and Fuzzy Confidence Functions

Camilo Franco[1(✉)], Nicolás Hernández[1,2], and Haydemar Núñez[2]

[1] Department of Industrial Engineering, University of the Andes, 111711 Bogota, Colombia
{c.franco31,nm.hernandez10}@uniandes.edu.co

[2] Department of Systems and Computer Engineering, University of the Andes, 111711 Bogota, Colombia
h.nunez@uniandes.edu.co

Abstract. In this paper we explore the use of fuzzy confidence functions for enhancing the performance of social recommenders with implicit preferences, focusing on K-Nearest Neighbors (KNN) collaborative filtering algorithms. Firstly, we measure the effects of including social relations for enhancing the performance of the algorithms with either explicit or implicit preferences, expecting to verify better results when social attributes are considered in the relevant neighborhood estimation. Secondly, it is proposed to enhance the social recomenders with implicit preferences by fuzzy confidence functions. An application is developed to measure the effects of including social relations and to illustrate our proposal on the fuzzy modeling of implicit preferences, recommending courses based on the students socio-demographic and academic information. As a result, the best recommendations are accomplished with socially-enhanced algorithms that make use of implicit preferences and fuzzy confidence functions, obtaining a FCP in test of 0.68.

Keywords: Recommendation systems · Collaborative filtering · K-nearest neighbors · Social closeness · Explainability

1 Introduction

Recommendation systems allow applying machine learning techniques in a personalized and useful fashion, building preference models with a great range of applications. Examining user-based collaborative filtering algorithms as a first-basic type of social recommender, building recommendations from similar user rating behavior, it is noted that the inclusion of relevant social information should only improve the systems' recommendations by accomplishing higher performance scores. From this standing point, the use of fuzzy confidence functions is explored for estimating implicit preferences and improving socially-enhanced recommendation systems.

V. Torra and Y. Narukawa (Eds.): MDAI 2021, LNAI 12898, pp. 118–130, 2021.
https://doi.org/10.1007/978-3-030-85529-1_10

There is an already important body of knowledge supporting the claim that social information allows enhancing the performance of intelligent systems (see, e.g., [5]). Perhaps the most popular implementations of recommender systems with Collaborative Filtering (CF) make use of Matrix Factorization (MF) algorithms, enhancing their predictive capability by incorporating the knowledge contained in social data sources, either through *social regularization* (see e.g. [13]), or directly including social analysis for identifying similar users [11,14].

A rather less developed body of literature can be found on K-Nearest Neighbors (KNN) collaborative filtering, making use of social and relevant neighbor information for estimating the users' preferences. One example can be found for recommending academic conferences with implicit-binary preferences [3], which builds the users' explicit ratings according to the number of papers published. Another example of enhanced-social KNN-CF can be found in [1], with the novelty of estimating the reliability of the neighborhood and implicit preferences by Dempster-Shafer theory, having significant effects on users holding insufficient relationships with their neighbors.

Developing socially enhanced KNN-CF algorithms, we include social information through a social similarity function, which takes into account the social background between users, and consider both implicit and explicit preferences. In this sense, an explicit framework allows interpreting users' preferences directly from an observed rating, which can be regarded (although it is not always properly specified) as a bipolar univariate, Likert-type scale, where 1 star reveals an absolute negative preference, 3 stars stand for absence of preference and 5 stars correspond with absolute positive preference (on this semantic issue, see e.g. [7]). On the other hand, implicit preferences refer to some binary knowledge on a user being interested in an item, just because there was some interaction with that item. But implicit preferences can be modeled in a more expressive and informative way, according to the *confidence* that users express on preferring a specific item (as suggested in [9]). Here we propose a fuzzy representation of such a confidence, gradually incorporating information on the implicit preference in a monotonically increasing manner.

There are multiple examples in literature dealing with techniques for estimating implicit-binary preferences from the users' behavior and the context of the user-item interaction (see e.g. [2,12]), but not necessarily taking into account the strength or degree of confidence of such an implicit preference (as stated above in the sense of [9]). Following the original proposal of [9], where it is acknowledged that beliefs or preferences are associated with varying confidence levels, we examine the use of a fuzzy representation of confidence for implicit feedback. Hence, a certain graduality is introduced, obtaining a smoothed estimation of the confidence in observing the implicit preference for an item, also contributing to the transparency and explainability of the system. For illustrating the use of fuzzy confidence functions, a case study is developed for academic course recommendation, where the proxy for confidence is given by the qualification or grade that the students achieve for the course.

This particular application for course recommendation is chosen not only to illustrate our proposal on fuzzy confidence functions, but also because it allows modelling the recommendation problem both from the perspective of having explicit preferences, taking as proxy of such preferences the grade obatined for a given course; and from the perspective of having implicit preferences, where the grade is used as the input of a fuzzy function for estimating the confidence for a given binary preference (see e.g. [4,6], for other applications for academic course recommendation). Hence, it can be tested if the recommendations can be enhanced by using fuzzy confidence functions, and at the same time, offering evidence on the relevance of including social information among users. As a result, the use of social information, together with the fuzzy modeling of confidence for estimating implicit preferences, obtains the best results for our application in course recommendation.

For presenting the complete methodology and the case study, the following paper is organized as follows. In Sect. 2 we present the data for our course recommendation application, as well as the social-enhanced KNN-CF algorithms that are going to be examined in this study and the evaluation metrics. Then, in Sect. 3 the fuzzy representation of confidence is introduced, necessary for computing the KNN-CF algorithms with implicit preferences. Finally, in Sect. 4, the results are presented, comparing the performance of the explicit/implicit KNN-CF recommenders with and without social information, and in Sect. 5, some final comments are given for future research.

2 Methodology

In the following section, the recollection of data for the course recommendation application is explained, as well as the socially-enhanced KNN-CF algorithms.

2.1 Data Recollection and Experimental Setting

Data was gathered from the historical information on students' characteristics and performance for courses between the first semester of 2016 and the second semester of 2018. This database has socio-demographic and academic information. For the former, it includes the anonymous ID of the students, along with their sex, age, city and region (department) from origin, the academic level (BA, MSc, PhD), and previous school (before beginning the university). Here, schools have an extra attribute consisting in their calendar, A or B, depending on the beginning and end of their academic year.

For the latter, information consisted on the course code, the grade of the student for that course, the semester in which the course was taken, the course department and faculty, and the credits for the course. In total there were 24862 students and 539130 courses taken by students with their respective grades.

Here it should be noted, as it will be better explained in Sect. 3, that the grade does not necessarily represent a rating as such, as it is not a subjective judgment of preference from the user, but rather an approximated degree of enjoying the course. For our explicit preference models we shall make the simplifying assumption that the grade is in itself the revealed rating between 0 and 5, but for our implicit preference approach, we shall take it as the proxy of confidence on the preference for the course, represented according to a given fuzzy membership function.

In order to evaluate the recommendation algorithms, we take 70% of the available ratings for training, and leave the remaining 30% for testing. For systems working with explicit preferences, the Root Mean Squared Error (RMSE) allows measuring the performance of the models regarding their predicted rating. However, with implicit preferences, a different type of measure is required, which allows grading the pertinence or accuracy of the binary recommendations. For doing this, the Fraction of Concordant Pairs (FCP) is going to be used [10], as it allows to take the preference value not on its cardinal value, but on its ordinal one. The FCP is defined by,

$$FCP = \frac{n_c}{n_c + n_d} \quad (1)$$

where the number of concordant items $n_c = \sum_u n_c^u$ is given for every user u, and pair of items (i, j), by $n_c^u = |\{(i, j)\hat{r}_{ui} > \hat{r}_{uj} \text{ and } r_{ui} > r_{uj}\}|$ where r_{ui} is the rating of user u to item i, and the number of discordant items n_d^u is analogously given for every user u, and pair of items (i, j), by $n_d^u = |\{(i, j)\hat{r}_{ui} > \hat{r}_{uj} \text{ and } r_{ui} < r_{uj}\}|$.

In this way, the FCP allows capturing when an item is assigned a high position in the recommendation ranking at least as good as the one actually assigned by the user. Furthermore, the FCP can be used to compare the algorithm's performance for both explicit and implicit feedback.

2.2 Social-Enhanced KNN-CF Algorithms

KNN-CF is a common approach to develop user-user recommendation systems, where unknown preferences can be predicted from similarly, like-minded, users, or what can be referred as the user neighborhood. Central to this approach is the estimation of weights for combining the members of the user neighborhood. Those weights are commonly computed by similarity functions, measuring the tendency of users to rate common items in a similar or concordant fashion.

Then, the objective of the KNN-CF algorithm consists in predicting the unobserved rating of user u for item i, given by r_{ui}, by means of the weighted average of the ratings from u's neighborhood, i.e., the k nearest neighbors of u. In this way, based on a given similarity measure $sim(u, v)$, let's define $N_i^k(u)$ as the set of all the k users which are most similar to u according to their preference for a given item i, and take the predicted value of r_{ui} as a weighted average of the ratings of neighboring users.

The *basic KNN* is modelled by Eq. (2).

$$\hat{r}_{ui} = \frac{\sum_{v \in N_i^k(u)} sim(u, v) \cdot r_{vi}}{\sum_{v \in N_i^k(u)} sim(u, v)}. \quad (2)$$

In case neighborhood information is totally or almost absent, i.e., where users do not have that much items in common, the baseline extension can be a better option, as in Eq. (3), taking into account the mean rating of user u for item i,

$$\hat{r}_{ui} = b_{ui} + \frac{\sum_{v \in N_i^k(u)} sim(u,v) \cdot (r_{vi} - b_{vi})}{\sum_{v \in N_i^k(u)} sim(u,v)} \tag{3}$$

where $b_{ui} = \mu + b_u + b_i$, and b_u and b_i indicate the observed deviations of user u and item i, respectively, from the total average rating μ.

It should be here noted that for explicit preferences, the rating r_{ui} will correspond in our course recommendation application to the grade obtained by student u in course i, while for our implicit preference implementation, the *rating* r_{ui} will be taken as a proxy of the confidence for the preference on a given course (according to the grade achieved by the student), as it will be explained in Sect. 3.

Lastly, the similarity function $sim(u,v)$ is here taken for every pair of users (u,v) as the Pearson correlation coefficient, which is computed by

$$\rho(u,v) = \frac{\sum_{i \in I_{uv}} (r_{ui} - \mu_u) \cdot (r_{vi} - \mu_v)}{\sqrt{\sum_{i \in I_{uv}} (r_{ui} - \mu_u)^2} \cdot \sqrt{\sum_{i \in I_{uv}} (r_{vi} - \mu_v)^2}} \tag{4}$$

where I_{uv} stands for the set of items in common between users u and v.

Extending the KNN-CF approach to a hybrid-social KNN-CF system, the similarity function has to be extended to make use of social attributes for properly identifying such a neighborhood. In the first place, social ties can be represented by a graph $G = (E, V)$, where the nodes in V are individuals (students) and the edges in E are the links or connections that exist among the nodes. Hence, a pair of nodes are linked together if there is an observed match between a given social dimension or attribute.

Let's recall that the social attributes considered in this application are the students' sex, age, city and region from origin, the academic level, and the previous school. All these attributes are directly measured, composing a set D of social factors. Therefore, for every pair (u,v), the coincidence in any social aspect $j \in D$ is computed here by a binary similarity $bsim_j(u,v)$, taking the value of 1 if there is a coincidence a 0 otherwise.

Aggregating the information that is gathered for all the social aspects in D, an overall social similarity $sim_{social}(u,v)$ is obtained for any neighboring pair (u,v), by

$$sim_{social}(u,v) = \frac{\sum_{j \in D} bsim_j(u,v)}{|D|}. \tag{5}$$

In this way, sim_{social} captures the proportion of social attributes in which the users make a social match, taken here as the approximate likelihood that they share the same preferences due to their social background. Therefore, both the user-item (sim_{user}, referring to similarities based only on preference over items) and the user-social similarities (sim_{social}) can now be combined together to potentially enhance the accuracy of KNN-CF recommenders. Here we take a convex combination between both sim_{user} and sim_{social}, given by

$$sim_{global}(u, v) = \alpha sim_{user}(u, v) + \beta sim_{social}(u, v) \tag{6}$$

where α and β represent the strength in which the preference or social based similarities are taken into account for computing the global similarity sim_{global}.

Following the Pearson similarity defined in Eq. (4), the social Pearson global similarity is here defined by

$$\rho_{global}(u, v) = \alpha\rho(u, v) + \beta sim_{social}(u, v). \tag{7}$$

After incorporating the social information into the KNN-CF by means of Eq. (7), it is still required to extend the algorithms to work with implicit preferences, given their special binary quality and the interpretation of the available rating (grade) for denoting a degree of confidence on the perceived binary feedback. For doing so we use soft, fuzzy, functions, as it will be explained next.

3 Fuzzy Confidence Functions

As previously mentioned, user-user CF algorithms are built from the interrelations of the users according to their interactions with the items. In this way, identifying users with a similar history of interactions or preferences over the same items, being those preferences explicit, as in a given rating for the items, or implicit, just by the occurrence of some interaction with the items, the CF system can estimate unknown relations between users and items.

Despite the explicit information being very useful for CF recommendation, implicit feedback is also relevant, given the potential unavailability of ratings or the reluctance of users to rate items. Besides, its recollection is easier than eliciting explicit ratings. Nonetheless, implicit preferences also have inherent limitations, such as being either positive or not-positive, without possibly expressing negative judgments (although this is also true for explicit frameworks unless the valuation structure is specified, as the usual five-star rating system can be interpreted according to the Likert scale, but also to a unipolar scale ranging from the absence of preference of one star, to the full preference of five stars), or the lack of expressiveness regarding the motives or opinion on the observed interaction (perhaps a mistake, a random selection, or even a disappointing experience with the item).

Whereas the explicit feedback offers an estimate for the value of preference, the implicit feedback offers a binary value that has to be weighted by its associated confidence [9]. In this sense, the rating variables r_{ui} in Eqs. (2)–(3), express an intensity of confidence when representing implicit feedback that should be properly modeled.

Firstly, let's introduce binary preferences p_{ui}, such that it takes the value of 1 if $r_{ui} > 0$, and 0 otherwise. In this way, if a user u interacted with an item i, it holds that $r_{ui} > 0$, and thus, there is evidence that u prefers i, i.e., $p_{ui} = 1$.

But if user u never interacted with an item i, it holds that there is a missing value for r_{ui}, or that $r_{ui} = 0$, and thus, it is taken as evidence that u does not prefer i, i.e., $p_{ui} = 0$.

Now, the verification of $p_{ui} = 1$ is subject to different degrees of confidence, which is here modeled according to the variable r_{ui}. In this way, the greater r_{ui}, the greater the confidence on user u preferring item i. Such a confidence c_{ui} can be modeled by the linear function [9],

$$c_{ui} = 1 + \lambda r_{ui} \tag{8}$$

where λ stands as a constant positive rate by which confidence on p_{ui} grows proportionally to r_{ui}, and it is here computed by $c_{ui} * p_{ui}$.

Therefore, confidence on a binary preference p_{ui} is understood according to the fuzzy set q_{ui}, expressing the degree in which preference of u over i can be affirmed (following fuzzy set theory [15]). Therefore, fuzzy confidence q_{ui} can be in general represented by its co-support and its core. The co-support of q_{ui} is taken as the subset of values for $r_{ui} = 0$, where there is absence of preference. And the core of q_{ui} refers to the subset of values for r_{ui} where confidence of preference is absolute. Then, there is a space in between the borders of the core and the co-support which can be resolved by a continuous and monotone transition (linear in the case of a triangular or trapezoidal fuzzy set). This is a first step towards trying a gradual transition for the model, but it is here noted that it should be possible to provide an appropriate axiomatic characterization of the minimal properties that are required for these fuzzy confidence functions.

Following this intuitive proposal, we can model such a transition by a convex function, following an exponential behavior,

$$q_{ui}^{exp} = \exp(r_{ui}), \tag{9}$$

or a concave transition, like with the logarithmic function,

$$q_{ui}^{log} = \log(r_{ui}) + 1, \tag{10}$$

where $\log(0) := 0$, or a both convex and concave transition, as given by the logistic function

$$q_{ui}^{S} = \frac{L}{1 + \exp^{-\gamma(r_{ui} - \bar{r})}}, \tag{11}$$

where L is the upper bound, γ is the growth rate and $\bar{r}$ is the function's middle value. All of these become hyper-parameters that need to be tuned in order to learn the users' confidence over their implicit preferences. The design for the different fuzzy representations are depicted in Fig. 1.

Following this approach, we implement the KNN-CF for both implicit and explicit feedback, testing the performance of the algorithms when using the social information together with the preference behavior of users.

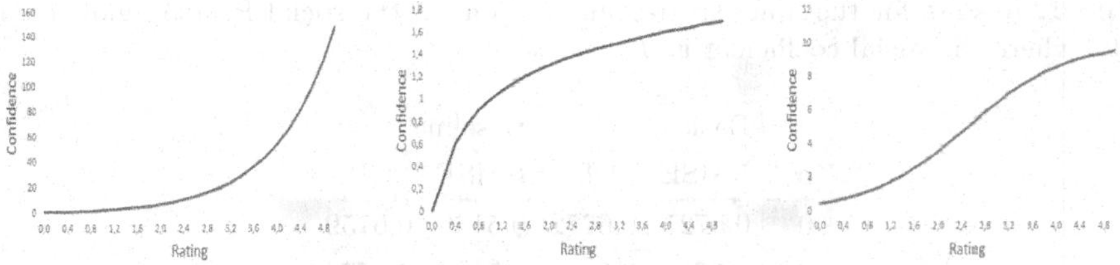

(a) (b) (c)

Fig. 1. Three plausible designs for fuzzy confidence following an (a) exponential (b) logarithmic or (c) logistic behavior.

4 Results

For the implementation of the experiments for the KNN-CF algorithms with implicit and explicit preferences, as well as with and without social information, we performed a broad search on the optimal number of neighbors. The search for the k neighbors took values from $\{2, 5, 7, 9, 15, 21, 51, 101\}$. Besides, the α parameter for computing the social Pearson global similarity in Eq. (7) is calibrated on the values $\alpha = \{0, 0.1, 0.2, ..., 0.9, 1\}$, where the social parameter β is such that $\beta = 1 - \alpha$. Hence, when $\alpha = 0$ the algorithm takes only social information into account, while for $\alpha = 1$ the algorithm is a pure KNN-CF algorithm, in the sense of Eqs. (2) or (3) together with Eq. (4), without any other social input.

4.1 Explicit Preferences

After estimating the explicit ratings r_{ui} according to Eq. (2) for the basic KNN-CF, and to Eq. (3) for the baseline KNN-CF, together with the social Pearson global similarity, we obtain the results presented in Table 1. Each entry for the respective α-value of the social Pearson similarity corresponds with its best result for the number of k nearest neighbors. For example, the best basic KNN-CF RMSE (0.5711) was obtained with $\alpha = 0.2$ and $k = 51$, which also obtains the best performance on the FCP (0.6778). As for the baseline KNN-CF, the best performance corresponds with $\alpha = 0.1$ and $k = 51$, with an RMSE of 0.5132 and a FCP of 0.6774.

On the other hand, the learning behavior for the KNN-CF algorithms, with and without social information (using the global social Pearson for the above mentioned α-values and the Pearson similarities, respectively), can be examined in Fig. 2. Such a behavior is depicted according to the number of nearest neighbors used to compute Eqs. (2)–(3).

In general, it can be seen that the best results arise when social data is taken into account, not only based on the performance metrics, but also on the sparsity of the estimation, as the use of the social Pearson similarity allows using a lower number of neighbors than with the Pearson similarity. From the previous

Table 1. Results for the different α-configurations of the social Pearson global similarity, where the social coefficient is $\beta = 1 - \alpha$

	Basic		Baseline	
α	RMSE	FCP	RMSE	FCP
0	0.5721	0.6772	0.5136	0.6759
0.1	0.5712	0.6778	0.5132	0.6774
0.2	0.5711	0.6778	0.5134	0.6771
0.3	0.5716	0.6757	0.5139	0.6758
0.4	0.5724	0.6754	0.5146	0.6747
0.5	0.5735	0.674	0.5155	0.6748
0.6	0.5745	0.674	0.5164	0.6747
0.7	0.5757	0.6729	0.5175	0.6741
0.8	0.578	0.6721	0.5196	0.6747
0.9	0.5817	0.671	0.5229	0.6731
1	0.5915	0.6728	0.523	0.6711

Table 1, it is evidenced that these best results are obtained when social data is incorporated with a high proportion. For the basic KNN-CF, it reaches its best performance with $\beta = 0.8$, if we focus on RMSE, while $\beta = 0.9$ accomplishes maximal performance based on the FCP. These results are confirmed under the baseline implementation, where the best performance (both on RMSE and FCP) is achieved with $\beta = 0.9$. In consequence, based on the RMSE, the baseline KNN-CF exhibits the best performance, with a RMSE $= 0.5132$, while based on the FCP, the basic KNN-CF achieves the best results, with a FCP $= 0.6778$. In this way, the neighborhood overpasses the baseline-aided KNN-CF rating estimation when we focus on the ranking results according to the FCP. Meanwhile, taking the course grades as explicit ratings, the RMSE is minimized by correcting the information given by the neighborhood with the respective baselines.

These are the results for the KNN-CF algorithms with explicit preference ratings, enhancing their performance by incorporating, besides the user preference behavior, their social background attributes. Now let's examine the performance of the algorithms with implicit preferences and their fuzzy confidence representation.

4.2 Implicit Preferences

Recalling the functions for estimating the confidence on the implicit preferences p_{ui}, given by the linear, exponential, logarithmic and logistic representation of Eqs. (8)–(11), the KNN-CF algorithms are implemented with the degrees of confidence that such representations allow taking into account. Hence, Eqs. (2)–(3) are computed with $r_{ui} = c_{ui}$ for linear confidence, as in Eq. (8), and with $r_{ui} = q_{ui}$ for exponential, logarithmic and logistic confidence, as in Eqs. (9)–(11).

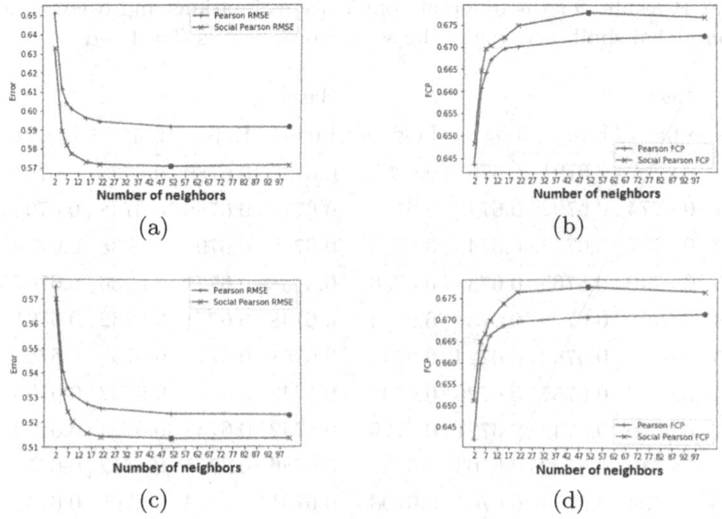

Fig. 2. RMSE and FCP behavior with different k neighbors for the (a,b) basic and (c,d) baseline KNN-CF recommenders. The Pearson and the social Pearson series correspond with the implementation of the respective KNN-CF algorithm with the Pearson and the social Pearson global similarity functions, respectively. The best models for each implementation are marked in bold.

In this way, for the linear estimation of confidence, the value of $\lambda = 40$ is fixed after an initial search for the optimal performance regarding the FCP, taking values for $\lambda \in \{1, 10, 20, 40, 80, 100, 160, 200\}$. This result agrees with the original proposal of [9], also suggesting that such a value of $\lambda = 40$ achieved good results.

On the other hand, the exponential and logarithmic estimation of confidence is rather direct, while for the logistic function there are various parameters to tune. After a discrete search, the best configuration with respect to the FCP performance was identified for $L = 110$, $\bar{r} = 4.4$, and $\gamma = 1.2$.

After estimating the confidence for implicit preferences, the basic and baseline KNN-CF algorithms were implemented. The results are presented in Table 2, showing the FCP performance for the different α-values of the Pearson global similarity computation. Analogously as the results presented in the previous Sect. 4.1, the values in Table 2 correspond with the best result for the number of k nearest neighbors.

Therefore, the best basic KNN-CF FCP was obtained with $\alpha = 0.1$ and $k = 101$, under the exponential fuzzy confidence estimation, achieving a FCP of 0.6792. On the other hand, the best baseline KNN-CF CFP performance also corresponds with $\alpha = 0.1$, but with $k = 51$ and the logistic fuzzy confidence estimation, with a FCP of 0.6795. Figure 3 presents the learning behavior for the respective KNN-CF algorithms using the $\alpha = 0.1$-social Pearson and the Pearson similarities, according to the number of nearest neighbors.

Table 2. FCP results for the different confidence estimations and α-configurations of the Pearson global similarity, where the social coefficient is $\beta = 1 - \alpha$

	Basic				Baseline			
α	Linear	Exp	Log	Logistic	Linear	Exp	Log	Logistic
0	0.6782	0.6791	0.675	0.6791	0.6759	0.6789	0.6738	0.6791
0.1	0.6774	0.6792	0.6747	0.679	0.6777	0.6786	0.6748	0.6795
0.2	0.6777	0.6774	0.6748	0.6783	0.6769	0.6793	0.6738	0.6782
0.3	0.6749	0.6765	0.6738	0.6756	0.6758	0.6771	0.6736	0.6767
0.4	0.6753	0.677	0.673	0.6754	0.6748	0.6771	0.6732	0.6761
0.5	0.6742	0.6769	0.6724	0.6743	0.6751	0.677	0.673	0.6752
0.6	0.6737	0.6757	0.6726	0.673	0.6748	0.676	0.6732	0.675
0.7	0.6729	0.6742	0.6721	0.6725	0.6742	0.6752	0.6724	0.6753
0.8	0.6719	0.6722	0.6703	0.672	0.6746	0.6756	0.6712	0.6751
0.9	0.6704	0.6718	0.6701	0.6708	0.6731	0.6754	0.6705	0.6751
1	0.6724	0.6752	0.6713	0.6734	0.6712	0.6742	0.67	0.6748

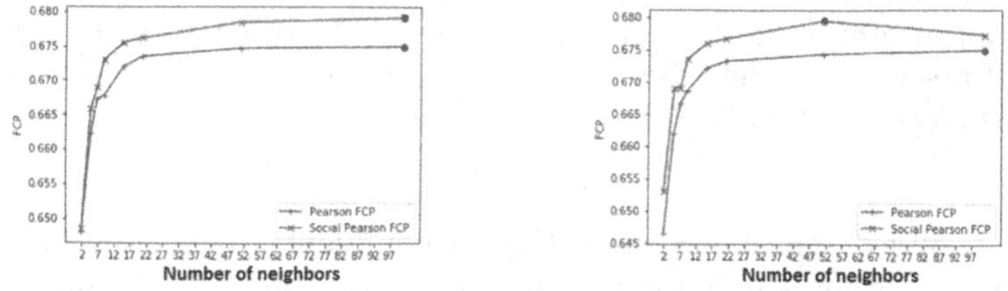

(a) (b)

Fig. 3. FCP behavior with different k neighbors for the (a) basic, exponential confidence and (b) baseline, logistic confidence KNN-CF recommenders.

As a result, again as with the explicit feedback scenario, it can be seen that the best results arise when social data is taken into account with a high proportion. This time, for both basic and baseline KNN-CF algorithms, the best FCP performance corresponds with $\beta = 0.9$. In absolute terms, the overall best prediction regarding all algorithm configurations, for either explicit or implicit feedback, corresponds with the KNN-CF baseline system and the fuzzy-logistic implicit preference procedure.

5 Final Remarks

In this paper we have implemented a study on basic and baseline socially-enhanced KNN-CF algorithms, examining its performance regarding the inclusion of the social background of users, besides the preferences over common items. The study has focused on recommending courses for students at the University of the Andes, accomplishing a FCP of 0.6795 for implicit preference prediction. This result was achieved by incorporating in a high proportion the social context, with the social parameter of $\beta = 0.9$ for the global social Pearson similarity function (with respect to the user-preference parameter of $\alpha = 0.1$), and by estimating the confidence on the observed implicit preferences by means of a fuzzy representation based on the logistic function.

For future research, it is projected to propose an axiomatic characterization for fuzzy confidence functions, and to explore recommendation systems making use of pairwise preference relations, bridging cognitive models and recommendation algorithms by means of fuzzy preference-aversion relations (as suggested in [8]).

References

1. Ahmadian, S., Joorabloo, N., Jalili, M., Ren, Y., Meghdadi, M., Afsharchi, M.: A social recommender system based on reliable implicit relationships. Knowl. Based Syst. **192**, 1–17 (2020)
2. Alahmadi, D., Zeng, X.: ISTS: implicit social trust and sentiment based approach to recommender systems. Expert Syst. Appl. **42**, 8840–8849 (2015)
3. Beierle, F., Tan, J., Grunert, K.: Analyzing social relations for recommending academic conferences. In: Proceedings 8th ACM International Workshop on Hot Topics in Planet, HotPOST 16. Paderborn, Germany, pp. 37–42 (2016)
4. Bhumichitr, K., Channarukul, S., Saejiem, N., Jiamthapthaksin, R., Nongpong, K.: Recommender Systems for university elective course recommendation. In: Proceedings 14th International Joint Conference on Computer Science and Software Engineering, JCSSE 2017. Nakhon Si Thammarat, Thailand, pp. 1–5 (2017)
5. Divyaa, L., Pervin, N.: Towards generating scalable personalized recommendations: integrating social trust, social bias, and geo-spatial clustering. Decis. Support Syst. **122**, 1–17 (2019)
6. Figueira, L., Cardoso, W.: Filtering graduate courses based on LinkedIn profiles. In: Proceedings 24th Brazilian Symposium on Multimedia and the Web, WebMedia 2018, Salvador, Brazil, pp. 141–147 (2018)
7. Franco, C., Rodríguez, J.T., Montero, J.: Building the meaning of preference from logical paired structures. Knowl. Based Syst. **83**, 32–41 (2015)
8. Franco, C., Rodríguez, J.T., Montero, J.: Learning preferences from paired opposite-based semantics. Int. J. Approx. Reason. **86**, 80–91 (2017)
9. Hu, Y., Koren, Y., Volinsky, C.H.: Collaborative filtering for implicit feedback datasets. In: Proceedings 2008 Eighth IEEE International Conference on Data Mining, ICDM 2008, Washington, USA, pp. 263–272 (2008)
10. Koren, Y., Sill, J.: OrdRec: an ordinal model for predicting personalized item rating distributions. In: Proceedings fifth ACM Conference on Recommender Systems, RecSys 2011, pp. 117–124 (2011)
11. Lai, Ch., Lee, Sh., Huang, H.: A social recommendation method based on the integration of social relationship and product popularity. Int. J. Hum. Comput. Stud. **121**, 42–57 (2019)
12. Li, G., Chen, Y., Zhang, Z., Zhong, J., Chen, Q.: Social personalized ranking with both the explicit and implicit influence of user trust and of item ratings. Eng. Appl. Artif. Intell. **67**, 283–295 (2018)
13. Ma, H., King, I., Lyu, M.: Learning to recommend with explicit and implicit social relations. ACM Trans. Intell. Syst. Technol. **2**, 1–19 (2011)

14. Wang, G., He, X., Ishuga, C.: HAR-SI: a novel hybrid article recommendation approach integrating with social information in scientific social network. Knowl. Based Syst. **148**, 85–99 (2018)
15. Zadeh, L.A.: Fuzzy logic and approximate reasoning. Synthese **30**, 407–428 (1975)

A Necessity Measure of Fuzzy Inclusion Relation in Linear Programming Problems

Zhenzhong Gao(✉) and Masahiro Inuiguchi(✉)

Osaka University, Osaka 560-8531, Japan
zhenzhong@inulab.sys.es.osaka-u.ac.jp, inuiguti@sys.es.osaka-u.ac.jp

Abstract. A programming problem with linear equality constraints can be generalised to the one with linear inclusions when coefficients are imprecisely given as possible ranges. In the problem with linear inclusions, the possible ranges of linear function values should always fluctuate within given ranges. In this paper, we investigate the programming problem with linear inclusions with coefficients being triangular fuzzy sets. To treat it, we introduce a new necessity measure, a linear extension of the one defined by the Dienes implication function, for the degree of the inclusions, and formulate a necessity measure maximisation problem. We propose a solution method based on the trade-off ratio and show that the maximisation problem becomes a regular linear programming problem in a particular condition. In general conditions, we also propose an algorithm utilising its properties and give a numerical example to demonstrate the solution procedure.

Keywords: Fuzzy linear programming · Necessity measure · h-level set · Trade-off ratio

1 Introduction

In applications, a linear programming problem (LPP) with coefficients imprecisely given by possible ranges can be modelled by linear inclusion relationships. Generally, one can study the set-inclusion constraints by building a maximum range that contains all possible ranges [9]. By utilising the set-inclusion relationships, Bard [1,2] constructed the LP problems with inexact coefficients described by crisp sets. Furthermore, in the view of interval linear programming, Hladík [3] also considered the same question and tried to solve it by shrinking or expanding the interval space.

However, using set-inclusion constraints is still too rough to describe an acceptable solution. Instead of giving a degree of trade-off ratio, it can only identify whether an inclusion relationship is valid or not. Hence, we need a more comprehensive estimation to assess an acceptable solution.

Supported by JSPS KAKENHI Grant Number JP18H01658.

V. Torra and Y. Narukawa (Eds.): MDAI 2021, LNAI 12898, pp. 131–140, 2021.
https://doi.org/10.1007/978-3-030-85529-1_11

Negoita [8] firstly considered such a problem by fuzzy sets called *robust programming*. He introduced fuzzy inclusive constraints expressed by inclusion relations between possible ranges and allowable ranges. Namely, the left-hand side fuzzy sets show the possible ranges with various level of estimation (from the narrowest to the widest) while the right-hand side fuzzy sets shows the allowable range for each different level of possible range estimation. However, he only concentrated on the inclusion relationships of discrete finite h-level sets, making the estimation still too rough. To treat the roughness, Inuiguchi and Tanino [7] applied a necessity measure approach to estimate the degree continuously. Furthermore, Inuiguchi et al. [5] tried several implication functions and modifier functions for more accurate estimation. However, the utilisation of modifier functions makes the model hard to calculate. To overcome the difficulty, Inuiguchi [4] proposed a simplified way to construct a necessity measure representing the decision maker's requests on fuzzy set-inclusive constraints in the setting of linear programming with uncertain coefficients.

Inspired by the tabular [4], we propose a fuzzy LPP to represent the original problem. Instead of considering the fuzziness representing the ambiguousness, we denote it as one's preference on the trade-off. For generality, we divide the constraints into hard and soft ones, where the hard is the one that should be fully satisfied, while the soft can be relaxed to a certain extent.

To represent the softness, we utilise symmetric triangular fuzzy numbers, enabling one to build a maximisation problem by the h-level sets. To treat the problem, we convert the set-inclusion constraints to a series of inequalities and form a non-linear model. We show that if only the right-hand-side coefficients contain softness, one can regard the model as a regular LPP by including the degree as one of the state variables. We also study the structure of the non-linear model generally and propose an algorithm linearly.

We organise the paper as follows. In Sect. 2, we give a brief preliminary about the fuzzy LP and NM. In Sect. 3, we propose the method for conversion and give the approach to solve it. In Sect. 4, we give a numerical example for illustrations, and finally, we briefly conclude our research with future work.

2 Preliminaries

2.1 Fuzzy LPP

Since we aim at the solution of an LPP, we regard it as a procedure to solve the following solution set:

$$\mathcal{S}(A, \boldsymbol{b}) := \{\boldsymbol{0} \leq \boldsymbol{x} \in \mathbb{R}^n : A\boldsymbol{x} = \boldsymbol{b}\} \tag{1}$$

where $A \in \mathbb{R}^{m \times n}$ and $\boldsymbol{b} \in \mathbb{R}^m$ are the coefficients in constraints.

Due to an objective function being trivial, we concentrate on the solution set of the corresponding fuzzy LPP. Since Inuiguchi et al. [6] have illustrated the fuzzy number in details, we only review the h-level set.

Definition 1 (h-level set). *An h-level set $[A]_h$ and a strong h-level set $(A)_h$ of a fuzzy subset A are crisp sets defined as below, respectively:*

$$[A]_h := \{r : \mu_A(r) \geq h\}, \quad (A)_h := \{r : \mu_A(r) > h\}, \tag{2}$$

where $\mu_A(r)$ denotes the membership function of A.

By Definition 1, it can be inferred that for any fuzzy subset A, $\forall 0 \leq h_1 \leq h_2 \leq 1$, $[A]_{h_2} \subseteq [A]_{h_1}$. In a sense, the higher h is, the more precise the information to describe A is. Therefore, one can regard h as the reliability degree to describe a fuzzy subset, and when treating h, it is preferable to have it as large as possible.

By converting the LPP (1) into a fuzzy one, we have the solution set as:

$$\mathcal{S}(\tilde{A}, \tilde{\boldsymbol{b}}) := \{\boldsymbol{0} \leq \boldsymbol{x} \in \mathbb{R}^n : \tilde{A}\boldsymbol{x} = \tilde{\boldsymbol{b}}\}, \tag{3}$$

where $\tilde{A} \subseteq \mathbb{R}^{m \times n}$ and $\tilde{\boldsymbol{b}} \subseteq \mathbb{R}^m$ denote the fuzzy coefficients.

Since $\mathcal{S}(\tilde{A}, \tilde{\boldsymbol{b}})$ becomes a fuzzy subset, we prefer a non-empty one with the greatest reliability degree, which is estimated by a *necessity measure.*

2.2 Necessity Measure

Possibility measure (PM) and necessity measure (NM) are techniques to measure the relation between two events by logical reasoning. Mathematically, given the information $r \in A$, a possibility (necessity) measure is to measure how possible (necessary) it can be for the condition $r \in B$.

Since PM is too weak, we only give a review of NM. Intuitively, an NM should follow the remark [5] below:

Remark 1. An NM should follow that

(i) $N_A(B) = 1$ iff $\text{cl}(A)_0 \subseteq [B]_1$,
(ii) $N_A(B) > 0$ iff $[A]_1 \subseteq \text{cl}(B)_0$,

where $\text{cl}(\cdot)$ denotes the closure of a set.

In Remark 1, the first condition means for all $x \in A$ in the weakest sense implies $x \in B$ in the strongest sense, and the second condition means for all $x \in A$ in the strongest sense implies $x \in B$ in the weakest sense. Since there exist multiple ways to define an NM, we review the original one by using membership functions.

Definition 2 (Necessity Measure). *An NM of a fuzzy subset B on another fuzzy subset A, which measures what extent $x \in B$ is certain when given $x \in A$ or what extent $x \in A$ implies $x \in B$, is defined as*

$$N_A(B) = \inf_{x \in X} I(\mu_A(x), \mu_B(x)), \tag{4}$$

where $I : [0,1] \times [0,1] \to [0,1]$ is an implication function (IF) and satisfies $I(0,0) = I(0,1) = I(1,1) = 1$ and $I(1,0) = 0$.

The most wide-spread type is the one using *Dienes* IF as $I^D(a,b) = \max\{1-a, b\}$ such that $N_A^D(B) = \inf_{x\in X} \max\{1-\mu_A(x), \mu_B(x)\}$. If one prefers to use the h-level set to represent the Dienes type, we have the result with its proof found in [6].

Proposition 1. *For the NM using Dienes IF, we have:*

$$N_A^D(B) \geq h \iff \mathrm{cl}((A)_{1-h}) \subseteq [B]_h \tag{5}$$

Proposition 1 suggests the way to use h-level sets to accomplish an NM, which is more comprehensive than using membership functions. For example, the necessity measure using Gödel IF [5] can be expressed more intuitively by h-level sets as $N_A^G(B) = h^* \iff \forall h < h^*,\ (A)_h \subseteq (B)_h$.

3 Model Conversion

3.1 Fuzzy LPP by Inclusion Relation

As Remark 1 shows the methodology of an NM by h-level sets, we notice that it is possible to have a more general expression. Instead of using a single h in an NM, we can separate it into two variables at both sides of the equation. Namely, we construct a fuzzy inclusion relationship of $[A]_{1-v} \subseteq [B]_w$ with two h-levels v and w. Hence, we have the following lemma[1].

Lemma 1. *An inclusion measure* $[A]_{1-v} \subseteq [B]_w$ *where* $v, w \in [0,1]$ *implies* $[A]_{1-v'} \subseteq [B]_{w'}$ *for all* $v' \leq v$ *and* $w' \leq w$.

Proof. It is apparent that $\forall v' \leq v,\ [A]_{1-v'} \subseteq [A]_{1-v}$ and $\forall w' \leq w,\ [B]_w \subseteq [B]_{w'}$, which gives out the final result. ∎

Unfortunately, such a manipulation constructs a multi-objective problem. Hence, we use weighted factors to convert it back to a single one, which gives out the following problem equivalent to Model (3).

$$\max_{0 \leq \boldsymbol{x} \in \mathbb{R}^n} \{\alpha v + \beta w : [A]_{1-v}\boldsymbol{x} \subseteq [\boldsymbol{b}]_w,\ v, w \in [0,1]\}, \tag{6}$$

where α and β are non-negative constants and $\alpha + \beta = 1$.

Moreover, for the simplification of using h-level sets in a fuzzy LPP, we prefer the fuzzy coefficients all being symmetric triangular fuzzy numbers, which are defined as

Definition 3 (Symmetric Triangular Fuzzy Number). *In this paper, we define a symmetric triangular fuzzy number as a fuzzy subset with symmetric and linear reference functions. Moreover, it could always be re-written by its h-level sets as*

$$[A]_h = [A^c - (1-h)A^r, A^c + (1-h)A^r] \tag{7}$$

[1] For the sake of simplicity, we always consider $[A]_0 = \mathrm{cl}(A)_0$ in this paper.

where A^c and A^r denote the centre and radius of $\mathrm{cl}((A)_0)$*, respectively. Mathematically, they equal to:*

$$A^c = \frac{1}{2}(\sup{(A)_0} + \inf{(A)_0}), \ A^r = \frac{1}{2}(\sup{(A)_0} - \inf{(A)_0})$$

3.2 Conversion Principles

Before continuing the procedure of solving the fuzzy problem, it is essential to have a discussion on the principle of conversion.

At first, if the original LPP is infeasible, the result of Problem (6) should always be *strictly less* than 1. If it equals to 1, then $v, w = 1$, which indicates there exists a feasible solution $\boldsymbol{x} \geq \mathbf{0}$ such that $[A]_0\boldsymbol{x} \subseteq [\boldsymbol{b}]_1$. According to Remark 1, it implies the necessity measure is 1, which contradicts the premise that the LP problem is infeasible.

Then, by the inclusion relation of two interval sets as $A \subseteq B \iff A^L \geq B^L$ & $A^U \leq B^U$, as well as the assumption such that A and B are symmetric triangular fuzzy subsets, we have the following conclusion that for all $i \in \{1, 2, \ldots, m\}$ and $j \in \{1, 2, \ldots, n\}$,

(i) The larger $\mathrm{cl}((\boldsymbol{b}_i)_0)$ is, the easier to have a solution.
(ii) The smaller $\mathrm{cl}((A_{ij})_0)$ is, the easier to have a solution.

Due to the complexity, one should not try to enlarge $\mathrm{cl}((A)_0)$. Alternatively, keeping A being constant may be a better choice for both model construction and calculation. Because of this, it is preferred to set $\beta > \alpha$, and when A is constant, we set $\alpha = 0$.

Hence, to use the fuzziness in representing the preference of a decision-maker, one should follow the several principles in the conversion.

(i) If a constraint is a hard one, keep the original coefficients constant and do not apply fuzziness on it.
(ii) If a constraint is a soft one, convert the coefficients into symmetric triangular fuzzy subsets.
(iii) Focus on $\boldsymbol{b}$ preferentially instead of A, and remember that the softer a constraint is, the larger $\mathrm{cl}((\boldsymbol{b})_0)$ should be.

After the conversion to the fuzzy problem in Model (6), we can continue to the solving approach.

4 Algorithm for the Fuzzy LP

Instead of considering the general condition with both fuzzy A and $\boldsymbol{b}$, it would be better to analyse the problem with only either of them at first. As it is indicated that focusing on $\boldsymbol{b}$ is preferential, we assume that only $\boldsymbol{b}$ contains fuzzy coefficients, which gives a constant A.

4.1 Special Case with Constant *A*

Assume a decision-maker has already set up fuzzy coefficients $\boldsymbol{b}$ with a constant matrix A, then Problem (6) equals to:

$$\begin{aligned} &\max\ w \\ &\text{s.t.}\ -A_s\boldsymbol{x} \leq -[\boldsymbol{b}_s]_w^L \\ &\qquad A_s\boldsymbol{x} \leq \ [\boldsymbol{b}_s]_w^U \\ &\qquad A_h\boldsymbol{x} = \ \boldsymbol{b}_h \end{aligned} \tag{8}$$

where A_s and A_h, $\boldsymbol{b}_s$ and $\boldsymbol{b}_h$ represent the soft and hard constraints, respectively.

Since each fuzzy entry in $\boldsymbol{b}$ is a symmetric triangular fuzzy number, one can write the h-level set by Definition 3 as $\forall\, h \in (0, 1]$,

$$[\boldsymbol{b}]_h = [\boldsymbol{b}^c - (1-h)\boldsymbol{b}^r, \boldsymbol{b}^c + (1-h)\boldsymbol{b}^r] \tag{9}$$

where $\boldsymbol{b}^c$ denote the only entry in $[\boldsymbol{b}]_1$ and $\boldsymbol{b}^r$ denotes the radius of $\mathrm{cl}((\boldsymbol{b})_0)$.

Consequently, we have the following model equivalent to Problem (8):

$$\begin{aligned} &\max\ w \\ &\text{s.t.}\quad A_h\boldsymbol{x} = \ \boldsymbol{b}_h \\ &\qquad -A_s\boldsymbol{x} \leq -\boldsymbol{b}_s^c + (1-w)\boldsymbol{b}_s^r \\ &\qquad A_s\boldsymbol{x} \leq \ \boldsymbol{b}_s^c + (1-w)\boldsymbol{b}_s^r \end{aligned} \tag{10}$$

By denoting w as a state variable, Problem (10) becomes a regular LPP. If it is infeasible, the tolerance given to $\boldsymbol{b}$ is still too narrow to make LPP (1) feasible even in the worst case.

4.2 General Case

After having the result with a constant A, we can proceed to study the situation with a fuzzy one. Similar to Problem (8), let us assume a decision-maker has set up α, β and the symmetric triangular fuzzy entries in A and $\boldsymbol{b}$, then the model becomes:

$$\begin{aligned} &\max\ \alpha v + \beta w \\ &\text{s.t.}\ A_h\boldsymbol{x} = \ \boldsymbol{b}_h \\ &\qquad (-A_s^c + vA_s^r)\boldsymbol{x} \leq -\boldsymbol{b}_s^c + (1-w)\boldsymbol{b}_s^r \\ &\qquad (A_s^c + vA_s^r)\boldsymbol{x} \leq \boldsymbol{b}_s^c + (1-w)\boldsymbol{b}_s^r \end{aligned} \tag{11}$$

Therefore, Problem (11) is equivalent to:

$$\max\{\alpha v + \beta w : (A^1 + vA^2)\boldsymbol{x} + \boldsymbol{b}^2 w = \boldsymbol{b}^1 + \boldsymbol{b}^2\} \tag{12}$$

where A^1, A^2, $\boldsymbol{b}^1$, $\boldsymbol{b}^2$ are conveniently defined from Problem (11).

Since Problem (12) is non-linear, we may have to apply some non-linear techniques for it. However, if we regard v or $\boldsymbol{x}$ as a constant variable at a specific step during the calculation, the system becomes linear. As an linear problem tends to be simpler than a non-linear one, solving the problem linearly is preferential. Hence, before applying non-linear algorithms directly, we prefer to do some analysis on Problem (12) at first.

Assume that we already have a pair of v^* and w^* at a specific step, and we want to improve the objective value. However, we cannot increase both v^* and w^* simultaneously because they are the intermediate solutions to Problem (12). Hence, the only way is to increase v^* with sacrificing w^* or the opposite direction. Since the situation with constant A gives the preference to have a larger w and $\alpha < \beta$, we usually have a large w^* and a small v^* at startup. Hence, let us consider the one by improving v^* and sacrificing w^*.

Let $\Delta v > 0$ and $\Delta w > 0$ denote the improvement such that $\alpha(v^* + \Delta v) + \beta(w^* - \Delta w) > \alpha v^* + \beta w^*$, which results in $\Delta w < (\alpha/\beta)\Delta v$. Moreover, to make the analysis more illustrative, we only consider the soft constraints and remove the slack variables in Problem (12). Then we have the constraints as:

$$(\boldsymbol{A}_s^1 + v\boldsymbol{A}_s^2)\boldsymbol{x} + \boldsymbol{b}_s^2 w \leq \boldsymbol{b}_s^1 + \boldsymbol{b}_s^2 \tag{13}$$

By adding Δv and Δw, Constraint (13) becomes:

$$(\boldsymbol{A}_s^1 + v^*\boldsymbol{A}_s^2)\boldsymbol{x} + \boldsymbol{b}_s^2 w^* + \Delta v A_s^2 x - b_s^2 \Delta w \leq \boldsymbol{b}_s^1 + \boldsymbol{b}_s^2 \tag{14}$$

To avoid the possibility that $\boldsymbol{x}$ may become infeasible, let $\Delta v, \Delta w \to 0^+$. Then inequalities in (14) becomes $\Delta v A_s^2 x \leq b_s^2 \Delta w$. Combined with $\Delta w < (\alpha/\beta)\Delta v$, it becomes the one as below:

$$\boldsymbol{A}_s^2 \boldsymbol{x} < \frac{\alpha}{\beta}\boldsymbol{b}_s^2 \tag{15}$$

Condition (15) implies that, if the current solution $\boldsymbol{x}$ does not satisfy it, then v^* and w^* at current step are already optimal to Problem (12) and cannot be improved. Due to A_s^2, $\boldsymbol{x}$ and b_s^2 being all non-negative, Condition (15) is usually hard to be achieved, especially with a large ratio α/β.

Therefore, we can have the algorithm as below:

Algorithm 1. - *Algorithm for Model (11)*

```
v <-- 0
value, x, w <-- Solve LP problem (12) with v
IF Condition (14) with x is not satisfied:
    OUTPUT x, v, w, value
    TERMINATE
ELSE:     #Apply bisection method
    v_0 <-- v; w_0 <-- w; x_0 <-- x; value_0 <-- value
    v_1 <-- 1
    value_1, x_1, w_1 <-- Solve LP problem (12) with v_1
    WHILE v_1-v_0 >= delta:
        v_c = (v_1+v_0)/2
        value_c, x_c, w_c <-- Solve LP problem (12) with v_c
```

```
    IF value_c <= value_1 AND value_1 == 1:
        OUTPUT x_1, v_1, w_1, value_1
    ELSE IF value_c > value_1:
        v_0 <-- v_c; w_0 <-- w_c
        x_0 <-- x_c; value_0 <-- value_c
    ELSE:
        v_1 <-- v_c; w_1 <-- w_c
        x_1 <-- x_c; value_1 <-- value_c
OUTPUT x_1, v_1, w_1, value_1
TERMINATE
```

5 Numerical Example

We propose the following example:

$$\begin{aligned} \text{Find } &\mathbf{0} \le \boldsymbol{x} \in \mathbb{R}^n \\ \text{s.t. } &3x_1 + 4x_2 = 7 \\ &9x_1 + 8x_2 = 16 \\ &x_1 - x_2 = 1 \end{aligned}$$

Since the number of constraints is over the variables, the system is obviously infeasible. At first, we only consider the right-hand-side vector being fuzzy one. Assuming Constraint 1 is a hard one, while Constraint 2 has higher priority than Constraint 3, we give 2 and 3 as $(16, 14, 18)_{LL}$ and $(1, -3, 5)_{LL}$, respectively. Hence we have the following system by introducing slack variables:

$$\max w \in [0, 1]$$
$$\text{s.t.} \quad \begin{bmatrix} 3 & 4 & 0 \\ -9 & -8 & 2 \\ 9 & 8 & 2 \\ -1 & 1 & 4 \\ 1 & -1 & 4 \end{bmatrix} \begin{bmatrix} x_1 \\ x_2 \\ w \end{bmatrix} + \begin{bmatrix} 0 \\ y_1 \\ y_2 \\ y_3 \\ y_4 \end{bmatrix} = \begin{bmatrix} 7 \\ -16+2 \\ 16+2 \\ -1+4 \\ 1+4 \end{bmatrix}$$

By the Simplex method, the solution is $w = 0.6935$ with the solution being $x_1 = 0.871, x_2 = 1.097$.

If one sets $\alpha = 0.05$ and $\beta = 0.95$, and considers fuzziness in A with $A_{21} = (9, 8.95, 9.05)_{LL}$, $A_{22} = (8, 7.98, 9.02)_{LL}$ and $A_{31} = (1, 0.999, 1.001)_{LL}$, then the Model becomes:

$$\max 0.05v + 0.95w$$
$$\text{s.t.} \quad \begin{bmatrix} 3 & 4 & 0 \\ -9 & -8 & 2 \\ 9 & 8 & 2 \\ -1 & 1 & 4 \\ 1 & -1 & 4 \end{bmatrix} \begin{bmatrix} x_1 \\ x_2 \\ w \end{bmatrix} + v \begin{bmatrix} 0 & 0 \\ 0.05 & 0.02 \\ 0.05 & 0.02 \\ 0.001 & 0 \\ 0.001 & 0 \end{bmatrix} \begin{bmatrix} x_1 \\ x_2 \end{bmatrix} + \begin{bmatrix} 0 \\ y_1 \\ y_2 \\ y_3 \\ y_4 \end{bmatrix} = \begin{bmatrix} 7 \\ 14 \\ 18 \\ 3 \\ 5 \end{bmatrix}$$

By Algorithm 1, we first solve $v^* = 0$ and $w^* = 0.6935$ with the solutions being $x_1 = 0.871, x_2 = 1.097$. Then, we check Condition (15), where in this example, $A_s^2 = [0.05\ 0.02;\ 0.001\ 0]$, $\boldsymbol{b}_s^2 = [2;\ 4]$, and $\alpha/\beta = 0.05/0.95$. Hence we find that $A_s^2 \boldsymbol{x} < (\alpha/\beta)\boldsymbol{b}_s^2$, which means we can do the improvement.

The following table is the result, which implies $v^* = 1$ and $w^* = 0.6861$, and the optimal solution is $x_1 = 0.854,\ x_2 = 1.109$.

v^*	w^*	$\alpha v^* + \beta w^*$	x_1	x_2
0	0.6935	0.6588	0.871	1.097
1	0.6861	0.7018	0.854	1.109
0.5	0.6898	0.6803	0.863	1.103

6 Conclusion

In this paper, we present a fuzzy model for an LPP with linear equalities. To get a desired solution, we utilise fuzziness to imply one's preference and priority, which results in an NM based on an h-level set. we obtain a non-linear programming problem, which can be linear if we only consider the right-hand-side part containing fuzziness. To solve the system, we first solve the linear case by standard LP techniques. Instead of applying non-linear tools directly, we analyse the structure of the system at first. The analysis suggests that we can still deal with it linearly, where we find an extra condition to identify whether it is possible to improve the optimal value. To illustrate our work, we give a numerical example in both situations.

For future progress, we separate our work into two sections. The first one is concerned with computational complexity. Since it still needs a bisection method to get the solution, we wonder if it is possible to have a direct one without iteration. As the space formed by w and v depends on the solved solution $\boldsymbol{x}$, we do not know whether it is always connected and convex. If it is, we can accomplish our goal with $v = 0$ and $v = 1$. However, if it is not, the algorithm may not give a globally optimised solution.

The second one is concerned with the solution set. Since we consider a fuzzy LPP to treat an infeasible LPP, we always have a fuzzy solution set. However, in this paper, we only solve one solution with the most significant reliability degree. Namely, we only solve one fuzzy solution in a fuzzy solution set, which is not enough for some situations. Hence, the result of the whole fuzzy solution set would become our goal of the following research and study.

References

1. Bard, J.F.: Inexact linear programming with generalized technological matrix sets. Eur. J. Oper. Res. **16**(1), 107–112 (1984)

2. Bard, J.F., Chatterjee, S.: Objective function bounds for the inexact linear programming problem with generalized cost coefficients. Comput. Oper. Res. **12**(5), 483–491 (1985)
3. Hladík, M.: Robust optimal solutions in interval linear programming with forall-exists quantifiers. Eur. J. Oper. Res. **254**(3), 705–714 (2016)
4. Inuiguchi, M.: Representation of the decision maker's preference on robust constraints in fuzzy linear programming. In: USB Proceedings of The 10th International Conference on Modeling Decisions for Artificial Intelligence, pp. 37–48 (2013)
5. Inuiguchi, M., Greco, S., Słowinski, R., Tanino, T.: Possibility and necessity measure specification using modifiers for decision making under fuzziness. Fuzzy Sets Syst. **137**(1), 151–175 (2003)
6. Inuiguchi, M., Ramík, J., Tanino, T.: Oblique fuzzy vectors and their use in possibilistic linear programming. Fuzzy Sets Syst. **135**(1), 123–150 (2003)
7. Inuiguchi, M., Tanino, T.: Necessity measures and parametric inclusion relations of fuzzy sets. Fundam. Inform. **42**, 279–302 (2000). https://doi.org/10.3233/FI-2000-423404
8. Negoita, C.: The current interest in fuzzy optimization. Fuzzy Sets Syst. **6**(3), 261–269 (1981)
9. Soyster, A.: Inexact linear programming with generalized resource sets. Eur. J. Oper. Res. **3**(4), 316–321 (1979)

Machine Learning

Mass-Based Similarity Weighted *k*-Neighbor for Class Imbalance

Anh Hoang([✉]), Toan Nguyen Mau, and Van-Nam Huynh

Graduate School of Advanced Science and Technology, Japan Advanced Institute of Science and Technology, 1-8 Asahidai, Nomi, Ishikawa 923-1292, Japan
{hoanganh,nmtoan91,huynh}@jaist.ac.jp

Abstract. Conventional distance-based or density-based classifiers like *k*-Nearest Neighbor algorithm face difficulty for class imbalance problem because they treat all neighbors equally though most of these instances belong to the majority class. Additionally, a varied density of data points existing in an imbalanced dataset poses another significant challenge for the distance-based classifiers. These two main challenges motivated us to investigate further a new mass-based similarity weighted *k*-neighbor approach for the imbalanced classification task. In our framework, the mass-based dissimilarity measurement is normed to obtain the similarity degree between two instances. Then the confidence of a neighbor instance is derived as the posterior probability while knowing the prior of the class label in training space and the likelihood in Bayes' theorem. We aim to add more importance to the neighbors, which have higher mass-based similarity weighted confidence than the others. Then, the weighted sum operator is used to combine the weighted confidences for predicting the class label. The experiments conducted on 60 imbalanced datasets show that the proposed approach outperforms the other 11 existing competitive methods in terms of precision-recall curves. We use the F1 score in our comparisons where we tested 12 models. We also use the Wilcoxon signed ranks test as a non-parametric statistical analysis to validate the experimental results.

Keywords: Imbalanced data · Imbalanced classification · Mass-based dissimilarity · *k*-nearest neighbor · Weighted sum · Aggregation function

1 Introduction

There has been vast literature on classification problems in knowledge discovery, data mining, and machine learning. Given a set of predefined categorical classes, classification means determining to which of these classes a specific instance belongs. A variety of classification approaches have been proposed and successfully applied in a wide range of application domains, e.g. C4.5 Decision Tree (DT) [1,2], Nave Bayes (NB) [3,4], *k*-Nearest Neighbor (*k*-NN) [5,6], Logistic Regression (LR) [7,8], Random Forest (RF) [9,10], Linear Support Vector

V. Torra and Y. Narukawa (Eds.): MDAI 2021, LNAI 12898, pp. 143–155, 2021.
https://doi.org/10.1007/978-3-030-85529-1_12

Machine (LinearSVM) [11,12], Gaussian RBF kernel SVM [13], Bagging algorithm [14,15], Decision Trees with AdaBoost [16,17], XGBoost [18,19], and Gaussian mixture model Proximity Weighted Evidential (mPEkNN) method [20]. However, most of these methods do not directly support imbalanced classification task. In an imbalanced dataset, the minority class has only a small portion of all the instances, while the majority class has a large percentage of all the objects. Besides, a varied density of data points is another significant challenge that distance-based or density-based classifiers cannot perform well.

To handle these class imbalance challenges, the preceding studies often focus on four group of methods: algorithmic modifications, resampling data space, cost-sensitive classification, and ensemble learning. Firstly, the algorithm-oriented approaches develop new algorithms or adapt existing ones for the class imbalance problem. Secondly, the resampling techniques, which preprocess the data to decrease the effectiveness of their class imbalance, include the over-sampling method like the Synthetic Minority Oversampling Technique or SMOTE for short [21], and under-sampling methods like the Tomek links algorithm [22]. Thirdly, the cost-sensitive learning solutions incorporate both the algorithmic and data-level methods for misclassification costs in the minority class. Lastly, the ensemble learning methods are conducted either by modifying the existing ensemble algorithms at the data level approaches or by embedding a cost-sensitive framework in the ensemble learning process. The limitation of these learning methods is that the objects from each class are treated the same, which means that the algorithms consume more resources to update learnable parameters for one class than another for imbalanced datasets.

In this paper, we propose a new mass-based similarity weighted confidence k-neighbor approach, so-called Sk-LMN, for the class imbalance problem. This approach can overcome the shortcomings of the distance-based or density-based classifiers for the imbalanced datasets. The experimental results show that Sk-LMN outperforms the other 11 competitive models tested on 60 imbalanced datasets in terms of the precision-recall (PR-AUC) metric. The F1 score is used for multiple comparisons 12 tested models as well. Besides, the Wilcoxon signed ranks tests are employed as non-parametric statistical analysis to validate the experimental results.

The remaining parts of this paper is organized as follows. Section 2 briefly introduces the mass-based dissimilarity measurement. Section 3 proposes the mass-based similarity weighted k-neighbor approach. Section 4 describes the experiments, whose comparison results are discussed. Section 5 draws the conclusion.

2 Mass-Based Dissimilarity Measurement

Distance metrics such as the Euclidean distance and Manhattan distance have been widely utilized for knowledge discovery, data mining, and machine learning. As a result, numerous classifiers based on distance functions have been introduced. However, these data-independent measures may have weaknesses because of the data-dependent properties of the dataset and the assumed distance axioms.

In this section, we briefly review the mass-based dissimilarity measurement [23,24] that alternates the algorithms based on distance function or geometric models. These distance-based measures may be directly replaced by mass-based measures. The remaining components of the original algorithms are the same. It should be noticed that the mass-based measures inherit the characteristics from research in psychology as follows: two points in a dense region of the space are less similar to each other than two points with the same inter-point distance in a sparse region, according to Tversky [25] and Krumhansl [26]. Therefore, for the class imbalance task, the shortcomings of the k-NN models and most of the neighbor-based classifiers are overcome by fundamentally changing their perspectives.

2.1 Definition

To understand how the mass-based dissimilarity measurement is developed, two concepts are introduced in each of the following definitions:

Definition 1. $S(x_t, x_i|H_j)$ is the smallest local space that covers two given data points x_t and x_i with respect to the hierarchical partitioning structure H_j as follows:

$$S(x_t, x_i|H_j) := \underset{S \in H_j, \{x_t, x_i\} \subseteq S}{\arg\max} \quad \text{depth}(S|H_j) \tag{1}$$

where depth$(S|H_j)$ is the depth or the path length of node S in H_j. This depth is equivalent to the number of separations required to isolate node S from the root node.

Definition 2. mass$(x_t, x_i|H_j)$ is the mass-based dissimilarity function between x_t and x_i, conditional on H_j. This is equivalent to the expectation of the probability that a random data point $z \in X$ belongs to the smallest space $S(x_t, x_i|H_j)$. That is found by Eq. (1) over $\mathcal{H}$. Where $\mathcal{H}$ is a set of H_j.

$$\text{mass}(x_t, x_i|H_j) := E_{\mathcal{H}}[P(z|z \in S(x_t, x_i|H_j))]$$

In practice, we have h finite hierarchical partitioning structures H_j for the dataset X. Thus, the mass$(x_t, x_i|H_j)$ is estimated as the average of the cardinality of $S(x_t, x_i|H_j)$ over all possible H_j.

$$\text{mass}_e(x_t, x_i) = \frac{1}{h}\sum_{j=1}^{h} \frac{|S(x_t, x_i|H_j)|}{|X|} \tag{2}$$

where $|X|$ is cardinality of the dataset X.

2.2 k-lowest mass$_e$ neighbors (k-LMN)

The context set of the query instance x_t is equivalently defined to the set of the k lowest mass-based dissimilarity neighbors around x_t, or k-LMN(x_t) for short:

$$k\text{-LMN}(x_t) = \{x_1, x_2, ..., x_k\}, \quad k \leq |X| \tag{3}$$

$$\text{where} \quad x_i = \underset{x \in X \setminus \{x_1, x_2, ..., x_{i-1}\}}{\arg\min} \quad \text{mass}_e(x_t, x), \quad i = 1, ..., k$$

In the other word, each x_i is a member of the k-LMN(x_t) that we used to define the set of neighbors of considering instance x_t.

3 Mass-Based Similarity Weighted *k*-neighbor (S*k*-LMN)

In this section, we propose the Sk-LMN approach for the class imbalance problem. Firstly, this approach is based on mass measurement instead of the distance-based or density-based functions. Thus, Sk-LMN can overcome the problem of the varied density of data points in the imbalanced datasets. Secondly, as uncertainty often exists in almost all datasets, the confidence of an instance plays an important role in the class imbalance problem where little information is available for the minority class. This confidence represents a conditional probability (Eq. 5) as the likelihood of a class label to which the instance belongs. There are several methods which also compute the conditional probability for classifying an instance, e.g. NB classifier. However, these methods cannot perform well for the imbalanced classification due to the poor estimation of the conditional density of the query instance associated with each class. Noticeably, Sk-LMN computes the conditional probability of neighbor instances belonging to the context set of the query instance rather than itself. Next, a simple weighted sum is used to aggregate the weighted confidence values provided by each individual neighbor of the query instance. The flowchart of the proposed approach is graphically illustrated in Fig. 1.

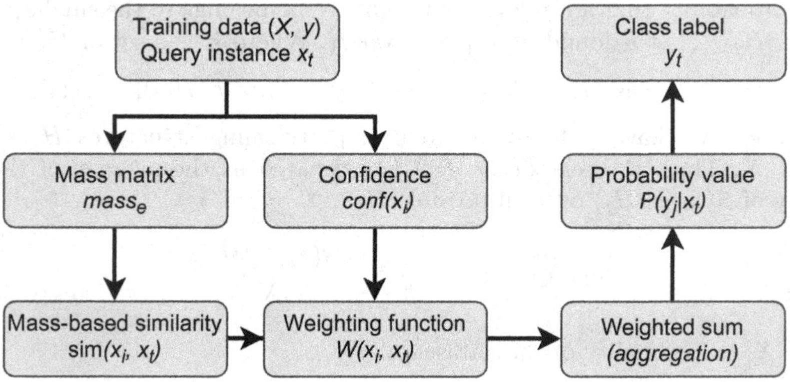

Fig. 1. Flowchart of the Sk-LMN approach.

For a new instance x_t, we find k lowest $mass_e$ neighbors (k-LMN) around it using the mass-based dissimilarity measurement as it was introduced in [27]. Let k-LMN(x_t) be a context set of the query instance x_t. Each member of the k-LMN(x_t), called x_i, assigns a weighted confidence value, which is computed by Eq. (4), supporting the prediction of the class label of x_t.

We observed that firstly a neighborhood instance will add more importance or larger weighted confidence value to class $y_j (1 \leq j \leq M)$ when this neighbor has a higher confidence that it belongs to y_j. A neighbor with a greater posterior probability should have a larger confidence than the one which is in the lower posterior probability area. Secondly, a neighbor will attach more importance or larger weighted confidence value to a specific class when the neighbor and the query instance are more similarity. We then formulate the weighting function that satisfies these two aforementioned observations as in Eq. (4).

$$W(x_i, x_t) = sim(x_i, x_t) \times conf(x_i) \tag{4}$$

where $conf(x_i)$ is the confidence of x_i represented by the posterior probability of class label y_i given x_i, and $sim(x_i, x_t)$ represents the mass-based similarity between x_i and x_t.

Algorithm 1. *train*Model(X, y)

Input: training data (X, y)
Output: $conf, mass_{max}$

Initialize array $conf$
$mass_{max} \leftarrow 0$
for $i = 1$ to $|X|$ **do**
    $conf[i] \leftarrow$ calculate confidence, Equation (5)
    for $j = i + 1$ to $|X|$ **do**
        $mass \leftarrow mass_e(x[i], x[j])$
        $mass_{max} \leftarrow \max(mass, mass_{max})$
    end for
end for
return $conf, mass_{max}$

3.1 Confidence Estimation

The confidence ($conf(x_i)$) that an instance x_i belonging to class y_i can be calculated by the following Eq. (5), regarding the Bayes' theorem.

$$conf(x_i) = P(y_i|x_i) = \frac{P(y_i) \times P(x_i|y_i)}{\sum_{j=1}^{M} P(y_j) \times P(x_i|y_j)} \tag{5}$$

where $y_i \in \{y_1, y_2, ..., y_M\}$ that is a finite set of M classes, $P(y_j)$ is the prior probability of y_j, and $P(x_i|y_j)$ is the likelihood probability. To calculate the likelihood, we use the Gaussian mixture model for estimating class-wise probability density function.

Algorithm 2. S*k*-LMN pseudo code

Input: training data (X, y), neighbor size k, query instance x_t
Output: class label y_t

```
conf(x_i), mass_max ← trainModel(X, y), from Algorithm (1)
s ← indices of k-LMN(x_t)
Initialize a list of W values
for i = 1 to k do
    index ← s[i]
    confidence ← conf[index]
    mass ← mass_e(x_t, x[index])
    similarity ← using Equation (6)
    weigh ← using Equation (4)
end for
Combine weighted confidence values using Equation (7)
ŷ_t ← predict class label, Equation (8)
return class label ŷ_t
```

3.2 Mass-Based Similarity Measurement

To obtain the similarity between two instances, the mass-based dissimilarity measurement introduced in [27] can be used. On one side, the similarity between two instances $(x_i, x_j \in X)$ can be maximum when (x_i, x_j) are in the same leaf node of the hierarchical partitioning structure. On the other side, the similarity will be minimum when the two data points are in the root node. To measure this similarity, normalization is applied as Eq. (6), so that $sim(x_i, x_j) \in [0, 1 - \frac{2}{N}]$, where N is the number of instances. Here, $mass_{max}$ is the maximum value of the estimated $mass_e(x_i, x_j)$ between two training instances.

$$sim(x_i, x_j) = 1 - \frac{mass_e(x_i, x_j)}{mass_{max}} \tag{6}$$

3.3 Weighted Sum Aggregation

Assume further that, for every neighbor instance x_i in the context set k-LMN(x_t), x_i assigns a numerical weighted confidence value $W(x_i, x_t)$ to support class y_j as its relative importance to the query instance x_t. We then used the weighted sum that is probably the best known and widely-used method for calculating the comprehensive evaluation or the total score of support in x_t. That is, for any query instance x_t we can compute the probability of x_t is assigned to class $y_j (1 \leq j \leq M)$ as,

$$P(y_j|x_t) = \frac{\sum_{x_i \in k\text{-LMN}(x_t), y_i = y_j} sim(x_t, x_i) \times conf(x_i)}{\sum_{1 \leq l \leq M} \sum_{x_i \in k\text{-LMN}(x_t), y_i = y_l} sim(x_t, x_i) \times conf(x_i)} \tag{7}$$

3.4 Decision Making

Equation (4) will return a larger weighted confidence value when a neighbor assigns more confidence and has more similarity to the query instance. To classify x_t, we use the weighted sum aggregation operator as in Eq. (7) to pool these discounted confidence values for each singleton class. According to this probability, we make the final decision using Eq. (8).

$$\hat{y}_t = \underset{1 \leq j \leq M}{\arg\max}\, P(y_j|x_t) \tag{8}$$

4 Experiential Results

The experimental study was conducted on 60 imbalanced datasets to compare the performances of the Sk-LMN approach with the other 11 competitive methods. Then, the Wilcoxon signed ranks test is employed as a non-parametric statistical analysis to validate the results.

4.1 Dataset Description

The imbalanced datasets were collected from the knowledge extraction based on the evolutionary learning (KEEL) website [28] to conduct experiments on a wide range of application domains, numbers of instances, numbers of features, and imbalance ratios. The imbalance ratio (IR) between the samples of the majority class and minority class of the datasets used in these experiments are in a wide range from 1.82 to 100.14. A dataset is highly imbalanced when the value of IR is higher than 100. We prepared these datasets for class imbalance tasks. The Table 1 presents the characteristics for 60 imbalanced datasets where **Idx.**, **#Inst.**, **#Ftr.**, and **IR** represent index of dataset, number of instances, features, and imbalance rate respectively.

4.2 Implementation Details and Evaluation Metrics

Sk-LMN is compared with other methods. The competitive models contain the conventional learning algorithms (C4.5 DT, NB, k-NN), logistic regression (LR), tree-based recent algorithms for imbalanced classification (RF), linear support vector machine (LinearSVM), SVM with RBF kernel (RBF_SVM), ensemble learning (Bagging, AdaBoost, and XGBoost), and recent evidential algorithm (mPEkNN).

There are many aspects and methods to evaluate the performance of a system for class imbalance problem. e.g. time, space, accuracy rate, F-series score, G-mean, Brier score, and the area under the curve (AUC) values. However, we consider the area under the precision-recall curves (PR-AUC values) as the most important factors due to its popularity in the literature. Moreover, we also include the F1 score of all testing models as well. These two metrics are used to assess the performance of the 12 competitive models.

Table 1. Descriptions of 60 imbalanced datasets.

Idx.	Dataset	#Inst.	#Ftr.	IR	Idx.	Dataset	#Inst.	#Ftr.	IR
1	Glass1	214	9	1.82	31	Glass-0-1-4-6_vs_2	205	9	11.81
2	Ecoli-0_vs_1	220	7	1.89	32	Glass-0-6_vs_5	108	9	12.50
3	Iris0	150	4	2.06	33	Ecoli-0-1-4-6_vs_5	280	6	13.74
4	Glass0	214	9	2.10	34	Shuttle-c0-vs-c4	1829	9	13.87
5	Haberman	306	3	2.78	35	Glass4	214	9	16.83
6	Vehicle2	846	18	2.88	36	Dermatology-6	358	34	16.90
7	Vehicle1	846	18	2.90	37	Winequality-white-9_vs_4	168	11	17.67
8	Vehicle3	846	18	2.99	38	Ecoli4	336	7	17.68
9	Vehicle0	846	18	3.25	39	Zoo-3	101	16	19.20
10	Ecoli1	336	7	3.36	40	Poker-9_vs_7	244	10	19.50
11	New-thyroid1	215	5	5.14	41	Shuttle-c2-vs-c4	129	9	20.50
12	Newthyroid2	215	5	5.32	42	Glass-0-1-6_vs_5	184	9	22.00
13	Segment0	2308	19	6.02	43	Shuttle-6_vs_2-3	230	9	22.00
14	Glass6	214	9	6.38	44	Glass5	214	9	25.75
15	Yeast3	1484	8	8.10	45	Winequality-red-4	1599	11	29.17
16	Ecoli3	336	7	8.60	46	Kddcup-guess_passwd_vs_satan	1642	38	29.98
17	Page-blocks0	5472	10	8.79	47	Yeast-1-2-8-9_vs_7	947	8	31.66
18	Yeast-0-3-5-9_vs_7-8	506	8	9.12	48	Abalone-3_vs_11	502	7	32.47
19	Yeast-0-2-5-7-9_vs_3-6-8	1004	8	9.14	49	Ecoli-0-1-3-7_vs_2-6	281	7	39.42
20	Ecoli-0-3-4_vs_5	200	7	9.53	50	Abalone-21_vs_8	581	7	40.50
21	Ecoli-0-6-7_vs_3-5	222	7	9.57	51	Yeast6	1484	8	41.40
22	Ecoli-0-1_vs_2-3-5	244	7	9.61	52	Kddcup-land_vs_portsweep	1061	38	49.52
23	Ecoli-0-2-3-4_vs_5	202	7	9.63	53	Abalone-19_vs_10-11-12-13	1622	7	49.69
24	Ecoli-0-2-6-7_vs_3-5	224	7	9.67	54	Poker-8-9_vs_6	1485	10	58.40
25	Ecoli-0-4-6_vs_5	203	6	9.68	55	Shuttle-2_vs_5	3316	9	66.67
26	Vowel0	988	10	9.98	56	Kddcup-buffer_overflow_vs_back	2233	38	73.43
27	Glass-0-1-6_vs_2	192	9	10.29	57	Kddcup-land_vs_satan	1610	38	75.67
28	Glass-0-4_vs_5	92	9	10.50	58	Poker-8-9_vs_5	2075	10	82.00
29	Ecoli-0-6-7_vs_5	220	6	10.58	59	Poker-8_vs_6	1477	10	85.88
30	Led7digit-0-2-4-5-6-7-8-9_vs_1	443	7	11.31	60	Kddcup-rootkit-imap_vs_back	2225	38	100.14

Besides, most of the classifiers have demonstrated beyond the binary classification as a multi-class problem that can simplify by the two-class task. Regularly, the minority class label is positive (or 1), and the majority class label is negative (or 0). In that case, the outcome of a classifier has been represented by a confusion matrix. This matrix has been used to calculate the F1 score, and AUC values. We have conducted the tenfold cross-validation test to evaluate the performance of 12 tested methods. As a result, these classifiers have ranked on each dataset in terms of the F1 score, and PR-AUC value, where a lower-ranked number indicates higher performance.

4.3 Results and Discussions

As illustrated in Fig. 2(a), we compared the 12 tested models in terms of the PR-AUC results from the tenfold cross-validation test on the 60 imbalanced datasets. It is worth noticing that the Sk-LMN model outperforms all the other tested models in both average values (**0.845**) and average ranks (**4.842**).

Fig. 2(b) shows the F1 score comparison results of the 12 models on the same tested datasets. The result shows that the proposed approach achieved the best average value (**0.738**) in the F1 score metric. However, the Sk-LMN model have

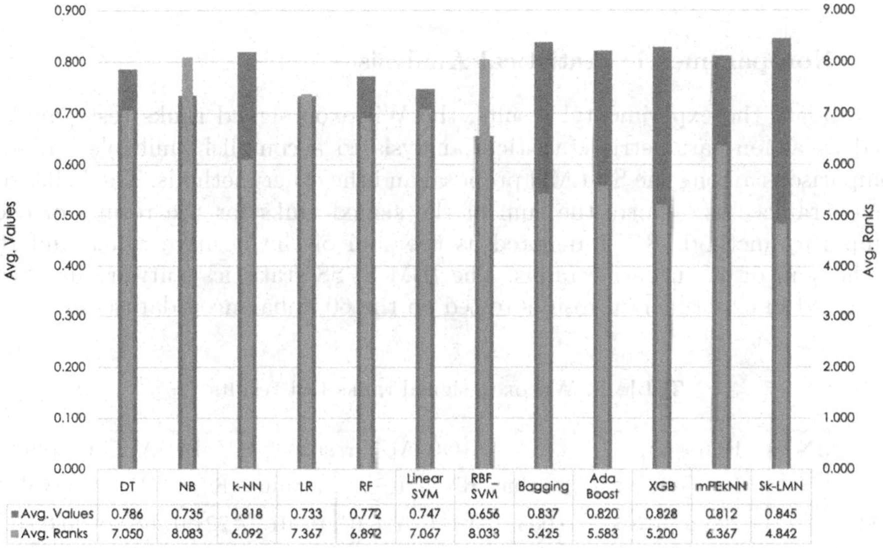
Avg. Values
Avg. Ranks
DT NB k-NN LR RF Linear SVM RBF_ SVM Bagging Ada Boost XGB mPEkNN Sk-LMN
Avg. Values 0.786 0.735 0.818 0.733 0.772 0.747 0.656 0.837 0.820 0.828 0.812 0.845
Avg. Ranks 7.050 8.083 6.092 7.367 6.892 7.067 8.033 5.425 5.583 5.200 6.367 4.842

(a) PR-AUC results.

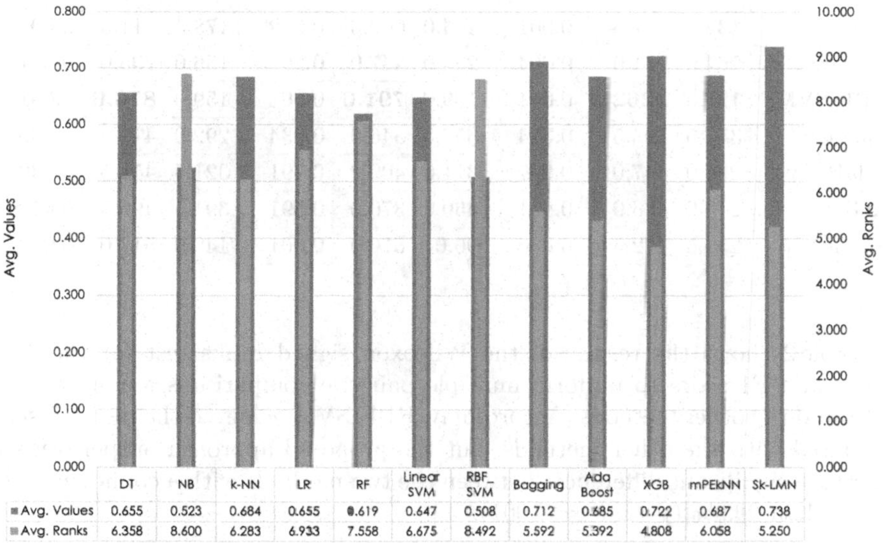
Avg. Values
Avg. Ranks
DT NB k-NN LR RF Linear SVM RBF_ SVM Bagging Ada Boost XGB mPEkNN Sk-LMN
Avg. Values 0.655 0.523 0.684 0.655 0.619 0.647 0.508 0.712 0.685 0.722 0.687 0.738
Avg. Ranks 6.358 8.600 6.283 6.933 7.558 6.675 8.492 5.592 5.392 4.808 6.058 5.250

(b) F1 results.

Fig. 2. Plots of results comparisons on 60 imbalanced datasets for 12 tested models. Note that a higher average value or a lower average rank indicates better performance.

reached the second-best average rank (**5.250**) while the best one (**4.808**) is the XGB method.

4.4 Non-parametric Statistical Analysis

To validate the experimental results, the Wilcoxon signed ranks test [29] was used as a non-parametric statistical analysis to accomplish multiple pairwise comparisons among the Sk-LMN proposal and the other methods. The Wilcoxon signed ranks test reports the sum of the signed-ranks for the results of each comparing method. R^+ is denoted as the sum of the positive ranks and R^- as the sum of the negative ranks. The IBM SPSS statistics software has been employed in experimental results tested on the 60 imbalanced datasets.

Table 2. Wilcoxon signed ranks test results.

Sk-LMN vs.	F1 results			ROC-AUC results			PR-AUC results		
	R^+	R^-	p-value	R^+	R^-	p-value	R^+	R^-	p-value
DT	226.5	676.5	0.005	136.0	725.0	0.001	221.0	682.0	0.004
NB	92.0	898.0	0.001	171.0	732.0	0.001	126.0	820.0	0.001
k-NN	309.0	511.0	0.174	105.0	598.0	0.001	190.0	513.0	0.015
LR	260.5	820.5	0.002	288.0	573.0	0.065	121.0	782.0	0.001
RF	132.5	857.5	0.001	284.0	662.0	0.022	178.5	811.5	0.001
LinearSVM	264.0	771.0	0.004	294.0	447.0	0.267	156.0	705.0	0.001
RBF_SVM	115.5	**1262.5**	0.001	199.0	**791.0**	0.001	159.0	**876.0**	0.001
Bagging	357.5	503.5	0.344	357.0	346.0	0.934	292.0	411.0	0.369
AdaBoost	294.0	447.0	0.267	371.0	409.0	0.791	321.5	458.5	0.339
XGB	470.0	476.0	0.971	450.0	370.0	0.591	391.5	496.5	0.613
mPEkNN	230.5	472.5	0.086	90.0	576.0	0.001	149.0	517.0	0.004

Table 2 shows the results of the Wilcoxon signed ranks test for the AUC values and F1 score to perform multiple pairwise comparisons among the Sk-LMN and the other methods. Accordingly, RBF_SVM achieved the best R^- score compared with the other methods, but the proposed approach outperforms it. There are significant differences between the two methods with a confidence level higher than 99.9% (p-value = 0.001).

5 Conclusion

This paper proposes an alternative approach for the imbalanced classification from perspectives of the mass-based measurement, neighbor-based algorithm, and information fusion. In Sk-LMN, we consider the confidence of a query instance's neighbor as the posterior probability that measures the uncertainty of its class label. To compute this confidence, we used the Gaussian mixture model for estimating the likelihood of the class label of the instance. Then, the similarity between the query instance and the neighbor instance has been used to weigh the confidence. The experimental study reveals that this weighted confidence method increases the likelihood of a minority class classification. The experiments conducted on 60 imbalanced datasets demonstrate that the proposed approach outperforms the other 11 competitive methods in terms of the PR-AUC metric. However, there is a limitation of this approach. We calculated on numerical feature values only. In future research, we plan to extend the Sk-LMN method for categorical features. We will also utilize the Dempster-Shafer theory of evidence by considering each neighbor as a piece of evidence to support the query instance. Then, we will use Dempster's rule of combination or any other combination rule to pool these pieces of evidence for classifying the query instance. The source code and datasets of the Sk-LMN project have been organized and available on Github at the following link: https://github.com/ImbOut/Sk-LMN.

References

1. Cieslak, D.A., Chawla, N.V.: Learning decision trees for unbalanced data. In: Daelemans, W., Goethals, B., Morik, K. (eds.) ECML PKDD 2008. LNCS (LNAI), vol. 5211, pp. 241–256. Springer, Heidelberg (2008). https://doi.org/10.1007/978-3-540-87479-9_34
2. Lee. J.-S., Auc4. 5.: Auc-based c4. 5 decision tree algorithm for imbalanced data classification. IEEE Access **7**, 106034–106042 (2019)
3. Murphy, K.P., et al. : Naive bayes classifiers. Univ. Br. Colum. **18**(60), 1–8 (2006)
4. Aridas, C.K., Karlos, S., Kanas, V.G., Fazakis, N., Kotsiantis, S.B.: Uncertainty based under-sampling for learning naive bayes classifiers under imbalanced data sets. IEEE Access **8**, 2122–2133 (2019)
5. Guo, G., Wang, H., Bell, D., Bi, Y., Greer, K.: KNN model-based approach in classification. In: Meersman, R., Tari, Z., Schmidt, D.C. (eds.) OTM 2003. LNCS, vol. 2888, pp. 986–996. Springer, Heidelberg (2003). https://doi.org/10.1007/978-3-540-39964-3_62
6. Zhang, S., Li, X., Zong, M., Zhu, X., Cheng, D.: Learning k for KNN classification. ACM Trans. Intell. Syst. Technol. **8**(3), 1–19 (2017)
7. Dreiseitl, S., Ohno-Machado, L.: Logistic regression and artificial neural network classification models: a methodology review. J. Biomed. Inform. **35**(5–6), 352–359 (2002)
8. De Caigny, A., Coussement, K., De Bock, K.W.: A new hybrid classification algorithm for customer churn prediction based on logistic regression and decision trees. Eur. J. Oper. Res. **269**(2), 760–772 (2018)
9. Svetnik, V., Liaw, A., Tong, C., Culberson, C., Sheridan, R.P., Feuston, B.P.: Random forest: a classification and regression tool for compound classification and qsar modeling. J. Chem. Inf. Comput. Sci. **43**(6), 1947–1958 (2003)
10. Paul, A., Prasad Mukherjee, D., Das, P., Gangopadhyay, A., Chintha, A.R., Kundu, S.: Improved random forest for classification. IEEE Trans. Image Process. **27**(8), 4012–4024 (2018)

11. Hsieh, C.-J., Chang, K.-W., Lin, C.-J. Keerthi, S.S., Sundararajan, S.: A dual coordinate descent method for large-scale linear SVM. In: Proceedings of the 25th international conference on Machine Learning, pp. 408–415, Helsinki, Finland, Springer (2008)
12. Chauhan, V.K., Dahiya, K., Sharma, A.: Problem formulations and solvers in linear SVM: a review. Artif. Intell. Rev. **52**(2), 803–855 (2019)
13. Ring, M., Eskofier, B.M.: An approximation of the gaussian RBF kernel for efficient classification with SVMs. Patt. Recogn. Lett. **84**, 107–113 (2016)
14. Roshan, S.V., Asadi, S.: Improvement of bagging performance for classification of imbalanced datasets using evolutionary multi-objective optimization. Eng. Appl. Artif. Intel. **87**, (2020)
15. Guo, L., Boukir, S., Aussem, A.: Building bagging on critical instances. Exp. Syst. **37**(2), (2020)
16. Hatwell, J., Gaber, M.M., Azad, R.M.A.: Ada-whips: explaining adaboost classification with applications in the health sciences. BMC Med. Inform. Decision Making **20**(1), 1–25 (2020)
17. Asim, K.M., Idris, A., Iqbal, T., Martínez-Álvarez, F.: Seismic indicators based earthquake predictor system using genetic programming and adaboost classification. Soil Dyn. Earthq. Eng. **111**, 1–7 (2018)
18. Ren, X., Guo, H., Li, S., Wang, S., Li, J.: A novel image classification method with CNN-XGBoost model. In: Kraetzer, C., Shi, Y.-Q., Dittmann, J., Kim, H.J. (eds.) IWDW 2017. LNCS, vol. 10431, pp. 378–390. Springer, Cham (2017). https://doi.org/10.1007/978-3-319-64185-0_28
19. Wang, C., Deng, C., Wang, S.: Imbalance-xgboost: leveraging weighted and focal losses for binary label-imbalanced classification with xgboost. Patt. Recogn. Lett. **136**, 190–197 (2020)
20. Kadir, M.E., Akash, P.S., Sharmin, S., Ali, A.A., Shoyaib, M.: A proximity weighted evidential k nearest neighbor classifier for imbalanced data. In: Lauw, H.W., Wong, R.C.-W., Ntoulas, A., Lim, E.-P., Ng, S.-K., Pan, S.J. (eds.) PAKDD 2020. LNCS (LNAI), vol. 12085, pp. 71–83. Springer, Cham (2020). https://doi.org/10.1007/978-3-030-47436-2_6
21. Chawla, N.V., Bowyer, K.W., Hall, L.O., Kegelmeyer, W.P.: Smote. synthetic minority over-sampling technique. J. Artif. Intell. Res. **16**, 321–357 (2002)
22. Devi, D., Purkayastha, B., et al.: Redundancy-driven modified tomek-link based undersampling: a solution to class imbalance. Patt. Recogn. Lett. **93**, 3–12 (2017)
23. Ting, K.M., Zhou, G.-T., Liu, F.T., Tan, J.S.: Mass estimation and its applications. In: Proceedings of the 16th ACM SIGKDD International Conference on Knowledge Discovery and Data Mining, pp. 989–998, New York, NY, USA, Association for Computing Machinery (2010)
24. Ting, K.M., Zhu, Y., Carman, M., Zhu, Y., Zhou, Z.-H.: Overcoming key weaknesses of distance-based neighbourhood methods using a data dependent dissimilarity measure. In: Proceedings of the 22nd ACM SIGKDD International Conference on Knowledge Discovery and Data Mining, pp. 1205–1214, Singapore, Springer (2016)
25. Tversky, A.: Features of similarity. Psychol. Review **84**(4), 327 (1977)
26. Krumhansl, C.L.: The interrelationship between similarity and spatial density: concerning the applicability of geometric models to similarity data. Am. Psychol. **5**, 445–463 (1978)
27. Hoang, A., Mau, T.N., Vo, D.V., Huynh, V.N.: A mass-based approach for local outlier detection. IEEE Access **9**, 16448–16466 (2021)

28. Triguero, J., et al.: Keel 3.0: an open source software for multi-stage analysis in data mining. Int. J. Comput. Intell. Syst. **10**(1), 1238–1249 (2017)
29. Wilcoxon, F.: Individual comparisons by ranking methods. In: Breakthroughs in statistics, pp. 196–202. Springer, New York (1992). https://doi.org/10.1007/978-1-4612-4380-9_16

Multinomial-Based Decision Synthesis of ML Classification Outputs

Alan J. Michaels(✉) and Lauren J. Wong

Hume Center for National Security and Technology, Virginia Polytechnic Institute and State University, Blacksburg, VA, USA
ajm@vt.edu

Abstract. With the explosion of interest in machine learning (ML)-based classification algorithms applied to streaming data sources, there is a decided benefit to the rapid decisions that a neural network (NN) can provide. However, there is also a desire to integrate successive real-time decisions on streaming data, which likely have temporal correlations, into an overall higher confidence result. This paper offers a predictive approach to aggregating the results of discrete classification outputs based upon that duration of temporal correlation, with specific examples for both an image processing (facial mask recognition) application and two radio frequency (RF) ML applications (specific emitter identification and automatic modulation classification). The decision aggregation technique employs a multinomial distribution representation of conditional decision probabilities, drawn from the confusion matrices of the classification problems, to show that an ML classifier possessing even a marginal ability to improve upon random guessing has the potential to drastically improve overall decision accuracy when operating on large continuous streams of data, boosting confidence in the resulting decision.

Keywords: Machine learning · Decision synthesis · NN · Classifier

1 Introduction

The use of feed-forward neural net (NN)-based ML classification techniques is widespread across image processing [1], natural language processing (NLP) [2], and RF applications [3,4]. Image processing is a dominant use-case for NN-based classification algorithms, with image recognition [5], full-motion video [6], and even niche applications like facial mask compliance [7] becoming widespread. NLP classification techniques vary from separation of speech components [8] and contextual sentiment analysis [9]. ML techniques applied to time-series RF signals (RFML) are used to perform automatic modulation classification (AMC) [10,11], device fingerprinting [12], specific emitter identification (SEI) [13,14], and for modeling RF interference [15].

In each scenario, an individual evaluation of the ML algorithm against a candidate input operates upon short captures of the observed signal or environment, which provides rapid correlation-based decisions [16]. Design of the NN

V. Torra and Y. Narukawa (Eds.): MDAI 2021, LNAI 12898, pp. 156–167, 2021.
https://doi.org/10.1007/978-3-030-85529-1_13

architecture often considers the minimum duration of the observable seeking classification (i.e. short and fixed signal captures or static images), yet often the incoming stimuli lasts far longer than a minimum (i.e. longer dynamic length signal captures or video). For example, an RF AMC algorithm can typically achieve solid performance using 100–1000 input symbols, yet most commercial transmissions (i.e. 802.11 g) are orders of magnitude longer in duration. Likewise, an image classification algorithm designed to detect the presence of a facial mask [17,18] can operate on virtually any instantaneous image capture of a given face, while the frame rate of common low-cost cameras enable capturing many temporally coherent images of the same face. When transitioning from laboratory training environments into real-time deployed environments, these classification techniques have been shown to degrade sufficiently that a single decision cannot be trusted [18]. However, subsequent NN-based decisions can be combined to achieve an operationally relevant high confidence decision, dependent upon how frequently the time-series data source is sampled (a driver of processing load), as examined in this work.

Additionally, most NN-based classification techniques employ an *a priori* known discrete set of possible outcome states, with the option for anomalous conditions to be categorized as a "none-of-the-above" or "unknown" category. For RFML in particular, many classification tools operate on extremely small amounts of data relative to expert-defined feature detectors/classifiers [19], making the direct performance comparison between unsupervised RFML-based signal classification solutions and traditional expert classification solutions inappropriate, due to the widely different observation timescales and amounts of input data. For example, the SEI work in [20] bases classification decisions on 128–222 symbols of data, while the comparative expert feature techniques used in [21] bases classification decisions upon 1330 symbols, nearly an order of magnitude increase in signal observations. As a result, direct comparison of the resulting accuracies from a single decision fails to fairly represent the differences.

This paper offers a mathematical model based on the multinomial distribution to normalize evaluations of different CNN classifiers so that *performance* can be based on a comparable amount of total input data. Multinomial distributions are widely used describe probabilities over many random trials drawn from discrete/disjoint sets. This same multinomial distribution is applied to training processes of CNN-based classifiers, with incorporation of stochastic pooling [22], probabilistic variants of dropout called *probout* [23], and computation of softmax loss [24]. Our use is independent of training, seeking instead to determine the most likely (similar to [23]) class of an observable input, based on an unordered string of classification outputs that are assumed correlated. Most underlying mathematical models assume *independent* observations, while temporal correlations of the input (a common object or signal) are likely in real life since we measure the input quicker than the scene/signal change. A notional example of this process is depicted in Fig. 1 comparing both image processing examples (short discrete captures) and short-term captures (frames) of a time-domain signal as is used in NLP and RF applications.

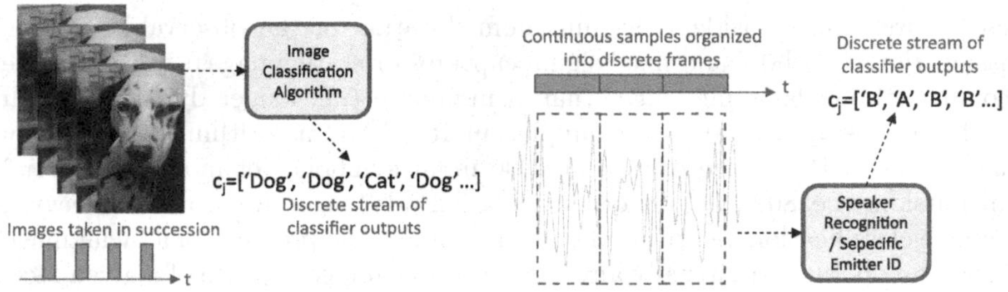

Fig. 1. Visualization of classifier decision aggregation across modalities for image processing (left) and NLP/RF applications (right).

In effect, our goal is to aggregate classification decisions from the NN into a higher accuracy overall decision, with predictive performance based upon the numerical estimation of the confusion matrix. Each possible class for the classification decision is treated as a distinct output state with quantifiable conditional probabilities, equivalent to one row of the confusion matrix, so that assumed independent classification decisions may be aggregated, taking advantage of the anticipated temporal correlations. The core framework for the proposed multinomial classification decision synthesis process is described in Sect. 2 including simplifying assumptions and expected variations in the model when the assumptions (statistical independence of successive inputs, distributions of output class decisions) deviate from reality. As concrete examples, the paper applies the multinomial decision framework to both an image processing-oriented binary facial mask recognition algorithm and two RFML-focused algorithms supporting SEI [20] and AMC [11] in Sect. 3; predictive performance bounds are provided, with discussion of limiting probability models. Overall conclusions and anticipated future use cases for the proposed model are provided in Sect. 4.

2 Multinomial-Based Decision Synthesis for ML Classification Problems

The framework presented in this section can be used in conjunction with any time-series classification decision algorithm where the set of possible output states is known *a priori* and successive decision outputs are likely to be reasonably coherent. In image processing, these assumptions are satisfied when multiple image frames capture the same object or individual with multiple successive frames. For RFML and NLP applications, these assumptions are satisfied when the duration of the observable signal exceeds the input frame (i.e. sample rate times frame size). Additionally, the decision synthesis framework given is applicable to any feed-forward NN architecture (i.e. MLP, CNN), with the chosen architecture assumed to be optimized to the desired task.

Consider the NN shown in Fig. 2 which is assumed for simplicity to have already undergone all chosen training processes, and whose decisions are selected

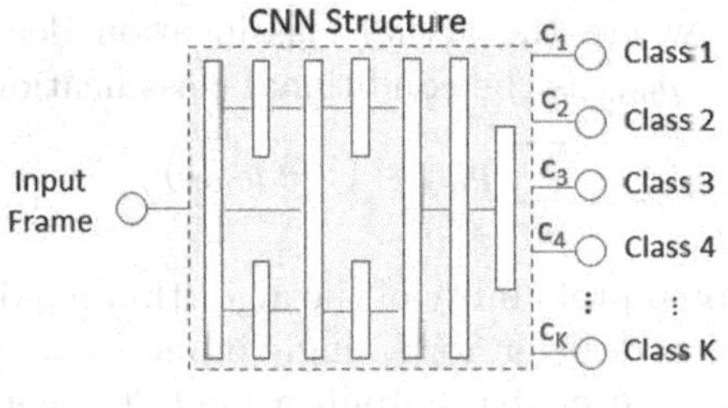

Fig. 2. Generalized CNN-based classification algorithm

among one of K possible output classes. The input to the NN is assumed to be a raw time series snapshot (i.e., a single image, a short RF capture), although optional data transforms may be incorporated without loss of generality. The assumed output of the network for each input frame is a classification decision representing a choice of one possible output state from a discrete selection of K possible classes. For our purposes, this classification decision is reinterpreted for a single input frame as a K-tuple of binary outputs $\{c_1, c_2, ..., c_K\}$, where $c_j \in \{0, 1\}$ and $\sum c_j = 1$. When a total of N input frames are processed by the NN, the stream of N outputs are accumulated over time, such that c_j terms represent a short-term count of classification occurrences.

2.1 Classification Decision Framework

If we consider N accumulated frames of output decisions in order to create an averaged decision over some chosen duration of N inputs, the order of those intermediate decisions does not matter, provided the input inputs are statistically independent. This independence will be assumed initially and subsequently addressed in the following section. Further, we assume the inputs only contain a single image/signal of interest, ignoring for the moment the scenarios of multiple faces or co-channel RF signals. As a result, we may consider the stream of output decisions using the multinomial distribution. The probability of selecting any specific combination $\{n_1, ..., n_K\}$, without replacement, of N outputs from the NN is shown in (1), where $\{p_{m1}, p_{m2}, ...p_{mK}\}$ represents the conditional probabilities of the NN algorithm producing a chosen output classification decision $\{1, ..., K\}$, given knowledge of the actual classification state m. These conditional probabilities $\{p_{m1}, p_{m2}, ...p_{mK}\}$ may also be viewed as a row of entries for the classifier's confusion matrix. Note that in the case of image processing, this confusion matrix must be pre-conditioned to be representative of image quality, and for RF/NLP problems, a similar consideration of quality such as a signal-to-noise ratio (SNR) must also be considered.

$$P(c_1 = n_1, ..., c_K = n_K | m) = \frac{N!}{n_1! \cdot ... \cdot n_k!} p_{m1}^{n1} \cdot ... \cdot p_{mk}^{nk} \tag{1}$$

Consider the case where the actual classification decision is state m; the conditional probability p_{mm} is the conditional classification accuracy, while

$$\sum_{j \neq m} p_{mj} = (1 - p_{mm})$$

represents the conditional probability of the algorithm making a specific pairwise error in the classification decision for a single input frame belonging to class m. Two extreme scenarios that enable simplification of the generalized multinomial model are (a) an assumption that the error probabilities are equiprobable and (b) an assumption that all of the error density is concentrated in a single false classification. In the first case, which is a best case scenario for classification accuracy, the probabilities for incorrect classification per incorrect state reduce to (2). In this case, (1) reduces to (3).

$$p_{mj} = \frac{(1 - p_{mm})}{K - 1} \forall j \neq m \tag{2}$$

$$P(c_1 = n_1, ..., c_K = n_K|m) = \frac{N!}{n_1! \cdot ... \cdot n_k!} p_{mm}^{n_m} \cdot (\frac{(1 - p_{mm})}{K - 1})^{N - n_m} \tag{3}$$

The second case focuses all of the error density into a single false state l, which is a worst case of a coherent error condition (i.e., errors are aggregated for a specific output state), collapsing the multinomial distribution of (1) into the binomial distribution (4).

$$\begin{aligned} P(c_1 = n_1, ..., c_K = n_K|m) &= P(c_m = n_m, c_l = N - n_m|m) \\ &= \frac{N!}{n_m! \cdot (N - n_m)!} \cdot p_{mm}^{n_m} \cdot (1 - p_{mm})^{N - n_m} \end{aligned} \tag{4}$$

In most classifiers, the error densities will lie in between these two extremes which offer bounding probabilities for estimating classifier performance. Anecdotally, the reasons for deviations between these two extremes have been application dependent. However, the error densities (off-diagonal entries of the confusion matrix) are typically split across a small number of incorrect state classifications, and tend to occur symmetrically, such that a larger value p_{mj} corresponds to a larger p_{jm}, suggesting a similarity between output classes.

2.2 Simplifying Assumptions and Operational Dependencies

The key simplifying assumptions made in the proposed framework are on statistical independence of the input[1] and the presence of a single observable. The assumption of statistical independence normalizes for the input sampling/framing processes, as well as any nonlinear variations occurring during

[1] We assume that the sampling process, such as capturing an image or measuring a signal, do not induce a temporal correlation to the decisions that is different from the actual object/signal of interest, such as might occur with plenoptic cameras or overlapping time windows as single frames.

the N input frames used to create an aggregate classification. The assumption that only one consistent observable (i.e. image or signal) is present for classification throughout the duration of the N input frames eliminates the case where classification decisions overlap in time, and is most accurate for slowly changing environments and/or signals This also assumes any nonlinear effects that transform the confusion matrix entries are also slowly changing such that they occur on a timeline greater than N input frames. Finally, an assumption is made that the chosen classes of the CNN-based classifier are disjoint (i.e., all probabilities add to 1), eliminating ordinal or hierarchical classification structures. In the case of image classification algorithms, these assumptions are generally met in facial identification, mask compliance analysis, and objection recognition settings. For RFML classification algorithms, these assumptions are reasonably accurate in the presence of randomly occurring background noise, while less so in the presence of periodic interferers.

Operational dependencies like image quality (as quantified by methods like Blind/Referenceless Image Spatial Quality Evaluator (BRISQUE) [25]) or signal quality (as quantified by SNR) also must be considered in the proposed decision framework by creating an ensemble of different confusion matrices, each conditioned upon a small range of the chosen quality metric.

2.3 Aggregating Classification Decisions

The goal of determining an aggregate classification decision on N input frames then translates to the subset of all possible multinomial selections since the decision metric applicable for a classification decision scenario is that of the classification chosen (class j, represented by c_j) most often; i.e.,

$$\max_j\{c_j\} = \begin{cases} j = m : \text{Accurate decision} \\ j \neq m : \text{Error} \end{cases}$$

Stated differently, an accurate decision occurs when $c_m > \max c_j, j \neq m$. Additionally, in the scenario where $c_m = \max c_j, j \neq m$, we may make a random guess based on the leading states, selecting the correct classification with a probability of $\frac{1}{K}$.

The challenge in most such scenarios is not in calculating the individual state probabilities, but rather efficiently selecting the proper subset of possible multinomial state combinations $\boldsymbol{c}$, meaning that the aggregate probability for selecting the correct overall classification decision from a set of N possible outputs states is the summation of multinomial state probabilities:

$$P(c_m \geq \max_{j\neq m} c_j) > \sum_{n_m > n_l \forall l \neq m} \binom{N}{\boldsymbol{n}} \prod_{l=1}^{K} p_l^{n_l} + \frac{1}{K} \sum_{n_m = \max_{n_l} \forall l \neq m} \binom{N}{\boldsymbol{n}} \prod_{l=1}^{K} p_l^{n_l}$$

Performing efficient subset selection is an ongoing area of research [26]; rather than focus on specific subset selection techniques, we focus on the applications of this model in the next section.

3 Decision Aggregation of Feed-Forward CNN Classifiers

To evaluate the efficacy of this multinomial framework against classifier frameworks in each of the common modalities as was shown in Fig. 1, we first consider a generic toy example that incorporates an arbitrary confusion matrix, representative of one environmental operating condition. Next, we apply the framework to a binary mask recognition task applied to real-time video streams with three different frame rates. This example illustrates the trade space between time and processing loads necessary to achieve an aggregate confidence level, and projects (4) to a binomial framework. Finally, we apply the full multinomial framework to two RFML tasks, SEI and AMC, across SNR ranges.

3.1 Toy Example

To provide a general example that demonstrates the technique in more detail, consider a feed-forward NN with 4 possible outputs ($K = 4$) and a test accuracy of 34% (i.e. $p_{11} = 0.34$), and the number inputs varying between 0 and 100 (i.e. $N \in \{1, 2, ..., 100\}$). Normally, a test accuracy of 34% is discounted as near useless, because, for each individual input frame, the correct classification accuracy is only marginally better than blind guessing (25%), so little confidence can be given to single output decisions. However, if we consider the following three scenarios such that the three error conditions are equiprobable (scenario 1), concentrated over two of the remaining three states (scenario 2), or varied over two of the remaining three states (scenario 3), the application of the proposed multinomial framework over varying values of N yields the aggregate classification accuracies shown in Fig. 3(a).

$$\begin{aligned}
\text{Scenario 1}: \{p_{11}, p_{12}, p_{13}, p_{14}\} &= \{34\%, 22\%, 22\%, 22\%\} \\
\text{Scenario 2}: \{p_{11}, p_{12}, p_{13}, p_{14}\} &= \{34\%, 32\%, 32\%, 2\%\} \\
\text{Scenario 3}: \{p_{11}, p_{12}, p_{13}, p_{14}\} &= \{34\%, 15\%, 22\%, 29\%\}
\end{aligned}$$

As expected, all three scenarios start with a classification accuracy of exactly $p_{11} = 0.34$. The oscillatory behavior for small numbers of output frames ($N < 10$) is due to the overall realization that error cases are in fact more likely than the correct classification, and thus the small population of outputs skewed probability that includes a higher density of ties ($c_m = \max_{j \neq m} c_j$) that are approximated as random guessing (25%). The difference between each of the three scenarios results from relative magnitudes of the correct classification and the highest density of error state probability. The extremely slow increase of Scenario 2 is representative of the near random guessing between three of the four states, while the quick convergence of Scenario 1 is indicative of the maximum separation between an assumed equiprobable error condition:

$$\frac{p_{mm}}{\max_{j \neq m} p_{mj}} = \frac{p_{mm}}{\frac{1}{N-1}(1 - p_{mm})} = \frac{(N-1)p_{mm}}{(1 - p_{mm})}$$

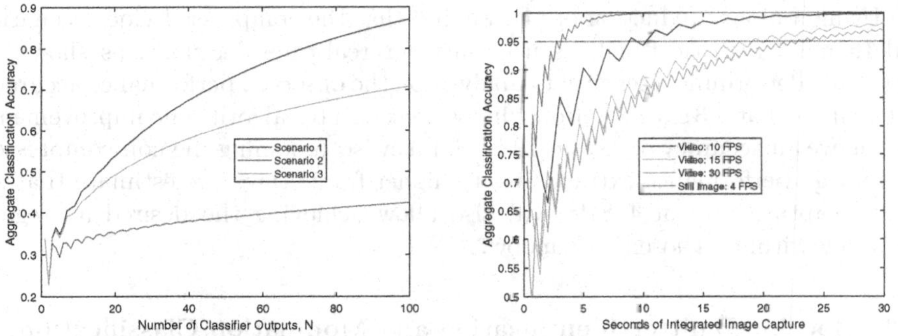

(a) (b)

Fig. 3. Comparison of three same-performance confusion scenarios for an arbitrary classification algorithm (a) and binomial reduction of aggregation method applied to a binary mask recognition problem (b).

The middle state, Scenario 3, has more evenly balanced error probabilities, indicative of more general training conditions for most classifiers, and converges quickly to higher classification accuracies, but not as quickly as the maximum possible rate based on Scenario 1. The rate of convergence for these three scenarios is fundamentally a ratio of large summations without common terms, making closed form reductions impractical, even when applying assumptions as described with Scenario 1. It should be noted that a final scenario, not shown in Fig. 3 occurs when the error density of any other state exceeds that of the correct classification - in such a scenario, the aggregate classification accuracy actually decays from p_{11}, making the proposed aggregation process counter-productive.

3.2 Binomial Reduction Applied to Mask Recognition Algorithm

Recent work in the area of binary mask recognition ('mask','no mask') has shown the ability to achieve a test accuracy of 99% accuracy using pristine images that are pre-processed such that the faces are centered within the each image [27]. Additional work [18] sought to generalize the algorithm to a wider range of ethnicities, more practical image qualities that may be encountered when using real-time webcams, facial images steered off camera boresight, and a series of intentional presentation attacks. While capable of achieving similar 99% accuracy on the original dataset, the generalized CNN used in [18] achieves closer to 72.91% accuracy when applied to still images, and progressively lower when applied to images derived from compressed video. However, in averaging 10 successive images from a compressed video stream with $\{10, 15, 30\}$ frames per second (FPS), the resulting accuracy is 75.67% (1 s total capture), 65.22% for 15 FPS (0.67 s), and 59.95% for 30 FPS was (0.33 s). This begs the operational question: how long must a person's face be observed in order to make a sufficiently accurate (i.e. 95%) decision?

Using a binary reduction of (4), we find that the compressed video scenarios fail to meet realistic timelines necessary for real-time decisions, as shown in Fig. 3(b). Performing a sensitivity analysis on the observed performance accuracy determines that a 3 s observation window necessitates algorithmic improvements to achieve an accuracy of 88% at 10 FPS. Likewise, assuming the non-compressed still image performance extrapolates to higher frame rates, we estimate that an image capture rate of 4 FPS will also allow achieving the desired aggregate accuracy within a chosen 3 s window.

3.3 Specific Emitter Identification and Modulation Classification

To further demonstrate the utility of the proposed multinomial model in synthesizing aggregate decisions in the RF domain, we provide exemplary use cases based on the SEI algorithm given in [20] and the AMC algorithm in [11]. In the CNN-based SEI algorithm of [20], a much smaller input frame size was considered than the traditional expert feature-based statistical IQ estimator algorithm given in [21], thought both contained 5 output classes ($K = 5$). Comparative results, normalized for the same amount of observation time, are shown in Fig. 4. Though the traditional expert feature-based algorithm achieved higher accuracy than the RFML-based algorithm when using only a single input capture, the expert feature-based algorithm utilized over an order of magnitude more symbols as input, making direct comparison between the two techniques unfair. Therefore, in [20], the authors utilized the multinomial approach given herein to aggregate decisions over 10 input captures, matching the equivalent amount of input observations. When the amount of data provided to each algorithm is approximately equivalent, the RFML-based algorithm outperforms the expert feature-based algorithm, especially at lower SNRs, showing a measurable benefit of the RFML-based approach.

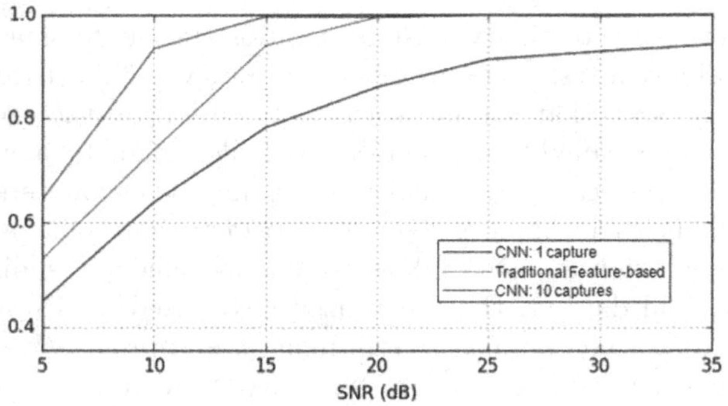

Fig. 4. Example use of the multinomial decision aggregation framework, comparing performance of RFML-based and traditional feature-based SEI algorithms.

Finally, when applying the same multinomial data aggregation models to an RFML-based AMC algorithm with four output classes ($K = 4$). The results, given in Fig. 5, show the proposed aggregation approach clearly improves the point-wise accuracy of the classifier when more input data is available than a single frame. At low SNRs, distinguishing between higher order QAM signals (i.e. 16QAM and 64QAM) is especially challenging. However, confusion between these two classes is notably lessened after the aggregation of just 10 input captures, and is negligible after the aggregation of 50 input captures.

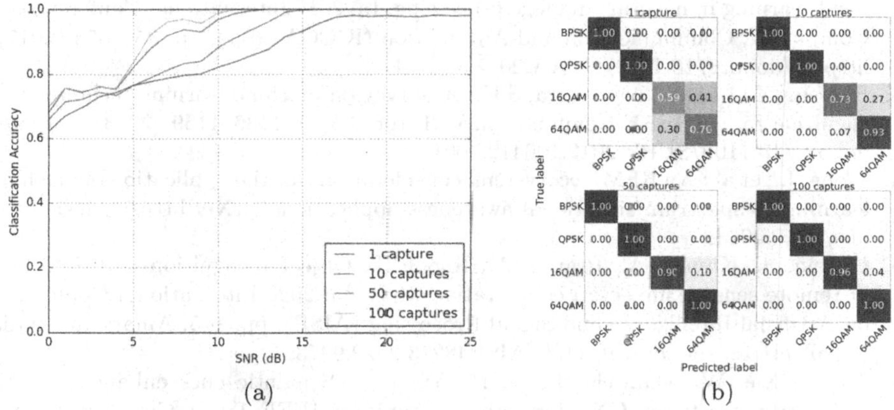

Fig. 5. Aggregate decisions for 4-state RFML-based AMC algorithm as functions of SNR (a) and confusion matrices (b) over $N = \{1, 10, 50, 100\}$ outputs.

4 Conclusions

This paper has focused on a framework for aggregating the outputs from any time-series classification decision algorithm using a multinomial distribution, shown to increase the overall classification accuracy when multiple consecutive inputs are observed and allow for direct comparison between techniques using differing input capture sizes. The rate of convergence for this output decision is primarily driven by the ratio of the classification accuracy of a single output decision and the most likely error case for that classification decision. While necessary to condition based upon the quality of the input, the process may be used to significantly improve classification accuracies of feed-forward NN-based classifiers that operate on low quality images or relatively short captures of RF signals, and may be extended across other different operating conditions (image frame rate, SNR, or other quality metric), if the confusion matrices for each condition are known. Future work is being pursued to relax the stated assumptions associated with temporal coherence, with the objective to better predict real-time performance in image classifiers and co-channel RF environments.

Acknowledgments. This work was partially supported by the Naval Surface Warfare Center, Crane Division, via the Naval Engineering Education Consortium under grant N00174-18-1-0005.

References

1. Abbas, Q., Ibrahim, M., Jaffar, M.: A comprehensive review of recent advances on deep vision systems. Artif. Intell. Rev. **52**, 39–76 (2019). https://doi.org/10.1007/s10462-018-9633-3
2. Sharma, A.R., Kaushik, P.: Literature survey of statistical, deep and reinforcement learning in natural language processing. In: 2017 International Conference on Computing, Communication and Automation (ICCCA 2017), pp. 350–354 (2017). https://doi.org/10.1109/CCAA.2017.8229841
3. Bkassiny, M., Li, Y., Jayaweera, S.K.: A Survey on machine-learning techniques in cognitive radios. IEEE Commun. Surv. Tutor. **15**(3), 1136–1159 (2013). https://doi.org/10.1109/SURV.2012.100412.00017
4. Wong, L., et al.: An RFML ecosystem: considerations for the application of machine learning to spectrum situational awareness applications. arXiv https://arxiv.org/abs/2010.00432
5. Fatima, S., Kumar, A., Pratap, A., Raoof, S.: Object recognition and detection in remote sensing images: a comparative study. In: 2020 International Conference on Artificial Intelligence and Signal Processing (AISP), pp. 1–5. Amaravati, India (2020) https://doi.org/10.1109/AISP48273.2020.9073614
6. Tu, Z., Xie, W., Dauwels, J., Li, B., Yuan, J.: Semantic cues enhanced multimodality multistream CNN for action recognition. IEEE Trans. Circ. Syst. Video Technol. **29**(5), 1423–1437 (2019). https://doi.org/10.1109/TCSVT.2018.2830102
7. Barabas, J., Zalman, R., Kochlan, M.: Automated evaluation of COVID-19 risk factors coupled with real-time, indoor, personal localization data for potential disease identification, prevention and smart quarantining. In: 2020 43rd International Conference on Telecommunications and Signal Processing (TSP), pp. 645–648. Milan, Italy (2020). https://doi.org/10.1109/TSP49548.2020.9163461
8. Nivetha, S.: A Survey on speech feature extraction and classification techniques. In: 2020 International Conference on Inventive Computation Technologies (ICICT), pp. 48–53. Coimbatore, India (2020). https://doi.org/10.1109/ICICT48043.2020.9112582
9. Ghorpade, T., Ragha, L.: Featured based sentiment classification for hotel reviews using NLP and Bayesian classification. In: 2012 International Conference on Communication, Information & Computing Technology (ICCICT), pp. 1–5. Mumbai, India (2012). https://doi.org/10.1109/ICCICT.2012.6398136
10. Romero, P., Dighe, K.: Fast and unsupervised classification of radio frequency data sets utilizing machine learning algorithms. Data Analyt. **2015**, 146–152 (2015)
11. Hauser, S.C., Headley, W.C., Michaels, A.J.: Signal detection effects on deep neural networks utilizing raw IQ for modulation classification. In: Military Communications Conference (MILCOM 2017), pp. 121–127. IEEE, MD, Baltimore (2017)
12. Merchant, K., Revay, S., Stantchev, G., Nousain, B.: Deep learning for rf device fingerprinting in cognitive communication networks. IEEE J. Select. Top. Sig. Process. **12**(1), 160–167 (2018)
13. Kim, K., Spooner, C.M., Akbar, I., Reed, J.H.: Specific emitter identification for cognitive radio with application to IEEE 802.11. In: IEEE GLOBECOM 2008–2008 IEEE Global Telecommunications Conference, pp. 1–5. New Orleans (2008)

14. Cain, I., Clark, J., Pauls, E., Ausdenmoore, B., Clouse, R., Josue, T.: Convolutional neural networks for radar emitter classification. In: 2018 IEEE 8th Annual Computing and Communication Workshop and Conference (CCWC), pp. 79–83. Las Vegas (2018)
15. Selim, A., Paisana, F., Arokkiam, J., Zhang, Y., Doyle, L., DaSilva, L.: Spectrum monitoring for radar bands using deep convolutional neural networks. In: GLOBECOM 2017–2017 IEEE Global Communications Conference, pp. 1–6. Singapore (2017)
16. LeCun, Y., Bengio, Y., Hinton, G.: Deep learning. Nature **521**, 436–444 (2015). https://doi.org/10.1038/nature14539
17. Wang, Z., et al.: Masked face recognition dataset and application. arXiv (2020). https://arxiv.org/abs/2003.09093
18. Billings, R.: On efficient computer vision applications for neural networks. Masters Thesis, Virginia Tech (2021)
19. Emam, A., Shalaby, M., Aboelazm, M., Bakr, H., Mansour, H.: A Comparative study between CNN, LSTM, and CLDNN models in the context of radio modulation classification. In: 2020 12th International Conference on Electrical Engineering (ICEENG), pp. 190–195. Cairo, Egypt (2020). https://doi.org/10.1109/ICEENG45378.2020.9171706
20. Wong, L.J., Headley, W.C., Michaels, A.J.: Specific emitter identification using convolutional neural network-based IQ imbalance estimators. IEEE Access **7**, 33544–33555 (2019). https://doi.org/10.1109/ACCESS.2019.2903444
21. Zhuo, F., Huang, Y., Chen, J.: Radio frequency fingerprint extraction of radio emitter based on I/Q imbalance. Procedia Comput. Sci. **107**, 472–477 (2017)
22. Zeiler, M.D., Fergus, R.: Stochastic pooling for regularization of deep convolutional neural networks. In: Proceedings of the International Conference on Learning Representations (ICLR) (2013)
23. Springenberg, J.T., Riedmiller, M.: Improving deep neural networks with probabilistic maxout units. CoRR. https://arxiv.org/abs/1312.6116
24. Gu, J., et al.: Recent advances in convolutional neural networks. Patt. Recogn. **77**, 354–377 (2018). ISSN: 0031-3203. https://doi.org/10.1016/j.patcog.2017.10.013
25. Mittal, A., Moorthy, A.K., Bovik, A.C.: No-reference image quality assessment in the spatial domain. IEEE Trans. Image Process. **21**(12), 4695–4708 (2012)
26. Bakir, S.: A subset selection procedure for multinomial distributions. J. Appl. Stat. **40**(7), 1608–1618 (2013)
27. Wang, Z., et al.: Masked face recognition dataset and application. arXiv preprint arXiv:2003.09093 (2020)

Quantile Encoder: Tackling High Cardinality Categorical Features in Regression Problems

Carlos Mougan[1(✉)], David Masip[2], Jordi Nin[3], and Oriol Pujol[4]

[1] Electronics and Computer Science, University of Southampton, Southampton, UK
C.Mougan-Navarro@southampton.ac.uk
[2] Centre Recerca Matematica, Universitat Autonoma de Barcelona, Barcelona, Spain
[3] Universitat Ramon Llull, ESADE, Barcelona, Spain
jordi.nin@esade.edu
[4] Department of Mathematics and Computer Science, Universitat de Barcelona, Barcelona, Spain
oriol_pujol@ub.edu

Abstract. Regression problems have been widely studied in machine learning literature resulting in a plethora of regression models and performance measures. However, there are few techniques specially dedicated to solve the problem of how to incorporate categorical features to regression problems. Usually, categorical feature encoders are general enough to cover both classification and regression problems. This lack of specificity results in underperforming regression models. In this paper, we provide an in-depth analysis of how to tackle high cardinality categorical features with the quantile. Our proposal outperforms state-of-the-art encoders, including the traditional statistical mean target encoder, when considering the Mean Absolute Error, especially in the presence of long-tailed or skewed distributions. Besides, to deal with possible overfitting when there are categories with small support, our encoder benefits from additive smoothing. Finally, we describe how to expand the encoded values by creating a set of features with different quantiles. This expanded encoder provides a more informative output about the categorical feature in question, further boosting the performance of the regression model.

Keywords: Statistical learning · Regression problems · Machine learning · Categorical features

1 Introduction

In the modeling stage of a machine learning (ML) prediction problem, there is the need of feeding the model with meaningful features that describe the problem in a relevant way. That is why the steps of data preparation and feature engineering

The author has contributed to this work while he was employed at European Central Bank.

V. Torra and Y. Narukawa (Eds.): MDAI 2021, LNAI 12898, pp. 168–180, 2021.
https://doi.org/10.1007/978-3-030-85529-1_14

are crucial in any ML project [2,10]. With the recent amount of available data, there is an inevitable increment in the variety of features. For the specific case of categorical variables this increment has two different effects: (a) the quantity of features is larger, and (b) the number of distinct values found in each feature (cardinality) increases [19]. When facing this second scenario, the problem of representing categorical features effectively and efficiently has a relevant effect on the performance of machine learning model.

Handling categorical features is a known and very common problem in data science and machine learning, given that many algorithms need to be fed with numerical data [23]. There are many well-known methods for approaching this problem [17]. However, depending on the kind of problem faced, namely classification or regression problems, some of the techniques for encoding categorical data are more suitable than others. This is particularly true when dealing with large-scale data where errors and outliers are more common and may hinder the computation of reliable statistical measures.

The most well-known encoding for categorical features with low cardinality is One Hot Encoding [1]. This produces orthogonal and equidistant vectors for each category. However, when dealing with high cardinality categorical features, one hot encoding suffers from several shortcomings [20]: (a) the dimension of the input space increases with the cardinality of the encoded variable, (b) the created features are sparse - in many cases, most of the encoded vectors hardly appear in the data -, and (c) One Hot Encoding does not handle new and unseen categories.

An alternative encoding technique is Label/Ordinal Encoding [3] which uses a single column of integers to represent the different categorical values. These are assumed to have no true order and integers are selected at random. This encoding handles the problem of the high dimensional encoding found in One Hot Encoding but imposes an artificial order of the categories. This makes it harder for the model to extract meaningful information. For example, when using a linear model, this effect prevents the algorithm from assigning a high coefficient to this feature.

Alternatively, Target Encoding (or mean encoding) [15] works as an effective solution to overcome the issue of high cardinality. In target encoding, categorical features are replaced with the mean target value of each respective category. With this technique, the high cardinality problem is handled and categories are ordered allowing for easy extraction of the information and model simplification. The main drawback of Target Encoding appears when categories with few (even only one) samples are replaced by values close to the desired target. This biases the model to over-trust the target encoded feature and makes it prone to overfitting. To overcome this problem several strategies introduce regularization terms in the target estimation [15]. A possibility is to use an estimator with additive smoothing, such as the M-Estimator, to estimate each category mean.

Although the techniques described before work for both regression and classification techniques, most of the literature focuses on the binary classification tasks even though the meaningful statistics of a binary classification are not well suited for other prediction tasks such as regression or multi-class classification. For example, when dealing with supervised learning regression tasks, calculating the target mean can give a misleading representation of the category due to the statistical properties of the mean if the data shows heavy tails. A reasonable change in this scenario would be the use of other summary statistics more suited to the target distribution. However, to the best of our knowledge, there is no previous research done in using other aggregation statistics other than the mean. In this paper, we study the use of the quantile as a better and more flexible summarizing statistic value on the regression tasks when measuring the Mean Absolute Error in high cardinality datasets. We study its effect in front of skewed and long-tailed distributions. We additionally introduce a regularization strategy to avoid over-representation of the encoded feature when the statistic is computed over a small target subset of data. Moreover, a richer extension of the studied encoder, namely target summaries, that consists of a discrete set of the basic quantile encoder with different hyperparameter values is introduced.

The strategy is evaluated in different regression scenarios including the presence of outliers, long-tailed, and skewed distributions. We summarize the main contributions of this paper as follows:

(i) We define the quantile encoder. This encoder improves the performance of a regression model when evaluated using the Mean Absolute Error. It is also remarkable that the Mean Absolute Error can be improved even when using a different target loss, such as least squares loss.
(ii) We show that the quantile encoder improves the performance of regression models when the distribution of the target is long-tailed.
(iii) Finally, we introduce the idea of summary encoder, an encoder designed for creating richer representations. This is built by leveraging information from different quantiles.

For the sake of reproducibility and to help with the development, experimentation, and testing of the methods used in this paper, an open-source python package containing all the implementations used for this paper is released. We refer to this package as *sktools* [7].

The rest of this article is organized as follows. Section 2 presents a formal definition of the proposed encoding. Section 3 shows the benchmarking datasets, experimental settings, results, and their corresponding discussion. Finally, in Sect. 4 the main conclusions of the paper are summarized and possible future work is presented.

2 Quantile Encoder

The problem that we aim to tackle is the improvement of the encoding of categorical variables in regression models. Target encoding with the mean is a valid approach, but not necessarily the most suitable. Target encoding can be easily generalized by replacing the mean with any other summarizing statistic. Thus, following a similar strategy to mean encoding, here we generalize the definition of Target encoding studying the use of the quantile as a summarizing encoding metric in the different categories.

Formally, Quantile encoding is defined as follows: Given a dataset $\mathcal{D} = \{(x_i, y_i)\}, i \in 1 \dots N$ with x_i a d-dimensional feature vector, y_i its corresponding label, we identify the j-th feature from sample x_i as $x_i^{(j)}$. For the following discussion we consider feature j-th a categorical variable with K_j different values. The quantile encoder replaces that feature as follows

$$\hat{x}_i^{(j)} = q_p(\{y_k\}), \quad \forall\ (x_k, y_k) \in \mathcal{D} \big|\ x_k^{(j)} = x_i^{(j)}, \tag{1}$$

where q_p is the quantile at p. Equation 1 assigns the p quantile of all targets that share the same categorical value for that feature.

A common issue when using target-based encodings such as mean encoding or the quantile encoding is not having enough statistical mass for some of the encoded categories. And, therefore, this creates features that are very close to the target label. Thus, they are prone to over-fitting. A possible solution is to regularize the target encoding feature using additive smoothing, as in [4,25]. To do so, we compute the quantile encoding using the following equation:

$$\tilde{x}_i^{(j)} = \frac{\hat{x}_i^{(j)} \cdot n_i + q_p(\{y\}) \cdot m}{n_i + m} \tag{2}$$

where,

- $\tilde{x}_i^{(j)}$ is the regularized Quantile Encoder applied to the value corresponding to element $x_i^{(j)}$.
- $\hat{x}_i^{(j)}$ is the non-regularized Quantile Encoder; the value corresponding to element $x_i^{(j)}$ as defined in Eq. 1.
- n_i is the number of samples sharing the same value as $x_i^{(j)}$.
- $q_p(\{y\})$ is the global p-quantile of the target.
- m is a regularization parameter, the higher m the more the quantile encoding feature tends to the global quantile. It can be interpreted as the number of samples needed to have the local contribution (quantile of the category) equal to the global contribution (global quantile).

The rationale of Eq. 2 is that, if a class has very few samples, $n_i \ll m$ then the quantile encoding will basically be the global quantile, $\tilde{x}_i^{(j)} \approx q_p(\{y\})$. If a class has a large number of samples, $n_i \gg m$ and $\tilde{x}_i^{(j)} \approx \hat{x}_i^{(j)}$ then the class quantile will have more weight than the global quantile. As a result, the Quantile Encoder transformation of categorical features has two different hyperparameters that can be tuned to increase, adjust, and modify the type of encoding. These are the following:

- m is a regularization hyperparameter. The range of this parameter is $m \in [0, \infty)$. For the specific case where $m = 0$, there is no regularization. The larger the value of m, the most the Quantile Encoder features tends to the global quantile.

- p is the value of the quantile of the target probability distribution. The range of this parameter is $p \in [0,1]$. When p is 0.5, we obtain the median encoder, as it applies $q_{0.5}$, the median of the target in each category.

2.1 Summary Encoder

A generalization of the quantile encoder is to compute several features corresponding to different quantiles per each categorical feature, instead of a single feature. This allows the model to have broader information about the target distribution for each value of that feature than just a single value. This richer representation will be referred to as *summary encoder*. Formally, it is defined as

$$\hat{x}_i^{(j)} = \{q_{p_m}(\{y_k\})\}, \quad \forall\,(x_k, y_k) \in \mathcal{D}|\; x_k = x_i^{(j)},\; m = 1 \ldots M \tag{3}$$

where M is the number of new features, and p_m are the values of the quantiles to use. This representation changes a single feature by a set of M quantiles according to the values in p_m.

3 Experiments

To perform an empirical evaluation of the proposed statistical encoding techniques, we developed a framework to ensure that all experiments described in this paper are fully reproducible. Original data, data preparation routines, code repositories, and methods are publicly available at [6]. Experiments have been organized into three groups: Firstly, we assess the performance of quantile encoder when compared with the state-of-the-art encodings, namely catboost, M-estimate, target, James-Steiner encoder, Generalized Linear Mixed Model Encoder, and ordinal encoder. To do this comparison we used the Mean Absolute Error. In the second group of experiments, we aim at showing the dependence of the encoding on the evaluation metric. To that effect, we study the performance of the quantile encoder when used with a least-squares loss model in terms of mean absolute error and mean squared error. Finally, we compare summary encoder with quantile and target encoders, the goal of these last experiments is to create more informative encoders to be able to boost regression algorithms performance.

3.1 Dataset Bench-Marking

The scenario we are addressing is characterized by datasets that display categorical features with high cardinality and skewed/long tail target distribution in a regression environment. Unfortunately, most of the machine learning benchmarking datasets do not display these common features of many real-life problems. Thus, following the open data for reproducible research guidelines described in [14] and for measuring the performance of the proposed methods, we have used a synthetic dataset for a more theoretical evaluation together with 4 open-source datasets for an empirical comparison. The selected datasets are:

- *The StackOverflow 2018 Developer survey* [21] is a data set with only five categorical features, namely country, employment status, formal education, developer type, and languages worked. The target variable is the annual salary of the user.
- *The StackOverflow 2019 Developer survey* [22] is a data set with a single categorical feature (country), few numerical features (working week hours, years of coding, and age), and a long-tailed target variable (salary).
- *Kickstarter Projects* [12] is a data set with crowd-funding projects where the goal is to predict what is the funding goal of each project. The categorical features are the crowd-funding type, the country, the state, and the currency.
- *Medical Payments* [13] is a data set with information about the price of a medical treatment. The dataset consists of 10 categorical features containing information about the state, city, zip code, country, physician type, physician country, the payment method, and its nature.

The datasets have undergone a minimal curation process, where miscellaneous features are removed and only the columns that are considered meaningful and informative for the modeling of the problem are kept.

For the synthetic dataset, we have used the Cauchy distribution in Eq. 4 to create a target distribution with long tails. The Cauchy distribution is parameterized by t and s, being t the location parameter and s the scale parameter as follows,

$$Cauchy(x; t, s) = \frac{1}{s\pi(1 + ((x - t)/s)^2)}. \tag{4}$$

Next,we have created two features, x_1 and x_2 accordingly,

$$c_i \sim U(0, 100), \quad x_1 \sim Cauchy(x; c_i, 1), \quad x_2 \sim Cauchy(x; c_i, 2),$$

where $\epsilon \sim N(0, 1)$ is sampled from a normal distribution and x_1 and x_2 are generated sampling from the Cauchy density function. Both features even if numerical are then treated as categorical variables. The corresponding regression target is generated as

$$y = x_1 + x_2 + \epsilon$$

3.2 Code and Reproducibility

For the sake of reproducibility, the code for the experiments has been encapsulated in a library, *sktools* [7]. It can be found https://sktools.readthedocs.io/. The library contains the implementations presented in this paper such as Quantile Encoder regularized with an additive smoothing and the Summary Encoder. Notebook and experiments are hosted on the following Github repository [6].

With respect to quantile encoder the default values for m and p are $m = 1$ and $p = 0.5$. Default values for the hyperparameters of the summarizing encoder are $m = 100$ and defines three encoding features containing the quantiles at 0.4, 0.5 and 0.6.

As a final estimator we have used a Generalized Linear Model with an reguralarization hyperparameter (*l1_penalty*) that we have optimized across the crossvalidation folds.

3.3 Comparison of all Encoding Methods

This first experiment consists of comparing the performance of the quantile encoder against state-of-the-art encoding techniques in terms of the Mean Absolute Error (MAE). To do so, we have used the methods implemented in the Category Encoders library [27] to evaluate them against our proposed encoding technique. This set of encoding techniques includes ordinal encoding, James-Stein Encoder [8,16,28], and several state of the art target encodings with different regularization values such as catboost [18], classical target encoder [15,27] and M-estimate target encoding [4] and generalized linear mixed models encoder [9,27]. Every experiment has been executed using 3 times repeated 4-fold cross-validation on the parameters of each method.

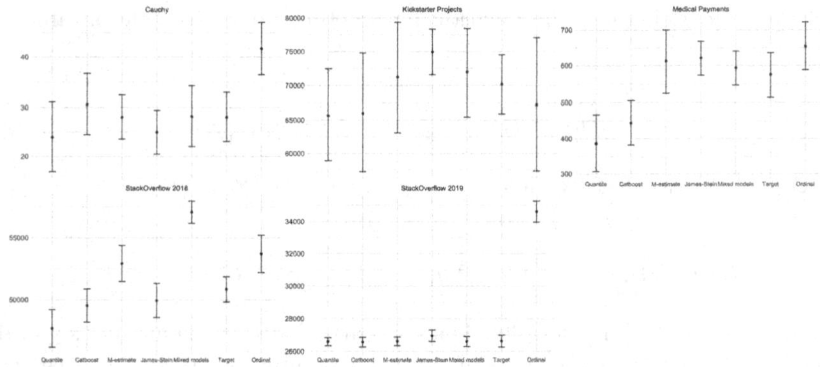

Fig. 1. Comparison between Quantile Encoder and several categorical encoding techniques using the cross validated MAE error.

Figure 1 shows the results of the comparison on the cross-validated sets for the aforementioned datasets. Observe that on average, the quantile encoder achieves the best scores, followed by catboost. In all datasets, Quantile Encoder performs in the worst-case scenario similar to other state-of-the-art techniques. However, in three out of five cases it produces a solid improvement of the Mean Absolute Error. As expected, ordinal encoder yields the worst performance in these experiments. One of the advantages of using the quantile to encode categorical features over the mean encoding is that it allows us to have one more tunable hyperparameter.

3.4 Encoding Dependence with Respect to the Evaluation Metric

When evaluating machine learning regression models, the following natural question arises *For which metrics does the encoding technique give an improvement with respect to the alternatives?*

The mean is the estimator that minimizes the Mean Squared Error (MSE), meanwhile for the Mean Absolute Error is the median [5,11]. This statement supports the hypothesis that the median encoder may improve the performance of any regression model when it is measured with the MAE. To provide empirical evidence we evaluate mean and quantile encoders in front of MAE and MSE evaluation metrics. It is important to highlight that MAE error has one advantage versus the MSE from an interpretation point of view, in the sense that MAE maintains the units of the quantities giving a more intuitive representation of the performance of the model to the user. We evaluate both encodings with both metrics using an Elastic Net [24,29] with default scikit learn hyperparameters ($alpha = 1.0$, $l1\_ratio = 0.5$) as estimator. Hyperparameters are optimized using a grid search with parameters $m \in \{0, 1, 10, 50\}$ and $quantile \in \{0.25, 0.5, 0.75\}$.

Table 1. (a) Comparison between Quantile Encoder and Target Encoder for different evaluation metrics, (b) Wilcoxon's test p-values.

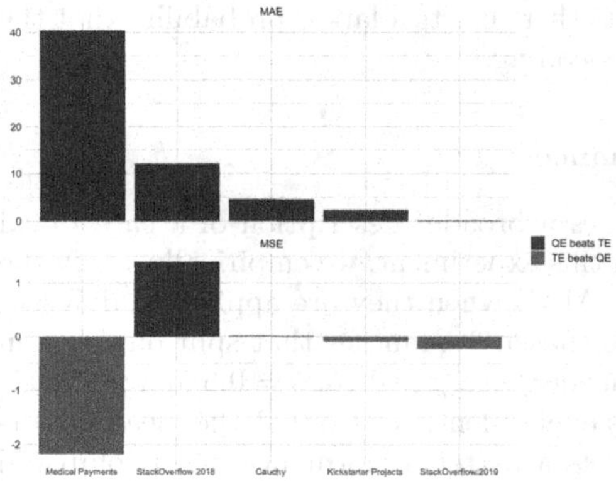

(a)

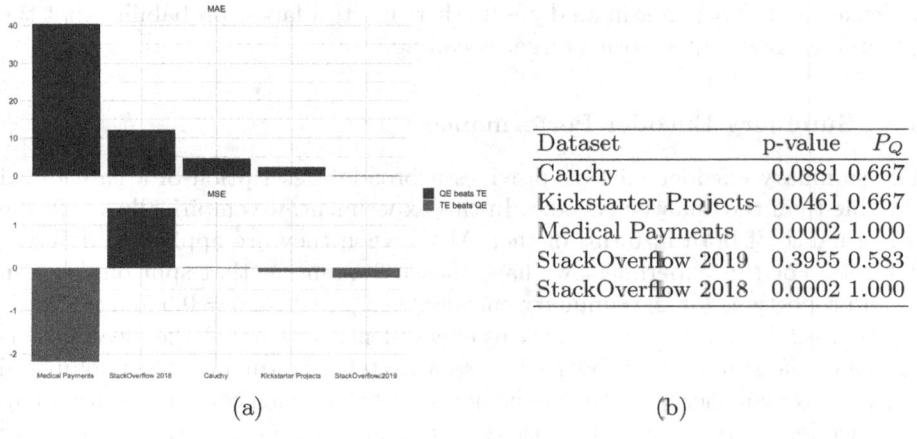

Dataset	p-value	P_Q
Cauchy	0.0881	0.667
Kickstarter Projects	0.0461	0.667
Medical Payments	0.0002	1.000
StackOverflow 2019	0.3955	0.583
StackOverflow 2018	0.0002	1.000

(b)

Figure in Table 3.4(a) shows the percentual difference between each encoding with respect to two metrics. In the upper part of the figure, we can see that the Quantile Encoder achieves better results than the Target Encoder for all datasets except for one when measuring the MAE metric. The encoding yields bigger percentual differences in medical payments and the StackOverflow datasets. In the lower part of the figure, we have the same plot using the MSE metric. Observe that mean encoder achieves better results in three out of five experiments. Observe that in this last case, percentual performance differences are smaller than in the case of MAE. Additionally, quantile encoder performs better than mean encoder in two of them. It is worth noting that the loss function of elastic net corresponds to a least-squares loss. Thus, this should benefit mean encoding and harm the performance of quantile encoder. However, results show the robustness of the quantile encoder even when in this adversarial case.

To verify the generalization of this observation, a quantile encoder is statistically validated on the selected data sets. The null hypothesis states that quantile encoder and mean encoder has the same performance when considering MAE. The p-value in Table 1(b) shows the results of the Wilcoxon test [26][1] on the MAE on 3 repetitions of 4-fold cross-validation. Observing Table 3.4(b) we see that the p-values in 3 out of 5 datasets are able to reject the null hypothesis at a significance level of 0.05. We observe that in the *cauchy* dataset the rejection level is found at a significance level of 0.10. Finally, in the 2019 StackOverflow dataset, we are not able to reject the null hypothesis. Despite this last result, the quantile encoder is not worse in any of the five datasets. This shows that the quantile encoder is a useful technique to encode categorical variables when optimizing the MAE. We additionally compute the probability of the quantile encoder outperforming the target encoder. This is shown in the column P_Q of Table 3.4. The value is computed by computationally estimating the empirical distribution of the difference of the performance values using kernel density estimation and integrating the area of the distribution where quantile encoder outperforms target encoding. The obtained results are in accordance with the Wilcoxon test. Note that in all datasets there exists a larger probability that the quantile encoder outperforms target encoding.

3.5 Summary Encoder Performance

The summary encoder method provides a broader description of a categorical variable than the quantile encoder. In this experiment, we empirically verify the performance of both in terms of their MAE when they are applied to different datasets. For this experiment we have chosen 3 quantiles that split our data in equal proportions for the summary encoder, *i.e.*, $p = 0.25$, $p = 0.5$ and $p = 0.75$.

Figure 2 depicts the results for this experiment. Notice that the mean performance of the summary encoder suggests a better performance when compared to the target encoder. The same behavior is observed when compared with quantile encoder in some cases. It must be noted that some extra caution needs to be taken when using the summary encoder as there is more risk of overfitting the more quantiles are used. This usage of the Quantile Encoding requires more hyperparameters as each new encoding requires two new hyperparameters, m and p, making the hyperparameter search computationally more expensive.

[1] The Wilcoxon test is a non-parametric statistical hypothesis test used to compare two repeated measurements on a single sample to assess whether their population means ranks differ.

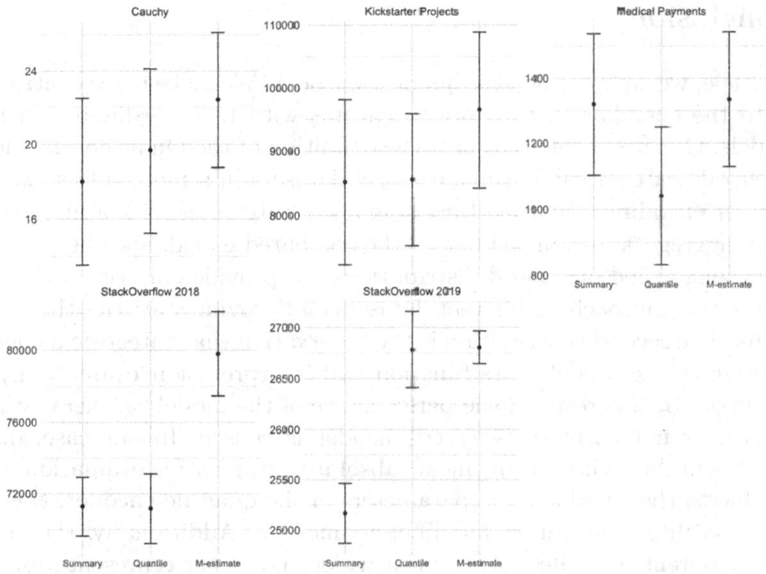

Fig. 2. Comparison between Summary, Quantile and Target encoders using the cross validated MAE error.

3.6 Discussion

The experiments show that quantile encoder represents better high cardinality categorical data in several scenarios. The observed improvements are:

- Quantile encoder is robust in front of outliers. On the contrary, target encoding is very sensitive to samples in the training set with extreme values.
- From an optimization point of view, the mean is the estimator that minimizes the MSE of a sample. On the contrary, and besides optimizing MAE, the use of quantile encoder is a sensible option for general use as it provides a highly tunable summary statistic suited to a broader set of metrics. Besides, from a regression point of view, MAE is a more intuitive metric that helps users interpreting the results.
- Finally, quantiles can be grouped to provide a much richer description of a categorical feature. For instance, we can run the percentiles 25, 50, and 75, which give much more information than just computing the mean. More features provide more information to the model. However, more features also increase the risk of overfitting and the problem starts gaining dimensionality. With the use of the Summary Encoder dimensionality does not become a hazard such as in the case of one-hot encoder. Nonetheless, the regularization techniques are to be considered to avoid overfitting in this case.

4 Conclusion

In this article, we have studied the quantile encoder. We make three contributions related to the encoding of categorical features with high cardinality in regression models. Our first contribution is the definition of the Quantile Encoder as a way to encode categorical features in noisy datasets in a more robust way than mean target encoding. Quantile Encoding maps categories with a more suitable statistical aggregation than the rest of the compared encodings when categories display in long-tailed or skewed distributions. To provide empirical evidence we benchmark the approach in different datasets and provide statistics that support our claims. The second contribution is the observation that categorical encodings are sensitive to the model's loss function and interpretation/evaluation performance metric. In this respect, the performance of the model can heavily change if a general or not correctly selected encoder is chosen. In our case, quantile encoder is suitable when using mean absolute error as an evaluation metric. Finally, due to the tunable hyperparameters of the quantile encoder, this shows a large versatility, being used for different metrics. Additionally, the concatenation of different quantiles allows for a wider and richer representation of the target category that results in a performance boost in regression models. To aid in the goal of open-source and reproducible research, we have released a toolkit, namely *sktools* [7], as an open-source Python package that provides a flexible implementation of the concepts introduced in this paper. For the summary encoder, we have used the M-estimate regularization technique, but further research can be done in the path of avoiding overfitting when creating a set of features out of a high-cardinality categorical feature. Strategies such as those found in leave-one-out encoding, or catboost encoder [18] could be considered to that effect.

Acknowledgements. This work was partially funded by the European Commission under contract numbers NoBIAS—H2020-MSCA-ITN-2019 project GA No. 860630.

This work has been partially funded by the Spanish project PID2019-105093GB-I00 (MINECO/FEDER, UE).

References

1. Bruin, J.: newtest: command to compute new test @ONLINE (2011). https://stats.idre.ucla.edu/stata/ado/analysis/
2. Burkov, A.: Machine Learning Engineering, 1 edn. Kindle Direct Publishing (2020)
3. Carey, G.: Coding categorical variables (2003). http://psych.colorado.edu/~carey/Courses/PSYC5741/handouts/Coding%20Categorical%20Variables%202006-03-03.pdf
4. Cestnik, B., Bratko, I.: On estimating probabilities in tree pruning. In: Kodratoff, Y. (ed.) EWSL 1991. LNCS, vol. 482, pp. 138–150. Springer, Heidelberg (1991). https://doi.org/10.1007/BFb0017010
5. Charles, J.G.: School of Statistics, University of Minnesota: Stat 5101 Lecture slides (2020). https://www.stat.umn.edu/geyer/f11/5101/slides/s4a.pdf

6. Masip, D., Mougan, C.: Quantile encoder experiments (2020). https://github.com/david26694/QE_experiments
7. Masip, D., Mougan, C.: Sktools:tools to extend sklearn, feature engineering based transformers (2020). https://sktools.readthedocs.io/
8. Efron, B., Morris, C.: Stein's paradox in statistics. Sci. Am. **236**, 119–127 (1977). https://doi.org/10.1038/scientificamerican0577-119
9. Gelman, A., Hill, J.: Data Analysis Using Regression and Multilevel/Hierarchical Models. Analytical Methods for Social Research. Cambridge University Press, Cambridge (2006). https://doi.org/10.1017/CBO9780511790942
10. Géron, A.: Hands-on machine learning with Scikit-Learn and TensorFlow : Concepts, Tools, and Techniques to Build Intelligent systems. O'Reilly Media, Sebastopol (2017)
11. Jaynes, E.T.: Probability Theory: The Logic of Science. Cambridge University Press, Cambridge (2003)
12. Kaggle: Kickstarter projects (2020). https://www.kaggle.com/kemical/kickstarter-projects. [Online; accessed 20-October-2020]
13. CMS.gov Centers for Medicare & Medicaid Services: Medical payments dataset (2020). Data retrieved from Center for Medicare and Medicaid Services, https://www.cms.gov/OpenPayments/Explore-the-Data/Dataset-Downloads
14. The Turing Way Community: The Turing Way: A Handbook for Reproducible Data Science (2019). https://doi.org/10.5281/zenodo.3233986
15. Micci-Barreca, D.: A preprocessing scheme for high-cardinality categorical attributes in classification and prediction problems. SIGKDD Explor. Newsl. **3**(1), 27–32 (2001)
16. Morris, C.N.: Parametric empirical bayes inference: theory and applications. J. Am. Stat. Assoc. **78**(381), 47–55 (1983)
17. Pargent, F., Bischl, B., Thomas, J.: A benchmark experiment on how to encode categorical features in predictive modeling. Master's thesis, School of Statistics (2019)
18. Prokhorenkova, L., Gusev, G., Vorobev, A., Veronika Dorogush, A., Gulin, A.: CatBoost: unbiased boosting with categorical features. arXiv e-prints arXiv:1706.09516 (2017)
19. Slakey, A., Salas, D., Schamroth, Y.: Encoding categorical variables with conjugate bayesian models for WeWork lead scoring engine (2019)
20. Slakey, A., Salas, D., Schamroth, Y.: Encoding categorical variables with conjugate bayesian models for WeWork lead scoring engine. arXiv e-prints arXiv:1904.13001 (2019)
21. Stackoverflow: Developer survey results 2018 (2018). https://insights.stackoverflow.com/survey/2018/
22. Stackoverflow: Developer survey results 2019 (2019). https://insights.stackoverflow.com/survey/2019/
23. Tutz, G.: Regression for Categorical Data. Cambridge Series in Statistical and Probabilistic Mathematics. Cambridge University Press, Cambridge (2011). https://doi.org/10.1017/CBO9780511842061
24. Wang, L., Zhu, J., Zou, H.: The doubly regularized support vector machine. Statistica Sinica **16**, 589–615 (2006)
25. Wikipedia contributors: Additive smoothing – Wikipedia, the free encyclopedia (2020). https://en.wikipedia.org/w/index.php?title=Additive_smoothing&oldid=937083796

26. Wilcoxon, F.: Individual comparisons by ranking methods. In: Kotz, S., Johnson, N.L. (eds.) Breakthroughs in Statistics. Springer Series in Statistics (Perspectives in Statistics). Springer, New York (1992). https://doi.org/10.1007/978-1-4612-4380-9_16
27. Will McGinnis: category encoders :a library of sklearn compatible categorical variable encoders (2020). https://contrib.scikit-learn.org/
28. Zhou, X.: Shrinkage estimation of log-odds ratios for comparing mobility tables. Sociol. Methodol. **45**(1), 320–356 (2015)
29. Zou, H., Hastie, T.: Regularization and variable selection via the elastic net. J. R. Stat. Soc. Ser. B **67**, 301–320 (2005)

Evidential Undersampling Approach for Imbalanced Datasets with Class-Overlapping and Noise

Fares Grina[1,2](✉), Zied Elouedi[1], and Eric Lefevre[2]

[1] Institut Supérieur de Gestion de Tunis, LARODEC, Université de Tunis, Tunis, Tunisia
fares.grina@isg.u-tunis.tn, zied.elouedi@gmx.fr

[2] Univ. Artois, UR 3926, Laboratoire de Génie Informatique et d'Automatique de l'Artois (LGI2A), 62400 Béthune, France
eric.lefevre@univ-artois.fr

Abstract. The class imbalance issue involves many real-world domains such as fraud detection, medical diagnosis, intrusion detection, etc. Most classification algorithms tend to perform poorly when the training dataset is class-imbalanced. This problem gets more challenging in the presence of other factors such as class-overlapping and noise. Among many methods, undersampling is a simple and efficient approach which re-balances the imbalanced dataset by removing majority samples. In this paper, we propose a novel method named Evidential Undersampling (EVUS), which is a re-sampling approach based on the theory of evidence. To avoid removing meaningful samples, each majority object is assigned a soft evidential label to gain more information about its location, then majority samples which are considered ambiguous or noisy by our framework, are eliminated from the training set. The conducted results with CART and SVM show that our proposal outperformed other well-known undersampling methods according to the AUC metric.

Keywords: Undersampling · Imbalanced data · Evidence theory · Classification

1 Introduction

In real-world classification tasks, instances are not always evenly distributed among classes. This form of distributions is typically referred to as *imbalanced* or *skewed*. In a binary class-imbalanced dataset, the class with the higher size is called the *majority* class, whereas the rare class is regarded as the *minority* class. Imbalanced classification has been reported in many domains such as medical diagnosis [4], fraudulent credit card detection [25], drug discovery [18], etc. From an application perspective, misclassifying a minority example is more critical than misclassifying a majority example [6]. For instance, failing to recognize a rare disease can be crucial. Class-imbalanced distributions can be due to a lot

V. Torra and Y. Narukawa (Eds.): MDAI 2021, LNAI 12898, pp. 181–192, 2021.
https://doi.org/10.1007/978-3-030-85529-1_15

of factors, including the domain's background (e.g. rare fraudulent transactions) or data collection (e.g. storage). This significantly deteriorates the performance of most classifier algorithms, since most of them assume an even distribution of the classes [14].

Over the years, many methods have been proposed to deal with imbalanced classification [12]. Generally, existent solutions can be categorized into two main groups: data-level [9], algorithm-level [14], as well as combinations of the two strategies. Algorithm-level methods involve modifying the classifier algorithm to adapt it for imbalanced datasets. It can also contain cost-sensitive solutions [1] and ensemble methods [29]. Data-level approaches typically change the class distribution of the training dataset by adding synthetic samples (oversampling), removing majority examples (undersampling), or both. One advantage is that data-level methods are independent of the used classifier. In other words, they are more flexible and do not require deep understanding of learning algorithms. Oversampling techniques are becoming more expensive in terms of complexity and memory as the amount of data is substantially increasing. In many classification cases, it is more effective to perform undersampling.

The most naive form of undersampling is random undersampling (RUS) [12], which randomly eliminates majority class objects to improve the class distribution of the training set. Nonetheless, it is possible that this method may remove potentially meaningful information from the dataset. To avoid this, many methods have been proposed to intelligently select unessential majority points for elimination. Some works used traditional filtering techniques, such as Edited Nearest Neighbors (ENN) [37], Nearest Neighbor Rule (CNN) [3], and Tomek Links (TL) [15]. These strategies discard majority points based on their nearest neighbors. Undersampling based on evolutionary algorithms were also proposed. In [10], the authors implemented evolutionary prototype selection to create an improved subset of the majority class. ACOSampling [40] is another evolutionary-based undersampling technique, which makes use of ant colony optimization (ACO) [7]. More recently, clustering-based methods were developed [21,22,27,35]. In [22], the authors used the k-means algorithm [16] to reduce the majority class size by only selecting the cluster centers for training. CBIS [35] is another clustering-based undersampling approach which uses clustering analysis to divide the majority class into sub-classes, and remove the unessential points in each sub-class.

Although studies have confirmed that a fair distribution of the classes typically performs better [5,14], it is important to note that class imbalance is usually not a concern when the classes are easily separable. However, real-world datasets tend to have class-overlapping regions. This issue can elevate the complexity of the classification, especially when allied with class imbalance. In addition, noise can also amplify the class imbalance problem [30], since rare instances and noise have similar characteristics. Thus, they may be treated as the same pattern.

In this paper, we propose a new undersampling method based on the evidence theory [31], which was recently used for oversampling [11]. The intuition of our proposed method is to improve the visibility of the minority class region in binary imbalanced datasets. To do that, we assign a soft evidential label to each majority class sample, in order to acquire information about their locations. Then, we eliminate the majority objects that are considered ambiguous (in overlapping regions), label noise (in the minority area), or outliers (far from both classes). This will not only improve the imbalance ratio, but also reduce the amount of overlap and noise in the training dataset. The considered evidential structure based on the theory of evidence is suitable for our objective, since it provides membership values towards classes, in addition to a belief mass assigned to meta-classes (both classes). This flexibility helps us develop precise rules to detect the unwanted samples for undersampling, with the possibility of tuning each rule individually.

The remainder of this paper will be divided as follows. The theory of evidence will be recalled in Sect. 2. Section 3 presents our idea, detailing each step. Experimental evaluation and discussion are conducted in Sect. 4. Our paper ends with a conclusion and an outlook on future work in Sect. 5.

2 Evidence Theory

The theory of evidence [8,31,33], also referred to as belief function theory or Dempster-Shafer theory (DST), is a flexible and well-founded framework for representing and combining uncertain information. The frame of discernment denotes a finite set of M exclusive possible events, e.g., possible class labels for an object in a classification problem. The frame of discernment is denoted as follows:

$$\Omega = \{w_1, w_2, ..., w_M\} \tag{1}$$

A basic belief assignment (*bba*) represents the amount of belief given by a source of evidence, committed to 2^Ω, that is, all subsets of the frame including the whole frame itself. Formally, a *bba* is represented by a mapping function $m : 2^\Omega \rightarrow [0, 1]$ such that:

$$\sum_{A \in 2^\Omega} m(A) = 1 \tag{2}$$

Each mass $m(A)$ measures the amount of belief allocated to a proposition A of Ω. A *bba* is called unnormalized if the sum of its masses is not equal to 1, and should be normalized under a closed-world assumption [32]. A focal element is a subset $A \subseteq \Omega$ where $m(A) \neq 0$.

The *Plausibility* function is another representation of knowledge defined by *Shafer* [31] as follows:

$$Pl(A) = \sum_{B \cap A \neq \emptyset} m(B), \quad \forall\ A \in 2^\Omega \tag{3}$$

$Pl(A)$ represents the total possible support for A and its subsets.

3 Evidential Undersampling Approach (EVUS)

EVUS starts by assigning soft labels to each majority point using the credal classification rule (CCR) introduced in [23]. Generally, it firstly consists of determining the centers of each class and meta-class (the overlapping region), then creating a *bba* based on the distance between the majority sample and each class center. The computed *bba* is later used for undersampling.

The remaining of this section will provide detailed descriptions of each step.

3.1 Determination of Centers

The simple approach to calculating class centers is to compute the mean value of the training data in the corresponding class. For the overlapping region, which is represented by a meta-class, the center is defined by the barycenter of the involved class centers as follows:

$$C_U = \frac{1}{|U|} \sum_{\omega_i \in U} C_i \tag{4}$$

where U represents the meta-class, ω_i are the classes involved in U, and C_i is the corresponding center.

3.2 Computing the Soft Labels

The evidential membership of each majority example is represented by a *bba* over the frame of discernment $\Omega = \{\omega_0, \omega_1, \omega_2\}$ where ω_1 and ω_2 represent respectively the majority and the minority class. The element ω_0 is included in the frame explicitly to represent the outlier, i.e., the unknown class.

Let x_s be a sample belonging to the majority class. Each class center represents a piece of evidence to the evidential membership of the majority sample. The mass values regarding the class memberships of x_s should depend on $d(x_s, C)$, i.e., the distance between x_s and the corresponding center of the class. The greater the distance, the lower the mass value. Hence, if x_s is more close to a specific class center, it means that x_s belongs very likely to the respective class. Thus, the initial (unnormalized) masses should be represented by decreasing distance based functions. To deal with anisotropic datasets, the Mahalanobis distance [24] is used in this work as recommended by [23].

The unnormalized masses are calculated as follows:

$$\hat{m}(\{\omega_i\}) = e^{-d(x_s, C_i)}, \quad i \in [1, 2] \tag{5}$$

$$\hat{m}(U) = e^{-\gamma\ \lambda\ d(x_s, C_U)}, \quad U = \{\omega_1, \omega_2\} \tag{6}$$

$$\hat{m}(\{\omega_0\}) = e^{t} \tag{7}$$

where $\lambda = \beta\ 2^{\alpha}$. A recommended value for $\alpha = 1$ can be used to obtain good results on average, and β is a parameter such that $0 < \beta < 1$. It is used to tune the number of objects committed to the overlapping region (see Sect. 3.3). The value of γ is equal to the ratio between the maximum distance of x_s to the centers in U and the minimum distance. It is used to measure the degree of distinguishability among the majority and minority classes. The smaller γ indicates a poor distinguishability degree between the classes of U for x_s. The outlier class ω_0 is taken into account in order to deal with objects far from both classes, and its mass value is calculated according to an outlier threshold t.

Finally, the previous unnormalized masses are normalized as follows:

$$m(A) = \frac{\hat{m}(A)}{\sum_{B \subseteq \Omega} \hat{m}(B)}, \quad \forall A \subseteq \Omega \tag{8}$$

3.3 Selecting Majority Samples for Elimination

Once basic belief assignments are created, the soft memberships are used to reject samples that are unwanted from the majority class. Rejection strategies in evidence theory are common in many applications [17,34]. The amount of information provided by evidential functions helps us determine whether a sample should be rejected or valid for classification.

As a result of bba creation, each majority object will have masses in 4 focal elements namely: $m(\{\omega_1\})$ for the majority class, $m(\{\omega_2\})$ for the minority class, $m(U)$ for the overlapping region U, and $m(\{\omega_0\})$ for the outlier class.

Overlapping Rejection. Ambiguous samples are usually located in regions where there is strong overlap between classes as seen in Fig. 1a. Consequently, this type of objects will have a high mass value in $m(U)$ in our framework. Thus, majority samples whose *bba* has the maximum mass committed to $m(U)$ are considered as part of an overlapping region, and are automatically discarded. Additionally, to avoid excessive elimination and allow tuning, it is possible to

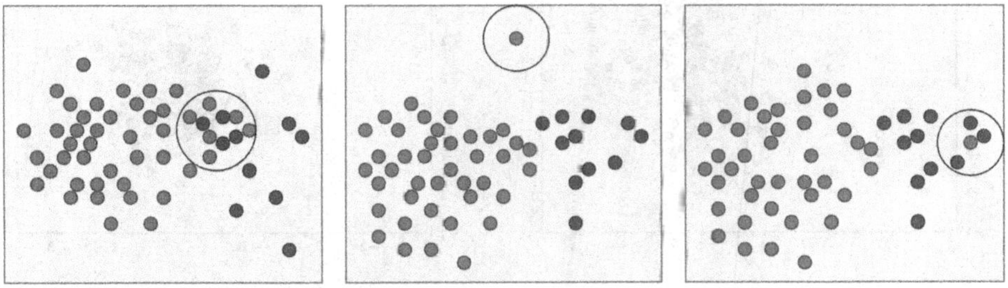

(a) Ambiguous samples in an overlapping area. (b) An outlier far from both classes. (c) A sample that could be characterized as label noise.

Fig. 1. Illustrations describing the different data difficulty factors that could worsen class imbalance. Green and red colors respectively represent the majority class and the minority one. (Color figure online) tune the parameter β. The bigger value of β will result in smaller number of objects committed to the overlapping region as seen in Fig. 2.

As for majority objects not in overlapping areas (i.e. the highest mass is not committed to $m(U)$), the object is necessarily committed to one of the singletons in Ω ($\{\omega_1\}$, $\{\omega_2\}$, or $\{\omega_0\}$). To make a decision of acceptance or rejection, the *plausibility* function defined in Eq. (3) is used. Each majority object x_s is assigned to the class with the maximum plausibility $Pl_{max} = max_{\omega \in \Omega} Pl(\{\omega\})$.

Label Noise. In EVUS, majority objects should normally have the maximum plausibility committed towards $m(\{\omega_1\})$ which represents the membership value towards the majority class. Accordingly, objects with Pl_{max} committed to $m(\{\omega_2\})$ signify that they are located in the minority region, as illustrated in Fig. 1c. In other words, this situation could be characterized by label noise, which is another data difficulty factor that amplifies the class imbalance issue [19]. In our undersampling framework, these types of majority objects are eliminated from the dataset.

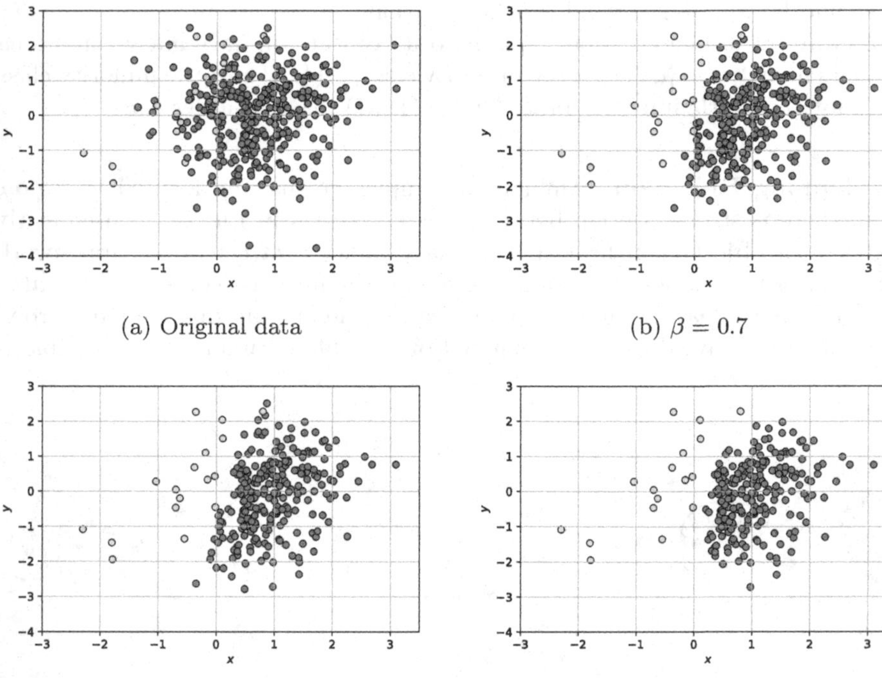

(a) Original data

(b) $\beta = 0.7$

(c) $\beta = 0.5$

(d) $\beta = 0.3$

Fig. 2. Undersampling made by EVUS on a synthetic imbalanced dataset with overlapping. Different tunings of the overlapping parameter β were tested.

Outlier Rejection. The final possibility occurs when Pl_{max} is committed to ω_0. This situation describes indecisive samples that are far from both classes and could be considered outliers as shown in Fig. 1b. In our framework, majority objects with the maximum plausibility committed towards ω_0 are eliminated from the training dataset. The parameter t in Eq. (7) can be used for tuning the outlier rejection, although $t = 2$ is recommended for good results on average. The bigger t results in smaller number of outliers, and it is recommended to take $t \in [2, 5]$.

4 Experimental Study

In this section, we describe our experimental setup and present the observed results.

4.1 Setup

Datasets. Binary imbalanced datasets were selected from the KEEL repository [2] to conduct the experimental study. Specifically, we have chosen a total of 20 datasets which vary in imbalanced ratios (1.87 to 129.44), number of instances (173 to 4174), number of features (6 to 41). The characteristics are further detailed in Table 1. The imbalance ratios (IR) are calculated as $\frac{\#majority}{\#minority}$. These variations allowed for comparisons in a range of different scenarios. In the case of *kr-vs-k-zero vs eight* and *kddcup-rootkit-imap vs back* datasets, categorical features were encoded as integers before applying undersampling. There was no further preprocessing done.

Baseline Classifiers. CART decision tree and Support Vector Machine (SVM) were chosen as baseline classifiers to conduct the comparisons. These learning methods are considered as one of the most used classifiers in class-imbalanced problems [12]. For all experiments, the implementations provided in the scikit-learn machine learning python library [28] were used, with the default parameters unchanged (RBF kernel was used for SVM).

Compared Methods. EVUS was compared against 5 other undersampling methods, in addition to baseline (BL). The approaches used are: random undersampling (RUS), Cluster Centroid undersampling (CC) [39], Condensed Nearest Neighbour editing (CNN) [13], One- Sided Selection (OSS) [20], and Near Miss undersampling (NM) [26]. The implementations provided by the python toolbox imbalanced-learn [21] were used for the compared methods.

Metric and Evaluation Strategy. The Area Under the ROC-Curve (AUC) was used as an evaluation measure. This metric provides a score to determine how well a classifier compensates its true positive and false positive rates. It has been shown to be a reliable assessment criterion for imbalanced classification problems [38]. The AUC scores were averaged through a 10-fold stratified cross validation to eliminate inconsistencies. It is worth mentioning that undersampling was performed only on the training set at each fold. Finally, statistical comparisons were carried out using the Wilcoxon's signed rank tests [36] to further evaluate the significance of the results.

Parameters. The following parameters were considered for EVUS: α was set to 1 as recommended in [23], the outlier tuning parameter t was fixed to 2 to obtain averagely good results, and we tested three different values for β in $\{0.3, 0.5, 0.7\}$ and selected the most performing value each time, since the amount of overlapping differs in each dataset. For the other reference methods, the recommended parameter in the respective papers were used.

Table 1. Description of the imbalanced datasets selected from the KEEL repository.

Datasets	Imbalance ratios (IR)	Features	Samples
wisconsin	1.860	9	683
glass0	2.060	9	214
vehicle3	2.990	18	846
ecoli1	3.360	7	336
yeast3	8.100	8	1484
ecoli-0-6-7 vs 3-5	9.090	7	222
yeast-0-3-5-9 vs 7-8	9.120	8	506
ecoli-0-2-6-7 vs 3-5	9.180	7	224
ecoli-0-1-4-7 vs 2-3-5-6	10.590	7	336
glass-0-1-4-6 vs 2	11.060	9	205
glass4	15.460	9	214
yeast-2 vs 8	23.100	8	482
winequality-red-4	29.170	11	1599
winequality-red-8 vs 6	35.440	11	656
kr-vs-k-zero vs eight	53.070	6	1460
winequality-white-3-9 vs 5	58.280	11	1482
poker-8-9 vs 6	58.400	10	1485
poker-8 vs 6	85.880	10	1477
kddcup-rootkit-imap vs back	100.140	41	2225
abalone19	129.440	8	4174

4.2 Results and Discussion

Table 2 presents the AUC scores obtained by CART and SVM on each imbalanced dataset after performing undersampling. The best scores are marked in bold. We can initially observe that undersampling improved the classification performance on all cases. Our proposed method achieves the best AUC scores in 13 out of 20 datasets for CART, and 11 out of 20 in the case of SVM. In 7 out of 20 datasets, EVUS performed better compared to the other methods for both classifiers. This can imply that the choice of the classifier did not affect much the end results. Furthermore, EVUS performed relatively better in cases when there are many borderline points in the dataset, i.e., when the overlapping between the classes is strong. By contrast, one can say that EVUS is not much of use in the cases of small number of borderline majority points.

Table 2. AUC results for KEEL datasets using CART and SVM.

Datasets	*CART*							*SVM*						
	BL	RUS	CC	CNN	OSS	NM	EVUS	BL	RUS	CC	CNN	OSS	NM	EVUS
wisconsin	0.935	0.945	0.939	0.910	0.914	0.936	**0.964**	0.971	0.972	0.971	0.968	0.971	0.967	**0.977**
glass0	0.753	0.758	0.747	0.762	0.767	0.673	**0.770**	0.500	**0.670**	0.656	0.500	0.500	0.485	0.667
vehicle3	0.708	0.720	0.709	0.675	0.672	0.649	**0.789**	0.500	0.606	**0.662**	0.648	0.517	0.639	0.619
ecoli1	0.826	0.842	0.848	0.852	0.876	0.804	**0.886**	0.861	0.902	0.889	**0.904**	0.875	0.838	**0.904**
yeast3	0.808	**0.897**	0.869	0.824	0.811	0.768	0.882	0.841	0.854	**0.923**	0.890	0.865	0.846	0.891
ecoli-0-6-7 vs 3-5	0.829	0.776	0.821	0.841	0.799	0.758	**0.852**	0.839	0.804	0.837	0.868	0.845	0.726	**0.886**
yeast-0-3-5-9 vs 7-8	0.659	0.614	0.544	0.616	**0.665**	0.546	0.663	0.597	0.728	0.676	0.597	0.607	0.538	**0.645**
ecoli-0-2-6-7 vs 3-5	0.822	0.790	0.797	0.825	0.809	0.748	**0.829**	0.839	0.827	0.862	**0.896**	0.842	0.670	0.867
ecoli-0-1-4-7 vs 2-3-5-6	0.777	0.756	0.814	0.822	0.819	0.735	**0.852**	0.840	0.869	0.902	0.850	0.815	0.674	**0.906**
glass-0-1-4-6 vs 2	0.571	0.680	**0.685**	0.590	0.560	0.601	0.596	0.500	**0.630**	0.585	0.500	0.500	0.473	0.600
glass4	0.838	0.850	0.716	0.743	0.698	0.875	**0.898**	0.500	0.788	0.843	0.500	0.500	0.573	**0.895**
yeast-2 vs 8	0.714	0.678	0.553	0.683	0.762	0.718	**0.764**	0.774	0.772	0.772	0.774	0.774	**0.816**	0.782
winequality-red-4	0.540	**0.595**	0.568	0.593	0.579	0.456	0.588	0.500	0.524	0.546	0.500	0.500	0.421	**0.556**
winequality-red-8 vs 6	0.612	0.695	0.713	0.759	0.637	0.454	**0.784**	0.500	0.662	0.546	0.500	0.500	0.356	**0.566**
kr-vs-k-zero vs eight	0.822	0.962	0.778	**1.000**	**1.000**	0.960	**1.000**	0.791	0.884	**0.959**	0.807	0.791	0.796	0.931
winequality-white-3-9 vs 5	0.630	0.644	0.558	0.595	0.612	0.542	**0.648**	0.500	**0.636**	0.550	0.500	0.500	0.526	0.585
poker-8-9 vs 6	0.527	0.543	0.630	0.563	0.543	**0.661**	0.573	0.500	0.525	0.497	0.500	0.500	0.531	**0.629**
poker-8 vs 6	0.549	0.511	**0.635**	0.496	0.547	0.541	0.548	0.500	0.522	0.515	0.500	0.500	**0.536**	0.525
kddcup-rootkit-imap vs back	0.992	0.997	0.561	**1.000**	0.911	0.560	**1.000**	0.997	**1.000**	0.998	0.550	0.973	0.999	**1.000**
abalone19	0.529	0.673	**0.690**	0.553	0.528	0.522	0.629	0.500	0.623	0.626	0.500	0.500	0.591	**0.665**

To assess the significance of the comparisons, Table 3 presents the statistical analysis made by Wilcoxon's signed ranks test. $R+$ represents the sum of ranks in favor of EVUS, $R-$, the sum of ranks in favor of the other compared methods, and p-values are calculated for each comparison. As shown in Table 3, all p-values are lower than 0.10.

Table 3. Wilcoxon's signed ranks test results comparing the AUC scores for both CART and SVM.

Comparisons	CART			SVM		
	$R+$	$R-$	p-value	$R+$	$R-$	p-value
EVUS vs RUS	173.5	36.5	0.009436	133.5	76.5	0.063069
EVUS vs CC	164.0	46.0	0.013321	154.0	56.0	0.03479
EVUS vs CNN	169.0	41.0	0.00015	177.0	33.0	0.000518
EVUS vs OSS	187.5	22.5	0.000107	210.0	0.0	<0.000001
EVUS vs NM	200.0	10.0	0.000041	198.0	12.0	0.000067

Thus, one can say that EVUS outperformed the compared techniques at a significance level of $\alpha = 0.10$.

5 Conclusions

Throughout this paper, we have proposed a new method called Evidential Undersampling (EVUS), based on the evidence theory. Majority samples are selected for elimination based on soft evidential labels, which provide us with more information about the point's location. This resulted in AUC improvements over well-known undersampling methods. The main motivation behind our proposed algorithm was to improve the visibility of the minority class regions, to avoid the misclassification of minority objects.

For future work, we intend to explore heuristic methods to further optimize the parameters β and t. This will provide a more adaptive behavior of the parameters based on the amount of overlap and noise present in the majority class.

References

1. Alaba, P.A., et al.: Towards a more efficient and cost-sensitive extreme learning machine: a state-of-the-art review of recent trend. Neurocomputing **350**, 70–90 (2019)
2. Alcala-Fdez, J., et al.: Keel data-mining software tool: Data set repository, integration of algorithms and experimental analysis framework. J. Mult.-Valued Log. Soft Comput. **17**, 255–287 (2010)
3. Angiulli, F.: Fast condensed nearest neighbor rule. In: Proceedings of the 22nd International Conference on Machine Learning (ICML 2005), pp. 25–32. Association for Computing Machinery, New York (2005)
4. Bridge, J., et al.: Introducing the gev activation function for highly unbalanced data to develop Covid-19 diagnostic models. IEEE J. Biomed. Health Inform. **24**(10), 2776–2786 (2020)
5. Chawla, N.V.: C4.5 and imbalanced data sets: investigating the effect of sampling method, probabilistic estimate, and decision tree structure. In: Proceedings of the ICML 2003 Workshop on Class Imbalances (2003)
6. Chawla, N.V., Japkowicz, N., Drive, P.: Editorial : Special issue on learning from imbalanced data sets. ACM SIGKDD Explor. Newslett. **6**(1), 1–6 (2004)
7. Colorni, A., Dorigo, M., Maniezzo, V.: Distributed optimization by ant colonies. In: Proceedings of the First European Conference on Artificial Life. vol. 142, pp. 134–142. Paris (1991)
8. Dempster, A.P.: A generalization of Bayesian inference. J. R. Statist. Soc. Ser. B (Methodol) **30**(2), 205–232 (1968)
9. Feng, Y., Zhou, M., Tong, X.: Imbalanced classification: an objective-oriented review. arXiv preprint arXiv:2002.04592 (2020)
10. García, S., Herrera, F.: Evolutionary undersampling for classification with imbalanced datasets: proposals and taxonomy. Evol. Comput. **17**(3), 275–306 (2009)
11. Grina, F., Elouedi, Z., Lefevre, E.: A preprocessing approach for class-imbalanced data using SMOTE and belief function theory. In: Analide, C., Novais, P., Camacho, D., Yin, H. (eds.) IDEAL 2020. LNCS, vol. 12490, pp. 3–11. Springer, Cham (2020). https://doi.org/10.1007/978-3-030-62365-4_1

12. Haixiang, G., Yijing, L., Shang, J., Mingyun, G., Yuanyue, H., Bing, G.: Learning from class-imbalanced data: review of methods and applications. Exp. Syst. Appli. **73**, 220–239 (2017)
13. Hart, P.: The condensed nearest neighbor rule. IEEE Trans. Inf. Theor. **14**(3), 515–516 (1968)
14. He, H., Garcia, E.A.: Learning from imbalanced data. IEEE Trans. Knowl. Data Eng. **21**(9), 1263–1284 (2009)
15. Ivan, T.: Two modification of CNN. IEEE Trans. Syst. Man Commun. SMC **6**, 769–772 (1976)
16. Kanungo, T., Mount, D.M., Netanyahu, N.S., Piatko, C.D., Silverman, R., Wu, A.Y.: An efficient k-means clustering algorithm: analysis and implementation. IEEE Trans. Patt. Anal. Mach. Intell. **24**(7), 881–892 (2002)
17. Kessentini, Y., Burger, T., Paquet, T.: A Dempster-Shafer theory based combination of handwriting recognition systems with multiple rejection strategies. Patt. Recogn. **48**(2), 534–544 (2015)
18. Korkmaz, S.: Deep learning-based imbalanced data classification for drug discovery. J. Chem. Inf. Model. **60**(9), 4180–4190 (2020)
19. Koziarski, M., Woźniak, M., Krawczyk, B.: Combined cleaning and resampling algorithm for multi-class imbalanced data with label noise. arXiv (2020)
20. Kubat, M., Matwin, S.: Addressing the curse of imbalanced training sets: one-sided selection. In: Proceedings of the Fourteenth International Conference on Machine Learning (ICML 1997), vol. 97, pp. 179–186 (1997)
21. Lemaître, G., Nogueira, F., Aridas, C.K.: Imbalanced-learn: a python toolbox to tackle the curse of imbalanced datasets in machine learning. J. Mach. Learn. Res. **18**(1), 559–563 (2017)
22. Lin, W.C., Tsai, C.F., Hu, Y.H., Jhang, J.S.: Clustering-based undersampling in class-imbalanced data. Inf. Sci. **409–410**, 17–26 (2017)
23. Liu, Z.G., Pan, Q., Dezert, J., Mercier, G.: Credal classification rule for uncertain data based on belief functions. Patt. Recogn. **47**(7), 2532–2541 (2014)
24. Mahalanobis, P.C.: On the generalized distance in statistics, In: Proceedings of the National Institute of Science, India, vol. 2, pp. 49–55 (1936)
25. Makki, S., Assaghir, Z., Taher, Y., Haque, R., Hacid, M.S., Zeineddine, H.: An experimental study with imbalanced classification approaches for credit card fraud detection. IEEE Access **7**, 93010–93022 (2019)
26. Mani, I., Zhang, I.: KNN approach to unbalanced data distributions: a case study involving information extraction. In: Proceedings of Workshop on Learning from Imbalanced Datasets. vol. 126 (2003)
27. Ofek, N., Rokach, L., Stern, R., Shabtai, A.: Fast-CBUS: a fast clustering-based undersampling method for addressing the class imbalance problem. Neurocomputing **243**, 88–102 (2017)
28. Pedregosa, F., et al.: Scikit-learn: machine learning in python. J. Mach. Learn. Res. **12**, 2825–2830 (2011)
29. Ribeiro, V.H.A., Reynoso-Meza, G.: Ensemble learning by means of a multi-objective optimization design approach for dealing with imbalanced data sets. Exp. Syst. Appl. **147**, 113232 (2020)
30. Sáez, J.A., Luengo, J., Stefanowski, J., Herrera, F.: SMOTE-IPF: Addressing the noisy and borderline examples problem in imbalanced classification by a re-sampling method with filtering. Inf. Sci. **291**(C), 184–203 (2015)
31. Shafer, G.: A Mathematical Theory of Evidence, vol. 42. Princeton University Press, Princeton (1976)

32. Smets, P.: The nature of the unnormalized beliefs encountered in the transferable belief model. In: Eighth Conference on Uncertainty in Artificial Intelligence, pp. 292–297. Elsevier (1992)
33. Smets, P.: The transferable belief model for quantified belief representation. In: Smets, P. (ed.) Quantified Representation of Uncertainty and Imprecision. HDRUMS, vol. 1, pp. 267–301. Springer, Dordrecht (1998). https://doi.org/10.1007/978-94-017-1735-9_9
34. Tong, Z., Xu, P., Denœux, T.: ConvNet and Dempster-Shafer theory for object recognition. In: Ben Amor, N., Quost, B., Theobald, M. (eds.) SUM 2019. LNCS (LNAI), vol. 11940, pp. 368–381. Springer, Cham (2019). https://doi.org/10.1007/978-3-030-35514-2_27
35. Tsai, C.F., Lin, W.C., Hu, Y.H., Yao, G.T.: Under-sampling class imbalanced datasets by combining clustering analysis and instance selection. Inf. Sci. **477**, 47–54 (2019)
36. Wilcoxon, F.: Individual comparisons by ranking methods. In: Kotz, S., Johnson, N.L. (eds.) Breakthroughs in Statistics. Springer Series in Statistics (Perspectives in Statistics). pp. 196–202. Springer, New York (1992)
37. Wilson, D.L.: Asymptotic properties of nearest neighbor rules using edited data. IEEE Trans. Syst. Man Cybernet. **3**, 408–421 (1972)
38. Xue, J.H., Hall, P.: Why does rebalancing class-unbalanced data improve AUC for linear discriminant analysis? IEEE Trans. Patt. Anal. Mach. Intell. **37**(5), 1109–1112 (2014)
39. Yen, S.J., Lee, Y.S.: Cluster-based under-sampling approaches for imbalanced data distributions. Exp. Syst. Appl. **36**(3), 5718–5727 (2009)
40. Yu, H., Ni, J., Zhao, J.: Acosampling: an ant colony optimization-based undersampling method for classifying imbalanced DNA microarray data. Neurocomputing **101**, 309–318 (2013)

Well-Calibrated and Sharp Interpretable Multi-Class Models

Ulf Johansson[1(✉)], Tuwe Löfström[1], and Henrik Boström[2]

[1] Department of Computing, Jönköping University, Jönköping, Sweden
{ulf.johansson,tuwe.lofstrom}@ju.se

[2] School of Electrical Engineering and Computer Science, KTH Royal Institute of Technology, Stockholm, Sweden
bostromh@kth.se

Abstract. Interpretable models make it possible to understand individual predictions, and are in many domains considered mandatory for user acceptance and trust. If coupled with communicated algorithmic confidence, interpretable models become even more informative, also making it possible to assess and compare the confidence expressed by the models in different predictions. To earn a user's appropriate trust, however, the communicated algorithmic confidence must also be well-calibrated. In this paper, we suggest a novel way of extending Venn-Abers predictors to multi-class problems. The approach is applied to decision trees, providing well-calibrated probability intervals in the leaves. The result is one interpretable model with valid and sharp probability intervals, ready for inspection and analysis. In the experimentation, the proposed method is verified using 20 publicly available data sets showing that the generated models are indeed well-calibrated.

1 Introduction

Interpretable predictive models make it possible to explain individual predictions, as well as discover and analyze underlying relationships in the data, without external explanation modules. We have previously argued, see e.g., [2], that models, and in particular interpretable models, that communicate *algorithmic confidence* are very informative. Obviously, for such models to earn a user's appropriate trust, rather than being outright misleading, the confidence measures must be well-calibrated [6].

Venn predictors, introduced in [9] are multi-probabilistic predictors with unique validity properties. Somewhat simplified, the multi-probabilistic predictions can be interpreted as a set of probability intervals, one for each label. The Venn predictors normally use an underlying classifier to divide all instances into a number of *categories*, based on a so-called *Venn taxonomy*. The relative frequency of each class label in a category is then used to estimate the probabilities

This research is partly funded by the Swedish Knowledge Foundation (DATAKIND 20190194).

V. Torra and Y. Narukawa (Eds.): MDAI 2021, LNAI 12898, pp. 193–204, 2021.
https://doi.org/10.1007/978-3-030-85529-1_16

for test instances falling into that category. Validity is obtained by including the test instance to be predicted in the calculation. However, as the true label is not known for the test instance, each possible label must be considered, resulting in as many label probability distributions as the number of possible labels.

An *inductive* Venn predictor [4] divides the training data into a *proper training set*, used to train the underlying model, and a *calibration set* used to calibrate the probabilities. For an extended introduction to inductive Venn predictors, see e.g., [3]. The main challenge for Venn predictors is to find a suitable Venn taxonomy, where too many categories lead to larger prediction intervals, and too few to a probabilistic model that is not sharp enough.

A solution to the challenge of finding a Venn taxonomy was introduced through *Venn-Abers predictors* [8]. In Venn-Abers, an optimized set of categories are found using isotonic regression. Venn-Abers is, however, in its basic version restricted to two-class problems. For multi-class problems, existing techniques use either one-vs-all or all-vs-all schemes, before applying Venn-Abers to each class and then aggregating the multi-probabilistic predictions into probability estimates, see e.g., [5]. Using this approach, there is no longer one predictive model used for the predictions, but a set of models. So, even if the algorithm is inherently capable of producing interpretable models, like decision trees or rule sets, it is no longer feasible to inspect or analyse the predictive models.

With this in mind, the overall purpose of this paper is to suggest a novel way of producing interpretable and sharp probabilistic decision trees utilizing Venn-Abers for multi-class. The key idea is that instead of generating full probability distributions over all labels in each leaf, we only derive prediction intervals for the labels that are actually predicted. With this strategy, it becomes possible to generate decision trees containing well-calibrated probability intervals in each leaf. It should be noted that these intervals convey not only information about the probability estimates, but also about how certain these estimates are. More specifically, the size of the intervals is dependent on how many calibration instances that fell into that leaf, with fewer instances leading to larger, i.e., more uncertain intervals.

2 Background

2.1 Probabilistic Prediction

Probabilistic predictors output a probability distribution over the possible classes. Informally, a probabilistic predictor is said to be *well-calibrated* if its confidences in the labels predicted correspond to the errors made. More formally, if p^{c_j} is the probability estimate for class j we expect:

$$p(c_j \mid p^{c_j}) = p^{c_j}. \tag{1}$$

2.2 Probability Estimation Trees

When decision trees are used as probabilistic predictions they are referred to as *probabilistic estimation trees* (PETs) [7]. While more sophisticated approaches are available, the estimates are in the basic setting based on the relative frequencies of the different labels in each leaf. If $g(i, j)$ is the number of instances belonging to class j that falls in the same leaf as instance i, the probability estimates are calculated using:

$$p_i^{c_j} = \frac{g(i, j)}{\sum_{k=1}^{C} g(i, k)} \tag{2}$$

2.3 Venn-Abers Predictors

Venn-Abers predictors use isotonic regression [10] to automatically optimize the taxonomy for two-class problems. The optimized taxonomy results in sharp predictions. As the Venn-Abers predictors are Venn predictors, they still inherit the validity guarantees.

For the Venn-Abers predictor to work, the underlying model must be a *scoring classifier*, i.e., the underlying model will output a *prediction score* $s(x_i)$ when applied to a test object x_i. A higher prediction score indicates a larger belief in the positive class, i.e., label 1. For a two-class scoring classifier, the prediction score can be used to obtain a prediction by comparing the score to a fixed threshold t, predicting label 1 if $s(x) > t$, and 0 otherwise. By using isotonic regression to fit an increasing function g using a number of prediction scores with known true targets, Venn-Abers predictors can let $g(s(x))$ be interpreted as the probability that the label for x is 1. An inductive Venn-Abers predictor producing a multi-probabilistic prediction is described in Algorithm 1 below.

Algorithm 1: Inductive Venn-Abers prediction

---

input : A learning algorithm A
A training set $\{z_1, \ldots, z_l\}$
Test object $x_l + 1$
output: $[low, high]$: the probability interval for $y_{l+1} = 1$

Divide the training set into a proper training set $\{z_1, \ldots, z_q\}$ and a calibration set $\{z_{q+1}, \ldots, z_l\}$;
Apply A to the proper training set to produce the scoring model H;
foreach $z_i = (x_i, y_i); i \in q+1, \ldots, l$ **do**
    `// for each calibration example`
    $s_i \leftarrow H(x_i)$;
end
$s_{l+1} \leftarrow H(x_{l+1})$;
Let g_0 be the isotonic calibrator for $\{(s_{q+1}, y_{q+1}), \ldots, (s_l, y_l), (s_{l+1}, 0)\}$;
Let g_1 be the isotonic calibrator for $\{(s_{q+1}, y_{q+1}), \ldots, (s_l, y_l), (s_{l+1}, 1)\}$;
Let the probability interval for $y_{l+1} = 1$ be $[low, high] \leftarrow [g_0(s_{l+1}), g_1(s_{l+1})]$;
return $[low, high]$

---

3 Method

In order to produce interpretable and sharp Venn predictors, we suggest applying Venn-Abers once for each test instance and possible label, while utilizing the same underlying decision tree and calibration set. More specifically, when considering a certain label, all other labels are regarded as belonging to the negative class. It must be noted that while this procedure requires $2C$ isotonic regressions for each test instance, where C is the number of classes, the standard approach trains C predictive models, and runs two isotonic regressions for each test instance. The centers of all intervals are compared and the class with the highest center is the leaf prediction. With this setting, we obtain well-calibrated probabilities for the labels predicted in each leaf, but the rest of the probability mass is not distributed over the other labels. The result is that we, after this calibration step, obtain one decision tree with well-calibrated prediction intervals in the leaves, i.e., an interpretable and informative model ready for inspection and analysis.

While a key property of the Venn-Abers predictor is the ability to output probability intervals (p_0, p_1), where the size is an indication of the confidence in the probability estimate, these intervals need to be aggregated into a single probability estimate when comparing to other techniques. Here, the recommendation, which is followed in this study, is to use a regularized value moved towards 0.5 [8].

$$p = \frac{p_1}{1 - p_0 + p_1} \tag{3}$$

In the experimentation, *scikit-learn* was used and we set the *DecisionTreeClassifier* parameter *min_weight_fraction_leaf* to 0.01, i.e., each leaf should contain at least 1% of the training instances. This will, of course, result in fairly small trees, i.e., encouraging comprehensibility. The testing protocol was 10×10-fold stratified cross-validation. For the VA models, the proper training set consisted of 2/3 of all the training instances, and the calibration set of 1/3. For the non-calibrated decision trees used for comparison, all training data was used for inducing the model.

In all experiments, 20 publicly available multi-class data sets from the UCI [1] repository are used. The data sets are described in Table 1 below, where *#class* is the number of classes, *#inst.* is the number of instances and *#attrib.* is the number of input attributes.

In the analysis, accuracy and area under the ROC curve (AUC) are used to measure the predictive performance. For the quality of the calibration, we employ log losses and the expected calibration error (ECE). The log loss is calculated using

$$\lambda_{log} = \begin{cases} -\log p & \text{if correct} \\ -\log(1-p) & \text{if incorrect} \end{cases} \tag{4}$$

where log is the binary logarithm and p the estimate for the predicted label. Here, the log loss function used avoids infinite results by clipping the probabilities making sure that they never are exactly 0 or 1.

Table 1. Data sets

Data set	#class	#inst	#attrib	Data set	#class	#inst	#attrib
balance	3	625	4	tae	3	151	5
cars	4	1728	6	user	5	403	5
cmc	3	1473	9	wave	3	5000	40
cool	3	768	8	vehicle	4	846	18
ecoli	8	336	7	whole	3	440	7
glass	6	214	9	wine	3	178	13
heat	3	768	8	wineR	6	1599	11
image	7	2310	19	wineW	7	4898	11
iris	3	150	4	vowel	11	990	11
steel	7	1941	27	yeast	10	1484	8

When calculating ECE, the probabilities for the predicted class are divided into M, in this study ten, equally sized bins, before taking a weighted average of the absolute differences between the fraction of correct (foc) predictions and the mean of the prediction probabilities (mop). Here, n is the size of the data set and B_m represents bin m.

$$ECE = \sum_{m=1}^{M} \frac{|B_m|}{n} \left| foc(B_m) - mop(B_m) \right| \tag{5}$$

4 Results

Before presenting the aggregated results over all data sets, we show a couple of induced VA-trees and a few sample calibration plots.

Figure 1 shows an induced VA-tree for the Image data set. To force the trees to be small enough for this analysis, the tree parameter *min_weight_fraction_leaf* was here set to 0.1, which requires the leaves to contain at least 10% of all instances. The seven classes are predicted by the seven leaf nodes with corresponding intervals for the calibrated probabilities. The sizes of the intervals are dependent on the number of instances falling into the leaves. Since we force the tree to be small, with a large number of instances in each leaf, all intervals will be fairly tight. With a fully-grown tree, the interval sizes could be expected to vary more, as a consequence of the much larger variation in the number of instances falling into each leaf node. Since these intervals are well-calibrated, we would expect, in the long run, the true error rate in each leaf node to be within the interval. So, for this particular tree, we would expect to be correct 61.1 – 62.2% of the time when predicting *window*, whereas we would be almost certain when predicting *path*, *sky* or *grass*.

In Fig. 2, a VA-tree for the Iris data set is shown. Even though we have used the same setting, forcing all leaves to contain at least 10% of all instances, the

```
saturation-mean <= 0.644
|   rawred-mean <= 99.500
|   |   region-centroid-col <= 159.500
|   |   |   intensity-mean <= 24.833
|   |   |   |   saturation-mean <= -1.844
|   |   |   |   |   saturation-mean <= -2.179
|   |   |   |   |   |    P(.847, .864) class: foliage
|   |   |   |   |   saturation-mean > -2.179
|   |   |   |   |   |    P(.611, .622) class: window
|   |   |   |   saturation-mean > -1.844
|   |   |   |   |    P(.809, .819) class: brickface
|   |   |   intensity-mean > 24.833
|   |   |   |    P(.909, .922) class: cement
|   |   region-centroid-col > 159.500
|   |   |    P(.989, 1.0) class: path
|   rawred-mean > 99.500
|   |    P(.989, 1.0) class: sky
saturation-mean > 0.644
|    P(.988, 1.0) class: grass
```

Fig. 1. Venn-Abers calibrated tree for the Image data set

probability intervals are much wider as a consequence of the smaller number of instances in the data set. In this particular tree, we see that the linearly separable class *Iris Setosa* has a high probability with the upper bound reaching 1.0. The two remaining classes have lower probabilities as a consequence of these leaf nodes making some errors on the calibration set.

```
petal length <= 0.800
|    P(.90, 1.0) class: Iris Setosa
petal length >  0.800
|   sepal width <= 4.850
|   |    P(.77, .85) class: Iris Versicolour
|   sepal width >  4.850
|   |    P(.83, .89) class: Iris Virginica
```

Fig. 2. Venn-Abers Iris

While the quality of the calibration varies between the data sets, it more or less always improves on the probability estimates from the trees. Starting with one of the most common patterns, Fig. 3 below shows a calibration curve where the Venn-Abers is able to substantially improve an already fairly good probabilistic model.

In these graphs, the top part is the actual reliability curve, while the lower part shows how the probability estimates for the test instances are distributed. In this example, it is interesting to see how the Venn-Abers lowers the most confident probability estimates from the overconfident tree.

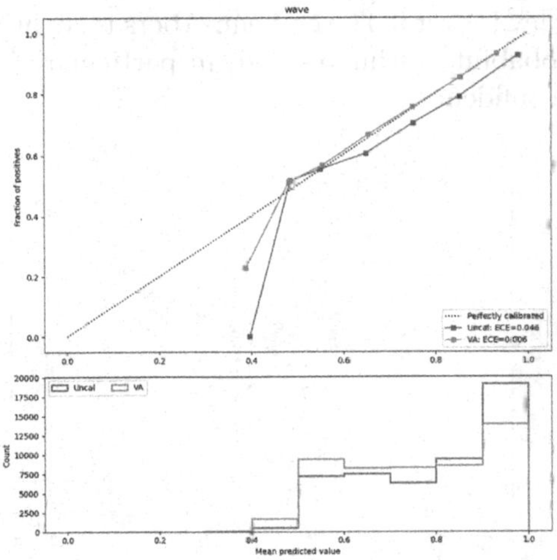

Fig. 3. Wave data set

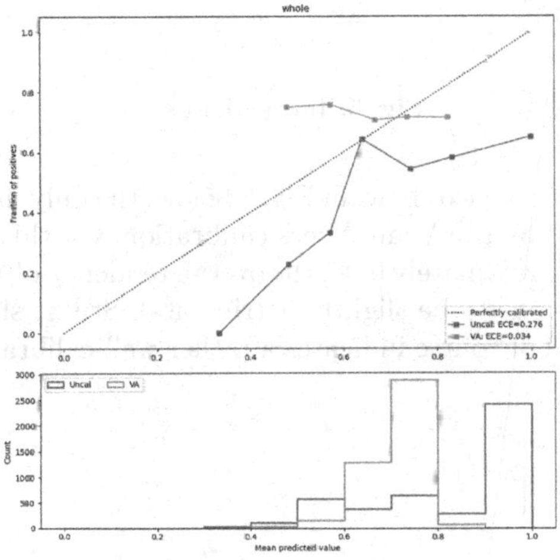

Fig. 4. Whole data set

The second observed pattern is the one where the calibration is the most successful; in these examples, a very poorly calibrated model is significantly improved. Figure 4 is one such example where the underlying model is extremely overconfident for all confidence levels. Interestingly enough, Venn-Abers produces very few predictions with high confidence. As will be presented later, when looking at the predictive performance, the calibration performed by Venn-Abers in this particular case actually also lead to significantly higher accuracy.

Figure 5 below shows another recurring pattern where the calibration marginally improves an already well-calibrated model. In this rather easy data set, where the accuracy is over 0.93, the Venn-Abers is slightly more conservative in the very high probability estimates and, in particular, for a few predictions with relatively low confidence.

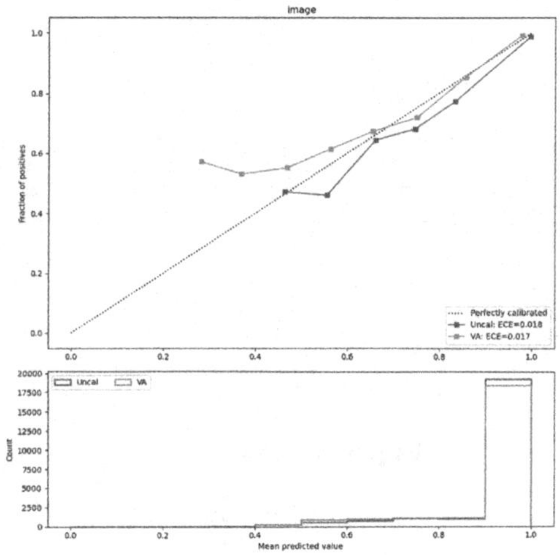

Fig. 5. Image data set

For completeness, we also show, in Fig. 6 below, the only data set (heat) where the ECE is increased by the Venn-Abers calibration. On this very easy data set, with accuracies of approximately 0.97, the overall tendency of Venn-Abers to lower the confidence turns out to be slightly detrimental. Still it should be noted that the ECE level of 0.02 of course indicates a rather well-calibrated model.

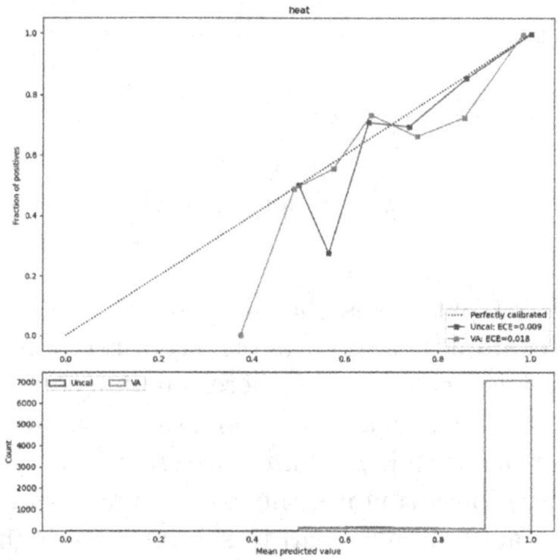

Fig. 6. Heat data set

We now look at aggregated results and start with predictive performances in Table 2 below.

Table 2. Predictive performance

	Accuracy		AUC			Accuracy		AUC	
	NoCal	VA	NoCal	VA		NoCal	VA	NoCal	VA
balance	.790	.804	.849	.780	user	.888	.872	.699	.774
cars	.933	.931	.914	.922	vehicle	.696	.686	.758	.760
cmc	.530	.517	.659	.666	vowel	.676	.660	.768	.747
cool	.921	.932	.956	.947	wave	.765	.761	.740	.728
ecoli	.834	.805	.722	.763	whole	.564	.716	.633	.499
glass	.697	.657	.589	.653	wine	.901	.906	.646	.616
heat	.970	.969	.939	.937	wineR	.592	.574	.622	.615
image	.931	.934	.904	.918	wineW	.521	.524	.591	.579
iris	.940	.941	.595	.650	yeast	.559	.552	.685	.655
steel	.706	.700	.760	.756	**Mean**	**.751**	**.752**	**.728**	**.724**
tae	.616	.591	.534	.521	**Mean rank**	**1.35**	**1.65**	**1.40**	**1.60**

Looking at the mean results, and in particular the mean ranks, we see that there is only a small loss in accuracy and ranking ability using the calibrated models. This is despite the fact that the uncalibrated tree models used a larger training set. Consequently, turning the underlying models into the more informative probabilistic classifiers by external calibration does not come at the expense of significantly lower predictive performance.

Turning to the calibration quality, the left part of Table 3 shows the differences between the average prediction confidence and the empirical accuracy. While this is a very crude metric, it clearly shows that the uncalibrated models are generally very overconfident; the mean difference is over nine percentage points taken over all data sets. For some data sets, like glass, tae and whole, the picture is actually significantly worse. The Venn-Abers, on the other hand, exhibits only small differences on most data sets and shows no systematic bias. This pattern is reinforced when looking at log losses, in the right part of Table 3, where the Venn-Abers lowers the log loss (often substantially) on every data set.

The fact that the benefit of calibration is substantial can also be seen when comparing ECE:s in the left part of Table 4. Actually, using Venn-Abers results in lower ECE on all data sets but one (Heat). On average, the reduction is approximately 75%. This should be seen as one of the main results of this study, showing that the suggested modification of Venn-Abers to multi-class problems works, even on these rather small data sets.

Table 3. Difference and log loss

	Difference		Log loss			Difference		Log loss	
	NoCal	VA	NoCal	VA		NoCal	VA	NoCal	VA
balance	.072	−.009	1.248	.406	user	.062	−.019	1.994	.324
cars	.000	−.009	.166	.153	vehicle	.142	.006	2.139	.523
cmc	.100	.021	.858	.636	vowel	.124	.003	2.063	.553
cool	.034	−.003	.197	.134	wave	.044	−.005	.818	.490
ecoli	.069	−.023	2.249	.423	whole	.275	−.005	6.882	.604
glass	.236	−.006	7.724	.610	wine	.075	−.041	2.383	.308
heat	.006	−.008	.137	.084	wineR	.106	.008	1.291	.663
image	.017	−.013	.330	.157	wineW	.062	.002	.694	.682
iris	.051	−.049	1.673	.234	yeast	.104	.006	.982	.650
steel	.071	−.002	1.218	.516	**Mean**	**.093**	**−.009**	**2.113**	**.442**
tae	.205	−.030	7.214	.693	**Mean rank**			**2.00**	**1.00**

The right part of Table 4 shows the tree sizes. With the parameter settings used, most trees are so small that they should be considered interpretable, sometimes even comprehensible. At the very least, it should be straightforward to understand the reasoning behind individual predictions. The fact that the calibrated trees are slightly smaller of course stems from the fact that they were induced using fewer training examples.

Table 5, finally, shows the average intervals produced by Venn-Abers, together with the empirical accuracies. With the exception of one single dataset (CMC)

Table 4. ECE and tree sizes

	ECE		Tree size			ECE		Tree size	
	NoCal	VA	NoCal	VA		NoCal	VA	NoCal	VA
balance	.075	.020	90.6	81.3	user	.069	.019	52.4	40.6
cars	.016	.009	45.6	48.5	vehicle	.142	.015	100.0	88.3
cmc	.111	.021	132.4	139.6	vowel	.124	.016	124.8	118.4
cool	.034	.024	38.6	34.5	wave	.046	.006	126.9	120.9
ecoli	.088	.026	53.8	44.4	whole	.276	.034	116.3	93.1
glass	.236	.039	71.5	49.5	wine	.075	.045	19.2	12.6
heat	.009	.018	39.2	36.7	wineR	.119	.020	145.4	140.8
image	.018	.017	50.2	48.8	wineW	.065	.012	146.2	149.1
iris	.051	.049	12.8	11.5	yeast	.105	.008	140.2	146.4
steel	.071	.008	125.2	120.1	**Mean**	**.100**	**.025**	**85.8**	**80.3**
tae	.264	.088	84.1	80.3	**Mean rank**	**1.95**	**1.05**	**1.80**	**1.20**

where the empirical accuracy is outside the average interval, the overall picture is that most intervals are both tight and cover the empirical accuracy. Most importantly, we again see that Venn-Abers predictors are able to provide (and communicate) both a confidence measure (given as the probability intervals) and an assessment of its certainty regarding the confidence; i.e., the width of the intervals.

Table 5. VA overall intervals

	Low	High	Accuracy		Low	High	Accuracy
balance	.781	.821	.804	tae	.529	.600	.591
cars	.918	.938	.931	user	.834	.901	.872
cmc	.525	.554	**.517**	vehicle	.677	.719	.686
cool	.924	.951	.932	vowel	.634	.718	.660
ecoli	.751	.844	.805	wave	.753	.766	.761
glass	.606	.722	.657	whole	.705	.723	.716
heat	.956	.982	.969	wine	.854	.921	.906
image	.916	.939	.934	wineR	.573	.596	.574
iris	.884	.948	.941	wineW	.522	.532	.524
steel	.686	.722	.700	yeast	.541	.584	.552

5 Concluding Remarks

We have in this paper suggested and evaluated a novel way of generating well-calibrated models for multi-class problems. The proposed approach utilizes Venn-Abers predictors in a way that only one decision tree is induced, thus the end result is an interpretable model with sharp and well-calibrated probability intervals in the leaves, ready for inspection and analysis. We argue that this is a highly informative model, with the extra benefit that the sizes of the probability intervals communicate a relative belief in the probability estimates.

References

1. Bache, K., Lichman, M.: UCI machine learning repository (2013)
2. Johansson, U., Löfström, T.: Well-calibrated and specialized probability estimation trees. In: Proceedings of the 2020 SIAM International Conference on Data Mining (SDM 2020), May 7–9, 2020. pp. 415–423. SIAM, Cincinnati (2020)
3. Johansson, U., Löfström, T., Boström, H.: Calibrating probability estimation trees using venn-abers predictors. In: Proceedings of the 2019 SIAM International Conference on Data Mining (SDM 2019), Calgary, Alberta, Canada, May 2–4, 2019. pp. 28–36 (2019)

4. Lambrou, A., Nouretdinov, I., Papadopoulos, H.: Inductive venn prediction. Ann. Math. Artif. Intell, **74**(1), 181–201 (2015)
5. Manokhin, V.: Multi-class probabilistic classification using inductive and cross Venn-Abers predictors. In: Proceedings of the Sixth Workshop on Conformal and Probabilistic Prediction and Applications. Proceedings of Machine Learning Research, vol. 60, pp. 228–240. PMLR, Stockholm, Sweden (2017)
6. Mueller, S.T., Hoffman, R.R., Clancey, W., Emrey, A., Klein, G.: Explanation in human-ai systems: a literature meta-review, synopsis of key ideas and publications, and bibliography for explainable AI (2019)
7. Provost, F., Domingos, P.: Tree induction for probability-based ranking. Mach. Learn. **52**(3), 199–215 (2003)
8. Vovk, V., Petej, I.: Venn-abers predictors. arXiv preprint arXiv:1211.0025 (2012)
9. Vovk, V., Shafer, G., Nouretdinov, I.: Self-calibrating probability forecasting. In: Advances in Neural Information Processing Systems. pp. 1133–1140 (2004)
10. Zadrozny, B., Elkan, C.: Obtaining calibrated probability estimates from decision trees and naive Bayesian classifiers. In: Proceedings of the 18th International Conference on Machine Learning, pp. 609–616 (2001)

Automated Attribute Weighting Fuzzy k-Centers Algorithm for Categorical Data Clustering

Toan Nguyen Mau(✉) and Van-Nam Huynh

School of Advanced Science and Technology, Japan Advanced Institute of Science and Technology, Ishikawa, Japan
{nmtoan91,huynh}@jaist.ac.jp

Abstract. Cluster analysis plays an important role in exploring the correlations in data by dividing datasets into separate clusters so that similar objects are located in the same cluster. Moreover, fuzzy cluster analysis can reveal the mixtures of clusters in datasets containing multiple distributions. Certainly, the outcome of clustering methods is approximately determined by the similarity definition. Thus, the similarity measurement is exceedingly important to the formation of fuzzy clusters. In fact, the similarity between two objects is mostly calculated by the mean of differences across multiple dimensions. However, the dissimilarity in some dimensions has little or no effect on the fuzzy clustering outcome. In this study, we explore such impacts for fuzzy clustering of data with categorical attributes. Accordingly, the impact of each attribute on each fuzzy cluster is calculated using an optimizer, and the overlapping dissimilar values are then adjusted by the corresponding weights. We propose to apply this approach to the Fk-centers clustering algorithm, and the experimental results show that our proposed method can achieve higher fuzzy silhouette scores than other related works. These results demonstrate the applicability of deploying of the proposed method in real-world application.

Keywords: Fuzzy clustering · Categorical data · k-representatives · k-centers

1 Introduction

Cluster analysis is one of the fundamental tasks in data mining, which aims to discover the natural groups of objects within a given dataset such that the objects belonging to the same cluster are similar and objects belonging to different clusters are dissimilar [1]. Cluster analysis can be categorized into different approaches: Hierarchical clustering techniques seek for a nested structure to describe the dataset while flat clustering techniques seek for the separated groups of similar objects [2]. Most flat clustering algorithms use the representations for reenacting the compactness of the clusters; these representations could be numeric vectors, frequencies, or both of them. In general, a data object belongs to only a

V. Torra and Y. Narukawa (Eds.): MDAI 2021, LNAI 12898, pp. 205–217, 2021.
https://doi.org/10.1007/978-3-030-85529-1_17

single cluster. However, this is an ill-suited assumption for the fuzzy spots on the boundaries between clusters. This problem can be handled by the assumption that a data object can simultaneously belong to multiple clusters with different membership degrees, an approach called fuzzy cluster analysis [3,4].

In addition, the similarity between objects can be differently defined for different types or characteristics of the data. The typical similarity measures such as Euclidean or Manhattan distances can work well for numeric data; however, these measures encounter many difficulties in handling categorical data [5,6]. On the other hand, the cluster representation is also an important aspect that decides the accuracy of the clustering task. Therefore, in regard to cluster analysis for special data types such as categorical data, the representations of data clusters can be accordingly modified. Consequently, k-means-like algorithms are one of the most widely used flat clustering techniques and have been extensively used for exploring the coherence of the data [2,7]. Several attempts have been made to define/redefine the representation of categorical data: Huang et al. use the highest categorical values on all attributes as the *mode* to acquire the representation [3,5], San et al. and Kim et al. use the representation called *representative* as the collection of probabilities of all categorical values on all attributes for each cluster [6,8], and Chen et al. propose using the weighted kernel-based *center* to resemble the cluster representation [9]. These flat clustering algorithms can be simply extended to handle the fuzzy analysis problem by using a parameter as the fuzzy degree; however, some definitions must be redefined to suit the fuzzy concept.

In previous work, we proposed a kernel-based clustering algorithm for fuzzy clustering of categorical data called Fk-centers, which is an extension of k-center algorithm. In this approach, the fuzzy frequency is estimated using the membership matrix. As a result, Fk-centers algorithm has the highest clustering effectiveness scores compared to other state-of-the-art approaches with an acceptable complexity [10]. However, the influence of categorical attributes on each cluster is ignored in the Fk-centers algorithm, which results in an algorithm convergence that is not absolute [11].

This research aims to enhance the previous work by increasing the performance for categorical data clustering. Technically, the contributions of categorical attributes are analyzed for each cluster, which can provide a better normalization of the overall objective function. The clustering effectiveness of the proposed method was verified using the fuzzy silhouette metric on common UCI categorical datasets.

2 Related Works

To highlight the significance of our proposed method, we selected the state-of-the-art categorical fuzzy clustering techniques as the competitors of our approach:

- Categorical encoding approaches: Fk-means directly converts unique categorical values into unique integer numbers [12]. FEk-means decomposes a categorical attribute into multiple numerical attributes yielding as many as

the number of unique categorical values of the original attribute [12]. Fuzzy space-based clustering (FSBC) algorithm computes the similarity matrix for all objects using an overlap measure and uses this matrix for clustering [13].
- FCentroids: FCentroids, as the first fuzzy clustering algorithm for categorical data, is an extension of the k-representatives algorithm [6,8].
- Single-objective genetic algorithm approaches: The SGA-Sep approach tries to maximize the average distances of objects in a cluster to others clusters. The SGA-Dist algorithm tries to minimize the average within-cluster distances. SGA-SepDist can express the cluster separation and cluster compactness at the same time by a fraction [14].
- Multi-objective genetic algorithm approaches: The MOGA algorithm defines cluster separation for all clusters and cluster compactness of each cluster as the two objectives [15]. In contrast, the NSGA-FMC algorithm defines equivalent objectives, but it uses the membership matrices as the chromosomes [16]. The MaOFCentroids algorithm also use fuzzy membership chromosomes, but it establishes many objectives at the same time and applies the NSGA-III selection technique [17].

3 Preliminaries

This research incorporates an attribute weighting technique into an existing k-means-like algorithm. First, we dicuss the Fk-centers algorithm, which is a superior algorithm for fuzzy clustering of a dataset with categorical attributes. Then, we introduce the selected automated method, which is based on the particle swarm optimizer.

3.1 Fuzzy Clustering for Categorical Data

Assume that the categorical dataset $X = \{\boldsymbol{x}_1, ..., \boldsymbol{x}_N\}$ containing N categorical objects needs to be clustered into k fuzzy clusters. Each categorical object $\boldsymbol{x}_i (1 \leq i \leq N)$ is a vector of D categorical values $\boldsymbol{x}_i = [x_{i1}, ..., x_{iD}]$; in other words, dataset X has D attributes. Each attribute has an independent finite set of all possible categorical values on this attribute, where $\mathcal{A}_d (1 \leq d \leq D)$ is the domain of the dth attribute.

There are several techniques that can calculate the dissimilarity of categorical objects, but in this study, the naive approach of the so-called overlap measurement is adopted to simplify the calculation. Technically, the overlap measure counts the number of the mismatch values in all attributes and uses it for the dissimilarity between two objects:

$$\mathrm{Dis}(\boldsymbol{x}_i, \boldsymbol{x}_j) = \sum_{d=1}^{D} \delta(x_{id}, x_{jd}), \quad \text{where } \delta(x_{id}, x_{jd}) = \begin{cases} 1, & \text{if } x_{id} \neq x_{jd} \\ 0, & \text{otherwise} \end{cases} \tag{1}$$

In terms of formulating the fuzzy clustering problem, each cluster has a representation denoted as $\mathbf{c}_j (1 \leq j \leq k)$, and these representations are usually the centers of the clusters, which minimizes the total distances from the objects to their nearest representations. The degrees of membership of object $\boldsymbol{x}_i (1 \leq i \leq N)$ are shared for k clusters, where U is called the membership matrix of all objects to all clusters:

$$U = [u_{i,j}| \quad 1 \leq i \leq N, 1 \leq j \leq k] \tag{2}$$

All the objects are treated equally and normalized:

$$0 \leq u_{i,j} \leq 1 \quad \text{and} \quad \sum_{j=1}^{k} u_{i,j} = 1 \tag{3}$$

Similar to a crisp clustering objective function, the objective function for fuzzy clustering is also designed to minimize the total distance from objects to their nearest clusters. However, the importance of proximal clusters is further strengthened by an exponent factor α:

$$\text{Minimize:} \quad \mathrm{P}(U, C) = \sum_{j=1}^{k} \sum_{i=1}^{N} u_{i,j}^{\alpha} \mathrm{Dis}(\boldsymbol{x}_i, \boldsymbol{c}_j) \tag{4}$$

In other words, α is the parameter defining the fuzzy degree of the clustering model with respect to $1 \leq \alpha < \infty$. With $alpha = 1$, the fuzzy clustering model degenerates to the crisp clustering model [17–19]. There are numerous techniques that can minimize the objective function in Eq. (4) such as genetic algorithms, expectation–maximization algorithms, and k-means-like algorithms. This study targets improving a k-means-like algorithm, which is discussed in the next section.

3.2 *k*-means Algorithm for Fuzzy Clustering (F*k*-means)

Most k-means-like fuzzy clustering algorithms start the processes with random representations, after which the algorithms sequential conduct the updating of the memberships and the centroids until they are converged. The degree of membership u_{ij} is the exponential inverse of the distance from object $\boldsymbol{x}_i$ to centroid $\mathbf{c}_j$:

$$u_{i,j} \cong \left(\sum_{m=1}^{k} \left(\frac{\mathrm{Dis}(\boldsymbol{x}_i, \boldsymbol{c}_j)}{\mathrm{Dis}(\boldsymbol{x}_i, \boldsymbol{c}_m)} \right)^{\frac{1}{\alpha-1}} \right)^{-1}, 1 \leq i \leq N, 1 \leq j \leq k \tag{5}$$

The centroids are then recalculated by the weighted mean of all objects with their degrees of membership to that clusters:

$$\boldsymbol{c}_j = \frac{\sum_{i=1}^{n} u_{i,j}^{\alpha} \boldsymbol{x}_i}{\sum_{i=1}^{n} u_{i,j}^{\alpha}}, \quad 1 \leq j \leq k \tag{6}$$

Different k-means-like algorithms may have different ways of formalizing the centroids and/or defining appropriate dissimilarity measures [18].

3.3 Previous Work (F*k*-centers Algorithm)

Recently, we have proposed to use a kernel representative (*fcenter*) for categorical data, which stores the frequencies of all categorical values. The fuzzy exponential probabilities of categorical values are calculated by their weighted contributions to different cluster following their membership degrees:

$$\mathrm{FPr}_{j,d}(v) = \frac{\sum_{\boldsymbol{x}_i \in X, x_{id}=v} u_{i,j}^{\alpha}}{\sum_{i=1}^{N} u_{i,j}^{\alpha}}, \quad 1 \le d \le D, 1 \le j \le k \tag{7}$$

The values of *fcenter* can be formulated as the weighted combination of the uniform distributions and the observed distributions:

$$\boldsymbol{c}_{jd} = [\{v, \lambda_j \frac{1}{|\mathcal{A}_j|} + (1-\lambda_j)\mathrm{FPr}_{j,d}(v)\} | \forall v \in \mathcal{A}_d], \quad 1 \le j \le k, 1 \le d \le D \tag{8}$$

where λ_j is the smoothing parameter for the jth cluster. The optimal value for λ_j can be learned using least squares cross-validation (LSCV) [20]. After obtaining the value of *fcenter*, the values must be normalized to become probability variables:

$$\begin{cases} \forall p \in \boldsymbol{c}_{jd}, 0 \le p \le 1 \\ \sum_{p \in \boldsymbol{c}_{jd}} p = 1 \end{cases}, \quad 1 \le j \le k, 1 \le d \le D \tag{9}$$

Because a *fcenter* is a vector of multiple distributions, the distance between an object to a *fcenter* can be calculated as follows:

$$\mathrm{Dis}(\boldsymbol{x}_i, \boldsymbol{c}_j) = \sum_{d=1}^{D} \sum_{\substack{v \in \mathcal{A}_d \\ v \neq x_{id}}} \mathrm{FPr}_{j,d}(v) = D - \sum_{d=1}^{D} \mathrm{FPr}_{jd}(x_{id}) \tag{10}$$

3.4 Attribute Weighting for Clustering

Lu et al. (2011) claimed that different dimensions can differently determine the values of cluster centroids, and the particle swarm optimizer can then be applied for weighting the importance of each dimension to each cluster [21]. In addition, these weights can also be calculated by investigating the scattering of categorical values in each cluster. For example, if only one categorical value of the dth attribute appears in the jth cluster then the contribution of the dth attribute to the jth cluster is high. In contrast, if all the categorical values of the dth attribute appear in the jth cluster with the same frequency, then the contribution of the dth attribute to the jth cluster is low.

Moreover, $w_{jd} (1 \le j \le k, 1 \le d \le D)$ is the weighting of the dth attribute for the jth cluster satisfying:

$$\begin{cases} 0 \le w_{jd} \le 1, & 1 \le j \le k, 1 \le d \le D \\ \sum_{d=1}^{D} w_{jd} = 1, & 1 \le j \le k \end{cases} \tag{11}$$

To minimize the objective function, the updated weighting can be recalculated by:

$$-\beta \log(\tilde{w}_{jd}) = 1 - \frac{\lambda_j^2}{|\mathcal{A}_j|} + (\lambda_j^2 - 1) \sum_{v \in \mathcal{A}_j} \mathrm{Pr}_{j,d}(v)^2 \tag{12}$$

where $\beta(> 0)$ is the parameter controlling the degree of convexity, and $\mathrm{Pr}_{j,d}(v)$ is the probability of value $c \in \mathcal{A}_d$ appearing in the jth cluster [9]. The weights are then normalized to satisfy Eq. (11):

$$w_{jd} = \frac{\tilde{w}_{jd}}{\sum_{d'=1}^{D} \tilde{w}_{jd'}}, \quad 1 \leq j \leq k, q \leq d \leq D \tag{13}$$

Overall, the weighting matrix $W = [w_{jd}]_{k \times D}$ is a variable that can be learned during the clustering process. The higher the value of w_{jd} is, the more support information that the dth attribute has for clustering of the jth cluster.

4 The Proposed Method

This paper proposes the WCFk-centers algorithm that applies attribute weighting to the Fk-centers algorithm. Figure 1 shows the overview of our proposed approach with the loop of three main processes: updating membership matrix U, *fcenters* C, and attribute weights W.

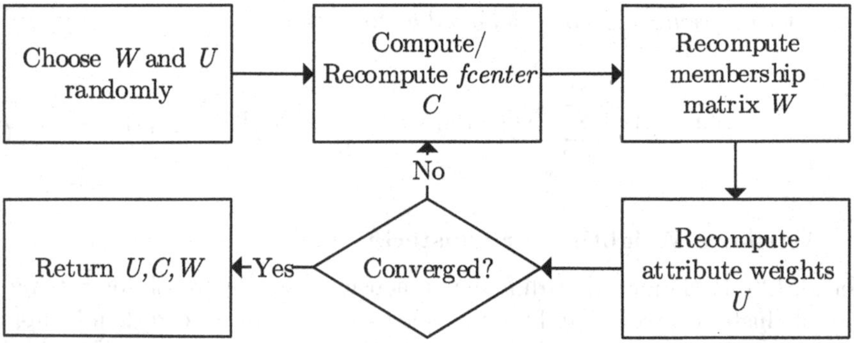

Fig. 1. Overview of the WFk-centers algorithm

According to the fuzzy membership, the fuzzy frequencies of categorical values can be estimated by Eq. (7). Similar to calculating smoothing parameters, the weight updating process can also use this fuzzy frequency estimation. The weight calculation in Eq. (12) can be redefined as follows:

$$-\beta \log(\tilde{w}_{jd}) = 1 - \frac{\lambda_j^2}{|\mathcal{A}_j|} + (\lambda_j^2 - 1) \sum_{v \in \mathcal{A}_j} \mathrm{FPr}_{j,d}(v)^2 \tag{14}$$

4.1 Objective Function

Our approach adds the weighting variables into the objective functions as,

$$\text{Minimize:} \quad \mathrm{P}(U, C, W) = \sum_{d=1}^{D} \sum_{j=1}^{k} \sum_{i=1}^{N} w_{jd} u_{ij}^{\alpha} \mathrm{Dis}(\boldsymbol{x}_i, \boldsymbol{c}_j) \tag{15}$$

Similar to other k-means-like algorithms, we also sequentially fix two of the three variables to find the current optimal values for the remaining variable. The algorithm stops when it converges or reaches a certain number of iterations called max_iter.

Algorithm 1. WCFk-center pseudocode

Require: Dataset X, N, k , iter_max
Ensure: The optimal $P(U, C, W)$
Assign random values for attribute weighting matrix W and a set of *fcenters* U such that the constraints on equations (3) and (11) are satisfied.
for iter = 0; iter $\leq$ iter_max; iter++ **do**
  Compute/recompute the smoothing parameters λ_j $(1 \leq j \leq k)$ for all clusters.
  Recompute the attribute weights W following equations (13) and (14).
  Compute/recompute the *fcenters* C following equation (8).
  Recompute the membership matrix U by equation (5). The new dissimilarity on equation (16) is used to update the membership matrix U.
  if The membership matrix U is unchanged **then**
    Break
  end if
end for
return U, C, and W

4.2 WCFk-centers Algorithm

Algorithm 1 summarizes our proposed method with the corresponding used formulations. In contrast to our previous work, this approach added the calculation of the attribute weights right after the smoothing parameter calculations. The updated attribute weights then directly affect the process of updating membership matrix U. Consequently, the distance measure in Eq. (10) is accordingly modified:

$$\mathrm{Dis}(\boldsymbol{x}_i, \boldsymbol{c}_j) = \sum_{d=1}^{D} w_{jd}(1 - \mathrm{FPr}_{jd}(x_{id})) \tag{16}$$

With the new dissimilarity measure, the contributions of attributes are allocated differently for each cluster such that some attributes may have higher priorities than others in the same cluster.

Certainly, WFk-centers is a local optimization approach, which may yield different outcomes with different initializations. To achieve high accuracy, several methods can be implemented to predict the potential initialization or the clustering process can be performed multiple times to select the best outcome. Indeed, these techniques take more time and resources.

5 Experiments and Results

The source code of our proposal is available at the Python repository URL: https://pypi.org/project/wcfkcenters/

5.1 Datasets and Testing Environment

Testing Datasets: We selected 10 common categorical datasets from the UCI repository [22] for analyzing our method's performance. These datasets are widely used by other research studies for clustering and classification problems. The number of instances, attributes, and classes of these datasets are given in Table 1. Among these datasets, Zoo and Tae have few numeric attributes and these are unused in our experiments.

Table 1. Common categorical datasets.

Name	#Items	#Attributes	#Classes
Soybean	47	21	4
Zoo	101	17	7
Tae	151	5	3
Hayes-roth	132	5	3
Connect	10,000	42	3
Chess	3,196	36	2
Mushroom	8,124	22	2
Splice	3,190	60	3
Tictactoe	958	9	2
Vote	435	16	2

Testing Environments: All of the experiments were carried out on a high-end computer with an Intel Xeon G-6240M 2.6 GHz (16 Cores x 4) CPU, and all programs were developed using Python programming language.

5.2 Evaluation Metrics

The performance metrics are divided into two categories: effectiveness and complexity.

For the effectiveness metric, we adopt the fuzzy silhouette (FSilhouette) evaluation metric. This score is the extension of the silhouette score to evaluate the accuracy of fuzzy clustering [4]. The FSilhouette evaluation metric takes into account the pairwise degrees of membership along with their distances:

$$\text{FSilhouette}(U, X) = \frac{\sum_{i=1}^{n} (\mu_{p,i} - \mu_{q,i})^{\alpha} s_i}{\sum_{i=1}^{n} (\mu_{p,i} - \mu_{q,i})^{\alpha}} \tag{17}$$

where $\mu_{p,i}$ and $\mu_{q,i}$ are the first and second largest values on the ith column in the membership matrix U, respectively, and s_i is the silhouette score of $\boldsymbol{x}_i$ [4].

For the complexity metric, the total running time of compared methods and their preprocessing time is analyzed.

5.3 Attribute Weighting Performance

Figure 2 shows the attribute weighting distributions of the proposed method. These weights are extracted after the final iteration of WCFk-centers for each testing dataset. The testing datasets have completely different weight distributions. On the Zoo and Chess datasets, the attributes seem not to have a large bias of clustering impacts for different attributes. Besides that, just a few attributes have an impact on clustering in Mushroom, Splice, and Tictactoe datasets.

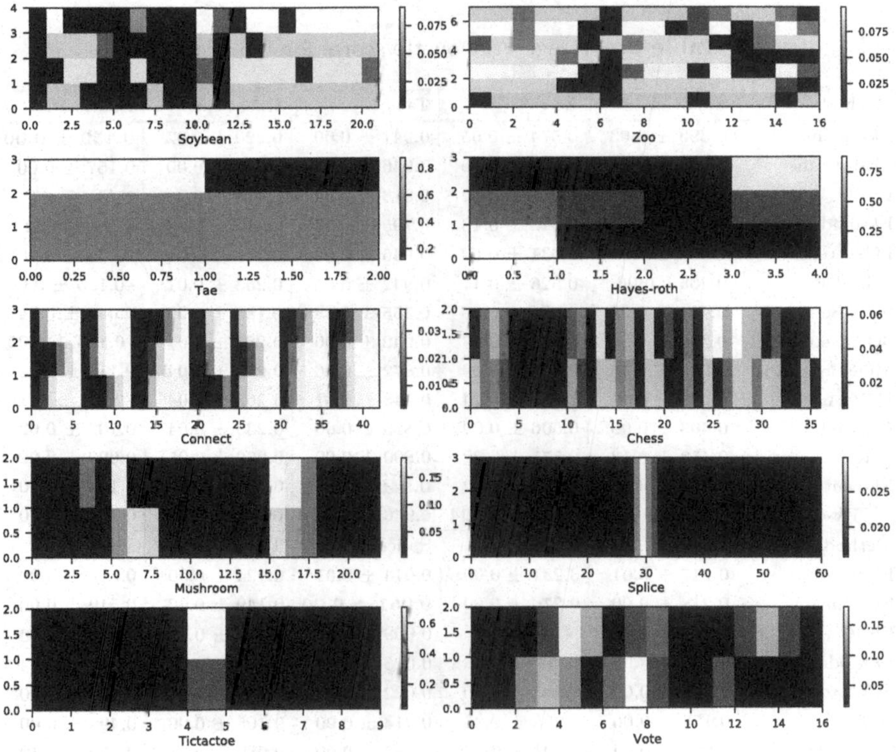

Fig. 2. Weighting distributions in UCI datasets; horizontal axis: attribute; vertical axis: cluster.

5.4 Fuzzy Clustering Performance Analysis

Table 2 shows the comprehensive comparisons of our method with other approaches in terms of fuzzy clustering effectiveness on the benchmark datasets. The encoding methods, such as Fk-means and FEk-means, have ordinary ranks in comparison to all methods. Fk-means has a massive amount of categorical information loss because the encoding order affects the similarity between categorical values. In terms of genetic algorithms, these methods need a significantly high number of generations to achieve convergence. Moreover, the complexities of these approaches are also affected by the size of chromosomes determined by the number of clusters k and the number of objects N. In this experiment, we set the maximum number of generations to 100, which is equal to the maximum number of iterations of other k-means-like algorithms. It is clear that with a higher number of generations, these genetic approaches achieve much better FSilhouette scores. However, this also means their processing time will significantly increase.

Table 2. Average FSilhouette scores for 128 runs

Method	Soybean	Zoo	Tae	Hayes-roth	Connect
Fk-means	0.393 ± 0.03	0.574 ± 0.05	0.941 ± 0.00	0.220 ± 0.02	**0.159 ± 0.00**
FEk-means	0.536 ± 0.00	0.565 ± 0.00	0.946 ± 0.00	0.354 ± 0.00	0.157 ± 0.00
FSBC	0.190 ± 0.01	0.258 ± 0.02	0.813 ± 0.04	0.175 ± 0.00	N/A
Fk-modes	0.456 ± 0.08	0.548 ± 0.13	0.795 ± 0.37	0.315 ± 0.03	0.146± 0.02
FCentroids	0.537 ± 0.00	0.631 ± 0.02	0.946 ± 0.00	0.363 ± 0.00	0.143 ± 0.00
SGA-Dist	0.455± 0.06	0.526 ± 0.11	0.942 ± 0.00	0.253 ± 0.01	–0.125 ± 0.06
SGA-Sep	0.201 ± 0.08	0.195 ± 0.18	0.555 ± 0.58	0.189 ± 0.02	0.046 ± 0.04
SGA-SepDist	0.383 ± 0.10	0.302 ± 0.17	0.942 ± 0.00	0.282 ± 0.01	–0.126 ± 0.05
MOFCentroids	0.378 ± 0.10	0.401 ± 0.09	0.872 ± 0.06	0.250 ± 0.04	0.110 ± 0.02
MOGA	0.288 ± 0.11	0.206 ± 0.24	0.548 ± 0.41	0.262 ± 0.04	–0.041 ± 0.10
NSGA-FMC	0.403 ± 0.09	0.506 ± 0.09	0.859 ± 0.06	0.235 ± 0.04	0.102 ± 0.02
MaOFCentroids	0.452 ± 0.13	0.575 ± 0.22	0.800 ± 0.09	0.269 ± 0.05	0.099 ± 0.02
Fk-centers	0.537 ± 0.00	0.639 ± 0.02	**0.946 ± 0.00**	0.363 ± 0.00	0.143 ± 0.00
WCFk-centers	**0.538 ± 0.00**	**0.650 ± 0.04**	0.946 ± 0.00	**0.368 ± 0.00**	0.153 ± 0.00
Method	Chess	Mushroom	Splice	Tictactoe	Vote
Fk-means	0.217 ± 0.01	0.231 ± 0.00	0.014 ± 0.01	0.124 ± 0.00	0.516 ± 0.00
FEk-means	0.218 ± 0.00	0.276 ± 0.00	**0.051 ± 0.00**	0.149 ± 0.00	0.519 ± 0.00
FSBC	0.038 ± 0.00	N/A	0.009 ± 0.00	0.028 ± 0.00	0.135 ± 0.00
FkModes	0.166 ± 0.04	0.209 ± 0.06	0.025 ± 0.00	0.132 ± 0.01	0.515 ± 0.04
FCentroids	0.246 ± 0.00	0.283 ± 0.00	0.032 ± 0.02	0.164 ± 0.00	0.527 ± 0.00
SGA-Dist	0.178 ± 0.06	0.203 ± 0.05	0.018 ± 0.00	0.105 ± 0.00	0.492 ± 0.00
SGA-Sep	0.094 ± 0.04	0.103 ± 0.04	0.012 ± 0.00	0.080 ± 0.01	0.159 ± 0.10
SGA-SepDist	0.048 ± 0.16	0.126 ± 0.08	0.014 ± 0.00	0.105 ± 0.00	0.478 ± 0.04
MOFCentroids	0.161 ± 0.02	0.220 ± 0.04	0.024 ± 0.01	0.101 ± 0.01	0.501 ± 0.01
MOGA	0.156 ± 0.13	0.123 ± 0.08	0.020 ± 0.00	0.114 ± 0.01	0.407 ± 0.14
NSGA-FMC	0.150 ± 0.03	0.212 ± 0.04	0.019 ± 0.00	0.096 ± 0.01	0.487 ± 0.02
MaOFCentroids	0.149 ± 0.03	0.202 ± 0.04	0.021 ± 0.00	0.095 ± 0.01	0.485 ± 0.05
Fk-centers	**0.246 ± 0.00**	0.283 ± 0.00	0.017 ± 0.04	0.164 ± 0.00	0.527 ± 0.00
WCFk-centers	0.232 ± 0.00	**0.287 ± 0.00**	0.035 ± 0.01	**0.164 ± 0.00**	**0.534 ± 0.00**

Markedly, *representative*-based approaches such as FCentroids and Fk-centers have a higher performance than other approaches, but our proposed method can outperform them for the FSilhouette score. In detail, FCentroids, Fk-centers, and WCFk-centers have average FSilhouette scores of 0.387 ± 0.00, 0.386 ± 0.01, and 0.391 ± 0.01, respectively.

5.5 Complexity Analysis

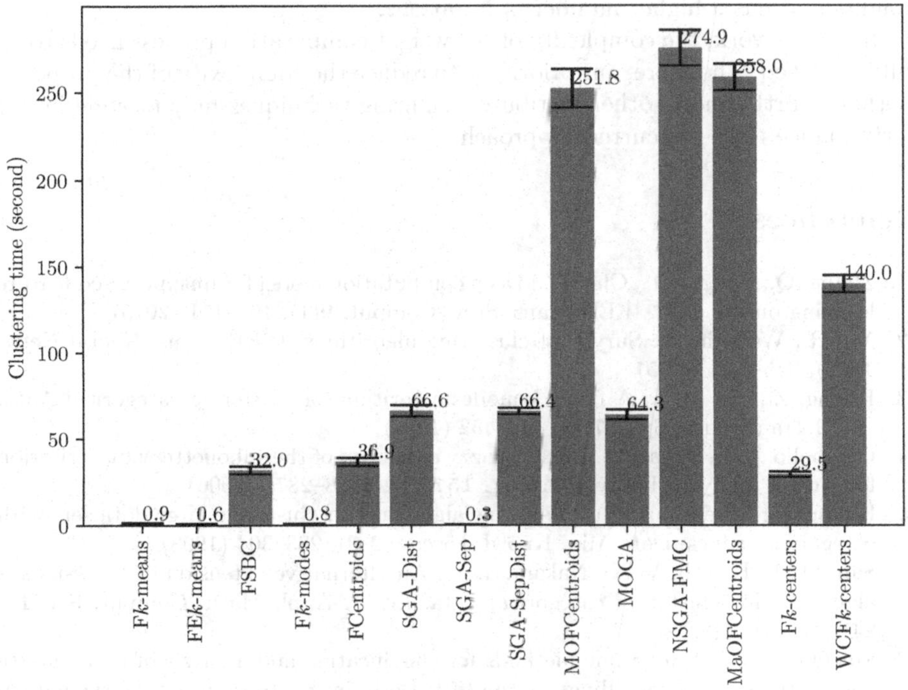

Fig. 3. Average clustering time for 10 UCI datasets for 128 runs

In Fig. 3, each dataset is clustered 128 times with different random initializations. The average clustering time and average standard deviations are calculated by taking the means of clustering time and the standard deviation of every method for every dataset. WCFk-centers employs a process to find the attribute weights in each iteration. As a result, WCFk-centers utilizes a compelling amount of processing time. This is the obvious result of the accuracy-complexity trade-off. On the other hand, because after each mutation process, the chromosomes need to be normalized, these genetic algorithms not only obtain worse accuracy but also take a significant amount of time to perform clustering.

6 Conclusion and Future Work

In this paper, WCFk-centers is proposed by applying a weighting technique to categorical attributes for the fuzzy clustering algorithm. The impact of each attribute on each cluster is clearly analyzed, which helps the fuzzy clustering algorithm achieve better presentations of clusters and reduces the importance of attributes that do not carry much clustering information. As a result, WCFk-centers can increase the degree of convergence compared to the original method. Therefore, our method needs extra processing time to calculate the attribute weights and has a higher number of iterations.

In future work, the complexity of the weight computation process needs to be fully analyzed. Therefore, our priority is to reduce the complexity of the proposed method. Furthermore, other attribute weighting techniques may provide better performance than the current approach.

References

1. Zhang, Q., Yang, L.T., Chen, Z.: Deep computation model for unsupervised feature learning on big data. IEEE Trans. Serv. Comput. **9**(1), 161–171 (2015)
2. Xu, R., Wunsch, D.: Survey of clustering algorithms. IEEE Trans. Neural Netw. **16**(3), 645–678 (2005)
3. Huang, Z., Ng, M.K.: A fuzzy k-modes algorithm for clustering categorical Aata. IEEE Trans. Fuzz. Syst. **7**(4), 446–452 (1999)
4. Campello, R.J., Hruschka, E.R.: A fuzzy extension of the silhouette width criterion for cluster analysis. Fuzzy Sets Syst. **157**(21), 2858–2875 (2006)
5. Huang, Z.: Extensions to the k-means algorithm for clustering large data sets with categorical values. Data Min. Knowl. Discov. **2**(3), 283–304 (1998)
6. San, O.M., Huynh, V.-N., Nakamori, Y.: An alternative extension of the k-means algorithm for clustering categorical data. Int. J. Appl. Math. Comput. Sci. **14**, 241–247 (2004)
7. MacQueen, J., et al.: Some methods for classification and analysis of multivariate observations. In: Proceedings of the fifth Berkeley Symposium on Mathematical Statistics and Probability, vol. 1, no. 14. Oakland, pp. 281–297 (1967)
8. Kim, D.-W., Lee, K.H., Lee, D.: Fuzzy clustering of categorical data using fuzzy centroids. Patt. Recogn. Lett. **25**(11), 1263–1271 (2004)
9. Chen, L., Wang, S.: Central clustering of categorical data with automated feature weighting. In: IJCAI, pp. 1260–1266 (2013)
10. Mau, T.N., Huynh, V.-N.: Kernel-based k-representatives algorithm for Fuzzy clustering categorical data. In: IEEE International Conference on Fuzzy Systems (2021, Under review)
11. Liu, H., Wu, J., Liu, T., Tao, D., Fu, Y.: Spectral ensemble clustering via weighted "k"-means: theoretical and practical evidence. IEEET Trans. Knowl. Data Eng. **29**(5), 1129–1143 (2017)
12. Potdar, K., Pardawala, T.S., Pai, C.D.: A comparative study of categorical variable encoding techniques for neural network classifiers. Int. J. Comput. Appl. **175**(4), 7–9 (2017)
13. Qian, Y., Li, F., Liang, J., Liu, B., Dang, C.: Space structure and clustering of categorical data. IEEE Trans. Neural Netw. Learn. Syst. **27**(10), 2047–2059 (2015)

14. Gan, G., Wu, J., Yang, Z.: A genetic fuzzy k-modes algorithm for clustering categorical data. Exp. Syst. Appl. **36**(2), 1615–1620 (2009)
15. Mukhopadhyay, A., Maulik, U., Bandyopadhyay, S.: Multiobjective genetic algorithm-based fuzzy clustering of categorical attributes. IEEE Trans. Evol. Comput. **13**(5), 991–1005 (2009)
16. Yang, C.-L., Kuo, R., Chien, C.-H., Quyen, N.T.P.: Non-dominated sorting genetic algorithm using fuzzy membership chromosome for categorical data clustering. Appl. Soft Comput. **30**, 113–122 (2015)
17. Zhu, S., Xu, L.: Many-objective fuzzy centroids clustering algorithm for categorical data. Exp. Syst. Appl. **96**, 230–248 (2018)
18. Dehariya, V.K., Shrivastava, S.K., Jain, R.: Clustering of image data set using k-means and fuzzy k-means algorithms. In: 2010 International Conference on Computational Intelligence and Communication Networks, pp. 386–391. IEEE (2010)
19. Ghosh, S., Dubey, S.K.: Comparative analysis of k-means and fuzzy c-means algorithms. Int. J. Adv. Comput. Sci. Appl. **4**(4), 36 (2013)
20. Li, Q., Racine, J.S.: Nonparametric Econometrics: Theory and Practice. Princeton University Press, Princeton (2007)
21. Lu, Y., Wang, S., Li, S., Zhou, C.: Particle swarm optimizer for variable weighting in clustering high-dimensional data. Mach. Learn. **82**(1), 43–70 (2011)
22. Frank, A., et al.: UCI machine learning repository, vol. 15, p. 22 (2011). http://archive.ics.uci.edu/ml

q-Divergence Regularization of Bezdek-Type Fuzzy Clustering for Categorical Multivariate Data

Yuchi Kanzawa(✉)

Shibaura Institute of Technology, Tokyo, Japan
kanzawa@shibaura-it.ac.jp

Abstract. In this paper, the q-divergence-regularized Bezdek-type fuzzy clustering approach is proposed for categorical multivariate data. Because the approach proposed here reduces to the conventional methods via appropriate control of the fuzzification parameters, it is considered as a generalization. Further, numerical experiments were conducted to show that the proposed method outperformed the conventional method in terms of clustering accuracy.

1 Introduction

The hard c-means (HCM) clustering algorithm [1] is generally used to partition objects into groups. Fuzzy clustering is an extension of this algorithm, where each object belongs to all or some clusters to varying degrees rather than being included in exactly one cluster alone. In the earliest fuzzy clustering method, fuzzy c-means (FCM) clustering, the linear membership weights of the HCM objective function were replaced with the powers of the memberships [2]. To discriminate this algorithm from the other alternatives that have since been proposed, this algorithm is referred to as the Bezdek-type FCM (BFCM) in this paper. Regularization of the HCM objective function is another fuzzy approach that is often used for cluster analysis. Miyamoto and Mukaidono incorporated a regularization term of the negative entropy of the membership [3] in the HCM objective function, thereby obtaining the entropy-based FCM (EFCM). However, the abovementioned algorithms tend to produce clusters of equal sizes, which is one of their disadvantages. Consequently, some objects in large clusters could possibly be misclassified in other smaller clusters if the cluster sizes are different. To solve this issue, some approaches have introduced variables to control the cluster sizes [4,5]. These methods using variables to control the cluster sizes corresponding to BFCM and EFCM are referred to as the revised BFCM (RBFCM) and revised EFCM (REFCM), respectively, in the present study. Furthermore, the RBFCM has been previously generalized in [6] and referred to as the GFCM.

Clustering of categorical multivariate data is one of the methods of summarizing the co-occurrence information comprising the mutual affinities among objects and items. For example, in the case of document-keyword frequency

V. Torra and Y. Narukawa (Eds.): MDAI 2021, LNAI 12898, pp. 218–230, 2021.
https://doi.org/10.1007/978-3-030-85529-1_18

information, the documents and keywords correspond to the objects and items, respectively. A multinomial mixture model (MMM) [7] is a probabilistic model used in clustering tasks for categorical multivariate data, where each component distribution is defined by a multinomial distribution. Honda et al. [8] proposed the Kullback–Leibler (KL) divergence-regularized fuzzy clustering model for categorical multivariate data induced by MMMs, which they referred to as the KLFCCMM. Furthermore, Kondo et al. extended the KLFCCMM algorithm by introducing q-divergence instead of KL divergence, resulting in the QFCCMM algorithm [9], and showed that the QFCCMM outperforms KLFCCMM in terms of clustering accuracy. The reason for this is because the q-divergence used in QFCCMM is a generalization of the KL divergence used in KLFCCMM. Thus, further generalization of the QFCCMM approach has the potential to result in a method that can produce more accurate clustering results.

In this work, we propose a fuzzy clustering algorithm for categorical multivariate data. First, we consider the Bezdek-type fuzzy clustering for categorical multivariate data induced by MMMs, i.e., BFCCMM, by replacing the object-cluster dissimilarities in the RBFCM objective function with those from the KLFCCMM objective function. Next, we show that the QFCCMM objective function can be interpreted as a regularization of the BFCCMM objective function by the q-divergence, where the fuzzification and q-divergence parameters have the same values. Then, we construct a new objective function by q-divergence regularization of the BFCCMM objective function, where the fuzzification and q-divergence parameters may have different values. The proposed method is referred to as the q-divergence-regularized Bezdek-type fuzzy clustering for categorical multivariate data induced by MMMs (QBFCCMM) because its objective function is obtained by q-divergence regularization of the BFCCMM method, Because the proposed QBFCCMM method can be reduced to the QFCCMM, KLFCCMM, and BFCCMM methods by controlling the values of the fuzzification parameters, the QBFCCMM method can be considered as a generalization of all these methods; therefore, the QBFCCMM method has the potential to yield more flexible clustering results than these conventional methods. In this study, we clarify the effects of the fuzzification parameters in the proposed method through numerical experiments with an artificial dataset. Furthermore, using a real dataset, we show that the proposed QBFCCMM outperforms the QFCCMM in terms of clustering accuracy.

The remainder of this paper is organized as follows. Section 2 introduces the notations used and some conventional algorithms. Section 3 describes the proposed algorithm. Section 4 presents the results of the numerical experiments to demonstrate the performance of the proposed algorithm. Finally, Sect. 5 presents the conclusions of this work.

2 Preliminaries

2.1 Divergence

Given two probability distributions P and Q, the KL-divergence of Q from P, $D_{\mathsf{KL}}(P||Q)$, is defined as

$$D_{\mathsf{KL}}(P||Q) = \sum_k P(k) \ln\left(\frac{P(k)}{Q(k)}\right). \tag{1}$$

The KL divergence is often used to derive fuzzy clustering [5,8] for vectorial and categorical multivariate data. The KL divergence can be extended using the q-logarithmic function

$$\ln_q(x) = \frac{1}{1-q}(x^{1-q} - 1) \quad (\text{for } x > 0) \tag{2}$$

as

$$D_q(P||Q) = \frac{1}{q-1}\left(\sum_k P(i)^q Q(k)^{1-q} - 1\right), \tag{3}$$

which is referred to as q-divergence [10]. In the limiting condition, as $q \to 1$, the KL divergence can be recovered.

2.2 Fuzzy Clustering for Vectorial Data

Let $X = \{x_k \in \mathbb{R}^p \mid k \in \{1, \cdots, N\}\}$ be a dataset of p-dimensional points. Then, the membership of x_k that belongs to the i-th cluster is denoted by $u_{i,k}$ ($i \in \{1, \cdots, C\}, k \in \{1, \cdots, N\}$), and the set of $u_{i,k}$ is denoted by u. The membership u has the constraint

$$\sum_{i=1}^{C} u_{i,k} = 1, \quad u_{i,k} \in [0, 1]. \tag{4}$$

The cluster center set is denoted by $v = \{v_i \mid v_i \in \mathbb{R}^p, i \in \{1, \cdots, C\}\}$. Further, the variable for controlling the cluster sizes is denoted by $\alpha = \{\alpha_i \in (0, 1)\}_{i=1}^{C}$, and it has the constraint

$$\sum_{i=1}^{C} \alpha_i = 1. \tag{5}$$

The HCM, BFCM, EFCM, RBFCM, and REFCM are subsequently obtained by solving the optimization problems

$$\underset{u,v}{\text{minimize}} \sum_{i=1}^{C} \sum_{k=1}^{N} u_{i,k} \|x_k - v_i\|_2^2, \tag{6}$$

$$\underset{u,v}{\text{minimize}} \sum_{i=1}^{C} \sum_{k=1}^{N} (u_{i,k})^m \|x_k - v_i\|_2^2, \tag{7}$$

$$\underset{u,v}{\text{minimize}} \sum_{i=1}^{C} \sum_{k=1}^{N} u_{i,k} \|x_k - v_i\|_2^2 + \lambda^{-1} \sum_{i=1}^{C} \sum_{k=1}^{N} u_{i,k} \log(u_{i,k}), \tag{8}$$

$$\underset{u,v,\alpha}{\text{minimize}} \sum_{i=1}^{C} \sum_{k=1}^{N} (\alpha_i)^{1-m} (u_{i,k})^m \|x_k - v_i\|_2^2, \tag{9}$$

$$\underset{u,v,\alpha}{\text{minimize}} \sum_{i=1}^{C} \sum_{k=1}^{N} u_{i,k} \|x_k - v_i\|_2^2 + \lambda^{-1} \sum_{i=1}^{C} \sum_{k=1}^{N} u_{i,k} \log\left(\frac{u_{i,k}}{\alpha_i}\right), \tag{10}$$

respectively, where $m > 1$ and $\lambda > 0$ are the fuzzification parameters.

2.3 Conventional Fuzzy Clustering Method for Categorical Multivariate Data

Consider a categorical multivariate dataset composed of N objects described by a set of quantitative variables $x_k^{(\ell)}$, with M items ($k \in \{1, \dots, N\}, \ell \in \{1, \dots, M\}$). The quantitative variables represent the co-occurrence relations among these objects and items.

The KLFCCMM and QFCCMM are obtained by solving the optimization problems

$$\underset{u,v,\alpha}{\text{maximize}} \sum_{i=1}^{C} \sum_{k=1}^{N} \sum_{\ell=1}^{M} u_{i,k} x_k^{(\ell)} \frac{1}{t} \left(\left(v_i^{(\ell)}\right)^t - 1 \right) - \lambda^{-1} \sum_{i=1}^{C} \sum_{k=1}^{N} u_{i,k} \ln\left(\frac{u_{i,k}}{\alpha_i}\right), \tag{11}$$

and

$$\underset{u,v,\alpha}{\text{maximize}} \sum_{i=1}^{C} \sum_{k=1}^{N} \sum_{\ell=1}^{M} (\alpha_i)^{1-m} (u_{i,k})^m x_k^{(\ell)} \frac{1}{t} \left(\left(v_i^{(\ell)}\right)^t - 1 \right) - \frac{\lambda^{-1}}{m-1} \sum_{i=1}^{C} \sum_{k=1}^{N} (\alpha_i)^{1-m} (u_{i,k})^m, \tag{12}$$

which are based on Eqs. (4), (5), and

$$\sum_{\ell=1}^{M} v_i^{(\ell)} = 1, \tag{13}$$

where $u_{i,k}$ is the membership of k-th object belonging to the i-th cluster, α_i is the variable controlling the size of the i-th cluster, $v_i^{(\ell)}$ is the ℓ-th item typicality for the i-th cluster, and $\lambda > 0, t$ are the fuzzification parameters.

3 Proposed Method

3.1 Basic Concepts

Since Eq. (11) can be equivalently written as

$$\underset{u,v,\alpha}{\text{minimize}} \sum_{i=1}^{C}\sum_{k=1}^{N} u_{i,k}\left(-\frac{1}{t}\sum_{\ell=1}^{M} x_k^{(\ell)}\left(\left(v_i^{(\ell)}\right)^t - 1\right)\right) + \lambda^{-1}\sum_{i=1}^{C}\sum_{k=1}^{N} u_{i,k}\ln\left(\frac{u_{i,k}}{\alpha_i}\right), \tag{14}$$

the KLFCCMM objective function is obtained by replacing $\|x_k - v_i\|_2^2$ by $-t^{-1}\sum_{\ell=1}^{M} x_k^{(\ell)}((v_i^{(\ell)})^t - 1)$. Therefore, we consider a Bezdek-type fuzzy clustering for categorical multivariate data induced by MMMs (BFCCMM), as

$$\underset{u,v,\alpha}{\text{minimize}} \sum_{i=1}^{C}\sum_{k=1}^{N}\sum_{\ell=1}^{M} (\alpha_i)^{1-m}(u_{i,k})^m \left(-x_k^{(\ell)}\frac{1}{t}\left(\left(v_i^{(\ell)}\right)^t - 1\right)\right). \tag{15}$$

by replacing $\|x_k - v_i\|_2^2$ in the RBFCM objective function with $-t^{-1}\sum_{\ell=1}^{M} x_k^{(\ell)}((v_i^{(\ell)})^t - 1)$. Furthermore, we note that the QFCCMM objective function is a q-divergence regularization of the BFCCMM objective function only if both the fuzzification parameter in Eq. (15) and q-divergence parameter in Eq. (3) have the same values. There is a potential to generalize the QFCCMM if the BFCCMM objective function is regularized by q-divergence with a different parameter value than the fuzzification parameter of the BFCCMM objective function, which may enable higher clustering accuracy.

Thus, we considered the fuzzification parameter m in the first and second terms in the QFCCMM objective function as being different, namely m_1 and m_2, respectively, and proposed a novel optimization problem as follows:

$$\begin{aligned}\underset{u,v,\alpha}{\text{minimize}} &\sum_{i=1}^{C}\sum_{k=1}^{N}\sum_{\ell=1}^{M} (\alpha_i)^{1-m_1}(u_{i,k})^{m_1} \left(-x_k^{(\ell)}\frac{1}{t}\left(\left(v_i^{(\ell)}\right)^t - 1\right)\right)\\ &+ \frac{\lambda^{-1}}{m_2 - 1}\sum_{i=1}^{C}\sum_{k=1}^{N} (\alpha_i)^{1-m_2}(u_{i,k})^{m_2},\end{aligned} \tag{16}$$

subject to Eqs. (4) and (5), where $m_1 > 1$, $m_2 > 1$, and $\lambda > 0$ are the fuzzification parameters. The clustering method obtained by solving this optimization problem is referred to as the q-divergence-regularized Bezdek-type fuzzy clustering for categorical multivariate data induced by MMMs (QBFCCMM) because its objective function is obtained by q-divergence regularization of the BFCCMM method, with different values for the q-divergence and fuzzification parameters.

Thus, the QBFCCMM method reduces to the QFCCMM method with $m_1 = m_2$, to the BFCCMM method with $\lambda \to +\infty$ or $m_2 \searrow 1$, and to the KLFCCMM method with $m_1 = m_2 \searrow 1$. Therefore, the proposed QBFCCMM method is a generalization of the QFCCMM, KLFCCMM, and BFCCMM methods. Hence, the proposed approach could potentially yield more flexible clustering results than other conventional clustering algorithms via control of the three fuzzification parameters.

3.2 Algorithm

The QBFCCMM method is obtained by solving the optimization problem given by Eqs. (16), (4), and (5), where the Lagrangian $L(u, v, \alpha)$ is defined as

$$L(u,v,\alpha) = \sum_{i=1}^{C}\sum_{k=1}^{N}\sum_{\ell=1}^{M}(\alpha_i)^{1-m_1}(u_{i,k})^{m_1}\left(-x_k^{(\ell)}\frac{1}{t}\left(\left(v_i^{(\ell)}\right)^t - 1\right)\right)$$
$$+\frac{\lambda^{-1}}{m-1}\sum_{i=1}^{C}\sum_{k=1}^{N}(\alpha_i)^{1-m_2}(u_{i,k})^{m_2} + \sum_{k=1}^{N}\gamma_k\left(1-\sum_{i=1}^{C}u_{i,k}\right)$$
$$+\eta\left(1-\sum_{i=1}^{C}\alpha_i\right) + \sum_{i=1}^{C}\zeta_i\left(1-\sum_{\ell=1}^{M}v_i^{(\ell)}\right) \quad (17)$$

with Lagrangian multipliers $(\gamma_1, \cdots, \gamma_N, \eta, \zeta_1, \cdots, \zeta_C)$. The necessary conditions for optimality are given as follows:

$$\frac{\partial L(u,v,\alpha)}{\partial u_{i,k}} = 0, \quad (18) \qquad \frac{\partial L(u,v,\alpha)}{\partial \gamma_k} = 0, \quad (21)$$
$$\frac{\partial L(u,v,\alpha)}{\partial v_i} = 0, \quad (19) \qquad \frac{\partial L(u,v,\alpha)}{\partial \eta} = 0, \quad (22)$$
$$\frac{\partial L(u,v,\alpha)}{\partial \alpha_i} = 0, \quad (20) \qquad \frac{\partial L(u,v,\alpha)}{\partial \zeta_i} = 0. \quad (23)$$

The optimal cluster center is obtained from Eq. (19) in a manner similar to those in the cases of the BFCCMM and QFCCMM methods:

$$v_i^{(\ell)} = \frac{\sum_{k=1}^{N}(u_{i,k})^{m_1}x_k^{(\ell)}}{\sum_{r=1}^{M}\sum_{k=1}^{N}(u_{i,k})^{m_1}x_k^{(r)}}. \quad (24)$$

Based on Eqs. (18) and (21), the optimal membership conditions are

$$f(u_{i,k}) \stackrel{\mathsf{def}}{=} m_1(\alpha_i)^{1-m_1}d_{i,k}(u_{i,k})^{m_1-1} + \frac{\lambda^{-1}}{m_2-1}m_2(\alpha_i)^{1-m_2}(u_{i,k})^{m_2-1} = \gamma_k, \quad (25)$$

and that given by Eq. (4), where

$$d_{i,k} = -\frac{1}{t}\sum_{\ell=1}^{M}\left(\left(v_i^{(\ell)}\right)^t - 1\right). \tag{26}$$

However, as it is difficult to explicitly obtain the optimal membership, we adopt the bisection method. If the γ_k value is given, then we obtain the optimal membership using the following algorithm:

Algorithm 1

STEP 1. Let the lower bound of $u_{i,k}$, $\underline{u_{i,k}}$, be 0. Let the upper bound of $u_{i,k}$, $\overline{u_{i,k}}$, be 1.

STEP 2. Set $\widehat{u_{i,k}} = (\underline{u_{i,k}} + \overline{u_{i,k}})/2$. If $\left|\overline{u_{i,k}} - \underline{u_{i,k}}\right|$ is sufficiently small, then terminate the algorithm and let the optimal $u_{i,k}$ be $\widehat{u_{i,k}}$.

STEP 3. If $m_1(\alpha_i)^{1-m_1}d_{i,k}(u_{i,k})^{m_1-1} + \frac{\lambda^{-1}}{m_2-1}m_2(\alpha_i)^{1-m_2}(u_{i,k})^{m_2-1} > \gamma_k$, let $\overline{u_{i,k}} = \widehat{u_{i,k}}$; otherwise, let $\underline{u_{i,k}} = \widehat{u_{i,k}}$. Go to STEP 2.

Note that there exists the unique solution of Eq. (25) because f is strictly increasing, $f(0) = 0$, and $f(u_{i,k}) \to +\infty$ with $u_{i,k} \to +\infty$.

The optimal γ_k value used in the STEP 3 in Algorithm 1 can also be obtained using the bisection method as follows. From $\alpha_i > 0$, $1 - m_1 < 0$, and $1 - m_2 < 0$, we have

$$\left(\min_{1\le j\le C}\{\alpha_j\}\right)^{1-m_1} \ge (\alpha_i)^{1-m_1} \ge \left(\max_{1\le j\le C}\{\alpha_j\}\right)^{1-m_1}, \tag{27}$$

$$\left(\min_{1\le j\le C}\{\alpha_j\}\right)^{1-m_2} \ge (\alpha_i)^{1-m_2} \ge \left(\max_{1\le j\le C}\{\alpha_j\}\right)^{1-m_2}, \tag{28}$$

from which the value of γ_k is bounded as

$$\begin{aligned}\gamma_k =& m_1(\alpha_i)^{1-m_1}d_{i,k}(u_{i,k})^{m_1-1} + \frac{\lambda^{-1}}{m_2-1}m_2(\alpha_i)^{1-m_2}(u_{i,k})^{m_2-1}\\ \ge& m_1\left(\max_{1\le j\le C}\{\alpha_j\}\right)^{1-m_1}\min_{1\le j\le C}\{d_{j,k}\}(u_{i,k})^{m_1-1}\\ &+\frac{\lambda^{-1}}{m_2-1}m_2\left(\max_{1\le j\le C}\{\alpha_j\}\right)^{1-m_2}(u_{i,k})^{m_2-1}\\ \ge& 0,\end{aligned} \tag{29}$$

$$\begin{aligned}\gamma_k =& m_1(\alpha_i)^{1-m_1}d_{i,k}(u_{i,k})^{m_1-1} + \frac{\lambda^{-1}}{m_2-1}m_2(\alpha_i)^{1-m_2}(u_{i,k})^{m_2-1}\\ \le& m_1\left(\min_{1\le j\le C}\{\alpha_j\}\right)^{1-m_1}\max_{1\le j\le C}\{d_{j,k}\}(u_{i,k})^{m_1-1}\\ &+\frac{\lambda^{-1}}{m_2-1}m_2\left(\min_{1\le j\le C}\{\alpha_j\}\right)^{1-m_2}(u_{i,k})^{m_2-1}\end{aligned}$$

$$\leq m_1 \left(\min_{1 \leq j \leq C} \{\alpha_j\} \right)^{1-m_1} \max_{1 \leq j \leq C} \{d_{j,k}\} + \frac{\lambda^{-1}}{m_2 - 1} m_2 \left(\min_{1 \leq j \leq C} \{\alpha_j\} \right)^{1-m_2} \quad (30)$$

because $(\alpha_i)^{1-m_1}$ and $(\alpha_i)^{1-m_2}$ decrease with respect to α_i, and $u_{i,k} \in [0, 1]$, which implies that there exists the unique $gamma_k$ satisfying Eqs. (25) and (4). Thus, the optimal γ_k value is obtained using the following algorithm.

Algorithm 2

STEP 1. Let the lower bound of γ_k, $\underline{\gamma_k}$, be 0. Let the upper bound of γ_k, $\overline{\gamma_k}$, be $m_1 \left(\min_{1 \leq j \leq C}\{\alpha_j\}\right)^{1-m_1} \max_{1 \leq j \leq C}\{d_{j,k}\} + \frac{\lambda^{-1}}{m_2 - 1} m_2 \left(\min_{1 \leq j \leq C}\{\alpha_j\}\right)^{1-m_2}$.
STEP 2. Set $\widehat{\gamma_k} = (\underline{\gamma_k} + \overline{\gamma_k})/2$. If $|\overline{\gamma_k} - \underline{\gamma_k}|$ is sufficiently small, then terminate the algorithm and let the optimal γ_k be $\widehat{\gamma_k}$.
STEP 3. Calculate $u_{i,k}$ using Algorithm 1.
STEP 4. If $\sum_{i=1}^{C} u_{i,k} > 1$, let $\overline{\gamma_k} = \widehat{\gamma_k}$; otherwise, let $\underline{\gamma_k} = \widehat{\gamma_k}$. Go to STEP 2.

Based on Eqs. (20) and (22), the optimal conditions of the variables controlling the cluster sizes are

$$g(\alpha_i) \stackrel{\text{def}}{=} (1 - m_1) \sum_{k=1}^{N} (\alpha_i)^{-m_1} (u_{i,k})^{m_1} d_{i,k} - \lambda^{-1} \sum_{k=1}^{N} (\alpha_i)^{-m_2} (u_{i,k})^{m_2} + \beta = 0$$

$$\Leftrightarrow (m_1 - 1) \left\{ \sum_{k=1}^{N} (u_{i,k})^{m_1} d_{i,k} \right\} (\alpha_i)^{-m_1} + \lambda^{-1} \left\{ \sum_{k=1}^{N} (u_{i,k})^{m_2} \right\} (\alpha_i)^{-m_2} = \beta. \quad (31)$$

and that given by Eq. (5). Because it is difficult to explicitly obtain the optimal variable controlling the cluster size, we adopt the bisection method. If the value of β is given, then we obtain the optimal variable controlling the cluster size using the following algorithm:

Algorithm 3

STEP 1. Let the lower bound of α_i, $\underline{\alpha_i}$, be 0. Let the upper bound of α_i, $\overline{\alpha_i}$, be 1.
STEP 2. Set $\widehat{\alpha_i} = (\underline{\alpha_i} + \overline{\alpha_i})/2$. If $|\overline{\alpha_i} - \underline{\alpha_i}|$ is sufficiently small, then terminate the algorithm and let the optimal α_i be $\widehat{\alpha_i}$.
STEP 3. If $(m_1 - 1) \left\{ \sum_{k=1}^{N} (u_{i,k})^{m_1} d_{i,k} \right\} (\alpha_i)^{-m_1} + \lambda^{-1} \left\{ \sum_{k=1}^{N} (u_{i,k})^{m_2} \right\}$ $(\alpha_i)^{-m_2} > \beta$, let $\overline{\alpha_i} = \widehat{\alpha_i}$; otherwise, let $\underline{\alpha_i} = \widehat{\alpha_i}$. Go to STEP 2.

Note that there exists the unique solution of Eq. (31) because g is strictly increasing, $g(\alpha_i) \to 0$ with $\alpha_i \to +\infty$, and $g(\alpha_i) \to +\infty$ with $g(\alpha_i) \searrow 0$.

The optimal β value used in STEP 3 in Algorithm 3 can also be obtained using the bisection method as follows. From $u_{i,k} \geq 0$, $m_1 > 1$, $m_2 > 10$, and

$\alpha_i \leq 1$, as well as the decreasing $(\alpha_i)^{-m_1}$ and $(\alpha_i)^{-m_2}$, there exists the unique β satisfying Eqs. (31) and (5), and we have the lower bound of β as

$$\begin{aligned}
\beta &= (m_1 - 1)\left\{\sum_{k=1}^{N}(u_{i,k})^{m_1} d_{i,k}\right\}(\alpha_i)^{-m_1} + \lambda^{-1}\left\{\sum_{k=1}^{N}(u_{i,k})^{m_2}\right\}(\alpha_i)^{-m_2} \\
&\geq (m_1 - 1)\min_{1\leq j\leq C}\left\{\sum_{k=1}^{N}(u_{j,k})^{m_1} d_{j,k}\right\}(\alpha_i)^{-m_1} + \lambda^{-1}\min_{1\leq j\leq C}\left\{\sum_{k=1}^{N}(u_{j,k})^{m_2}\right\}(\alpha_i)^{-m_2} \\
&\geq (m_1 - 1)\max_{1\leq j\leq C}\left\{\sum_{k=1}^{N}(u_{j,k})^{m_1} d_{j,k}\right\}(\alpha_i)^{-m_1} + \lambda^{-1}\max_{1\leq j\leq C}\left\{\sum_{k=1}^{N}(u_{j,k})^{m_2}\right\}(\alpha_i)^{-m_2} \\
&\geq (m_1 - 1)\max_{1\leq j\leq C}\left\{\sum_{k=1}^{N}(u_{j,k})^{m_1} d_{j,k}\right\} + \lambda^{-1}\max_{1\leq j\leq C}\left\{\sum_{k=1}^{N}(u_{j,k})^{m_2}\right\},
\end{aligned} \tag{32}$$

while the upper bound of β can only be obtained using the following algorithm instead of analytically.

Algorithm 4

STEP 1. Let the lower bound of β, $\underline{\beta}$, be $(m_1 - 1)\max_{1\leq j\leq C}\left\{\sum_{k=1}^{N}(u_{j,k})^{m_1} d_{j,k}\right\} + \lambda^{-1}\max_{1\leq j\leq C}\left\{\sum_{k=1}^{N}(u_{j,k})^{m_2}\right\}$. Set the candidate of the upper bound of β, $\overline{\beta} > \underline{\beta}$.

STEP 2. Obtain $\{\widehat{\alpha_i}\}_{i=1}^{C}$ from Algorithm 3 with $\beta = \overline{\beta}$. If $\sum_{i=1}^{C}\widehat{\alpha_i} > 1$, set $\overline{\beta} \leftarrow \kappa\overline{\beta}$ with $\kappa > 1$, and go to STEP 2. If $\sum_{i=1}^{C}\widehat{\alpha_i} < 1$, $\overline{\beta}$ is the upper bound of β. Then, terminate this algorithm.

Using the lower bound of β in Eq. (32) and the upper bound of β obtained from Algorithm 4, the optimal β value is obtained using the following algorithm.

Algorithm 5

STEP 1. Let the lower bound of β, $\underline{\beta}$, be that given in Eq. (32). Let the upper bound of β, $\overline{\beta}$, be that obtained from Algorithm 4.

STEP 2. Set $\widehat{\beta} = (\underline{\beta} + \overline{\beta})/2$. If $|\overline{\beta} - \underline{\beta}|$ is sufficiently small, then terminate the algorithm and let the optimal β be $\widehat{\beta}$.

STEP 3. Calculate $\{\alpha_i\}_{i=1}^{C}$ using Algorithm 3.

STEP 4. If $\sum_{i=1}^{C}\alpha_i > 1$, let $\underline{\beta} = \widehat{\beta}$; otherwise, let $\overline{\beta} = \widehat{\beta}$. Go to STEP 2.

Based on the above discussion, we propose the following algorithm for the QBFCCMM clustering method:

Algorithm 6 (QBFCCMM)

STEP 1. Given the number of clusters C and fuzzification parameter (m_1, m_2, λ), where $m_1 > 1$, $m_2 > 1$, and $\lambda > 0$, let the set of initial membership be u.

STEP 2. Obtain v using Eq. (24).
STEP 3. Calculate β using Algorithms 4 and 5, and obtain the variable controlling the cluster size using Algorithm 3.
STEP 4. Calculate γ_k using Algorithm 2, and obtain the membership using Algorithm 1.
STEP 5. Check the stopping criterion for (u, v, α). If the criterion is not satisfied, go to STEP 2.

4 Numerical Experiment

In this section, we present some numerical examples to investigate the fuzzification property of the proposed method using an artificial dataset as well as its clustering accuracy using a real dataset.

The first example involves an artificial dataset with three clusters, wherein each cluster comprises 50 points in a two-dimensional simplex, as shown in Fig. 1. We observe that for all combinations of the fuzzification parameter values, appropriate clustering results are obtained using the proposed method.

Figures 2–5 show the fuzzy classification functions (FCFs) for the first cluster obtained using the proposed method with $(m_1, m_2, \lambda, t) = (1.1, 1.1, 10, 10^{-5})$, $(m_1, m_2, \lambda, t) = (1.05, 1.1, 10, 10^{-5})$, $(m_1, m_2, \lambda, t) = (1.5, 1.001, 10, 10^{-5})$, and $(m_1, m_2, \lambda, t) = (1.5, 3, 10, 10^{-5})$, respectively. Comparing Figs. 2 and 3, we observe that the larger the fuzzification parameter value m_1, the fuzzier is the FCF. From Figs. 4 and 5, we observe that the larger the fuzzification parameter value m_2, the fuzzier is the FCF. Furthermore, we note that the fuzzification effect of m_1 is stronger than that of m_2. Figures 6 and 7 show the FCFs for the first cluster obtained using the proposed method, with $(m_1, m_2, \lambda, t) = (1.2, 1.2, 10, 10^{-5})$ and QFCCMM method, with $(m, \lambda, t) = (1.2, 10, 10^{-5})$, respectively, from which we can confirm that the proposed method with $m_1 = m_2$ produces the same result as that of the QFCCMM method.

The second example involves a real dataset referred to as "Cora" [11], which is composed of 2708 scientific publications that are categorized into seven different classes. Each publication in the dataset is represented by a 0- or 1-valued word vector indicating the having or not having the matching word from a dictionary containing 1432 unique words. In the experiment, this dataset was clustered using the proposed algorithm and the QFCCMM method. The cluster number C was correspondingly set as the number of classes for the dataset. The fuzzification parameter λ for the two algorithms was set from $\lambda \in \{10^{0\times5+1}, 10^{1\times5+1}, \cdots, 10^{3\times5+1}\}$. The fuzzification parameters m_1 and m_2 for the QBFCCMM and m for QFCCMM were set from $m \in \{1+10^{-1}, 1+10^{-2}, \cdots, 1+10^{-4}\}$. The fuzzification parameter t for the QFCCMM and QBFCCMM was fixed at 10^{-5}. The initialization was performed such that the initial object memberships assigned according to the actual class labels. The clustering results were evaluated using the adjusted Rand index (ARI) [12], which ARI takes a value in $[-1, 1]$, with higher values being preferred. The highest ARI value for QFCCMM was 0.7302, whereas that for QBFCCMM was 0.7431. Thus, the proposed method outperformed the QFCCMM in terms of clustering accuracy.

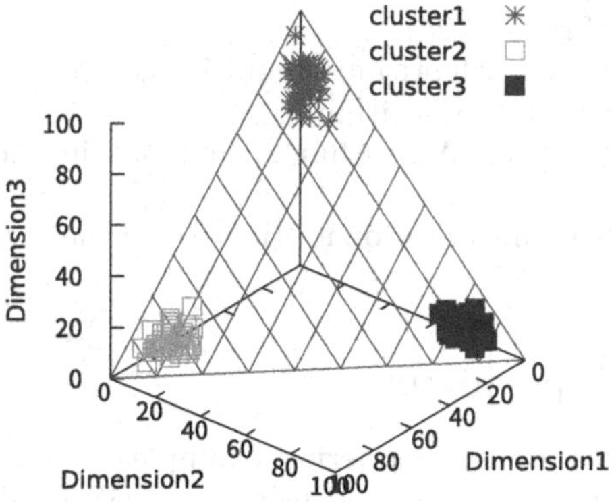

Fig. 1. Artificial dataset used for numerical verification of proposed method.

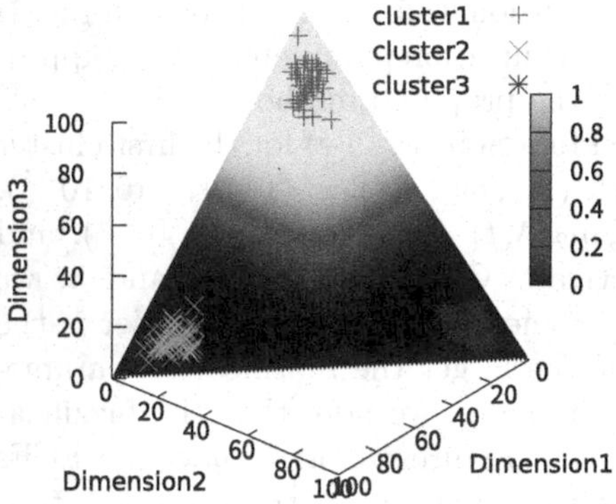

Fig. 2. FCF of the proposed method with $(m_1, m_2, \lambda, t) = (1.05, 1.1, 10, 10^{-5})$.

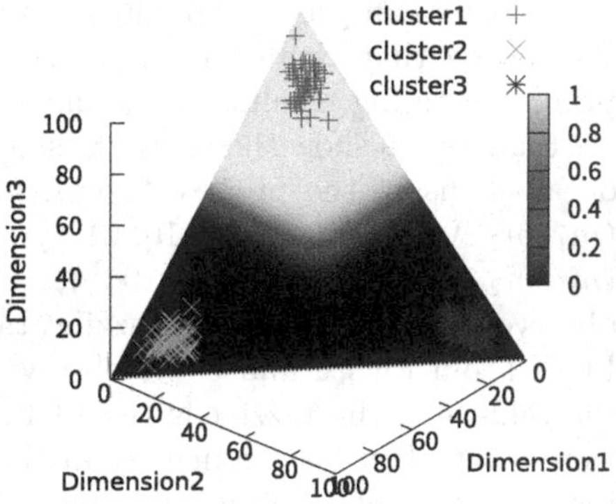

Fig. 3. FCF of the proposed method with $(m_1, m_2, \lambda, t) = (1.1, 1.1, 10, 10^{-5})$.

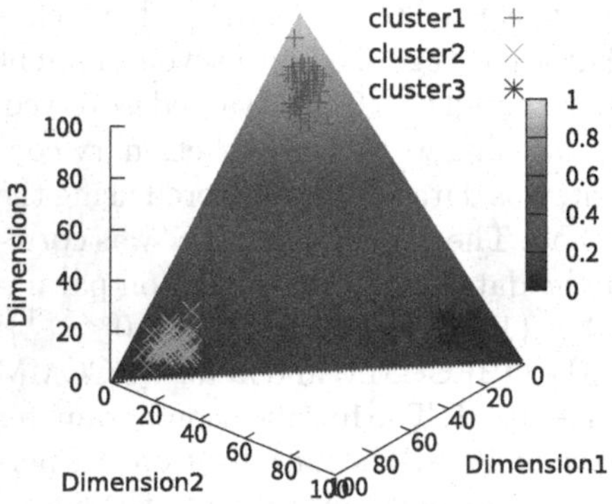

Fig. 4. FCF of the proposed method with $(m_1, m_2, \lambda, t) = (1.5, 1.001, 10, 10^{-5})$.

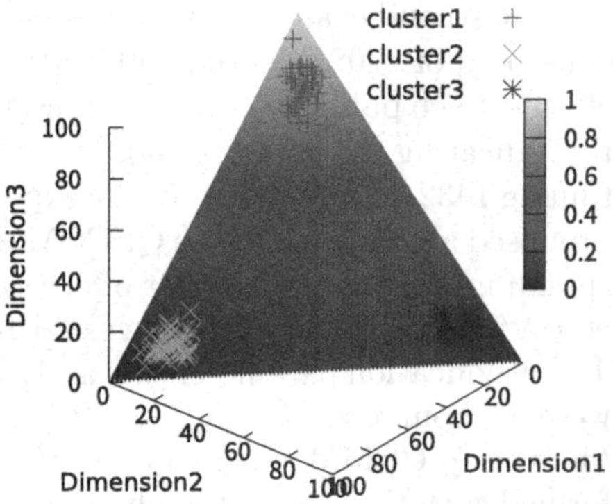

Fig. 5. FCF of the proposed method with $(m_1, m_2, \lambda, t) = (1.5, 3, 10, 10^{-5})$.

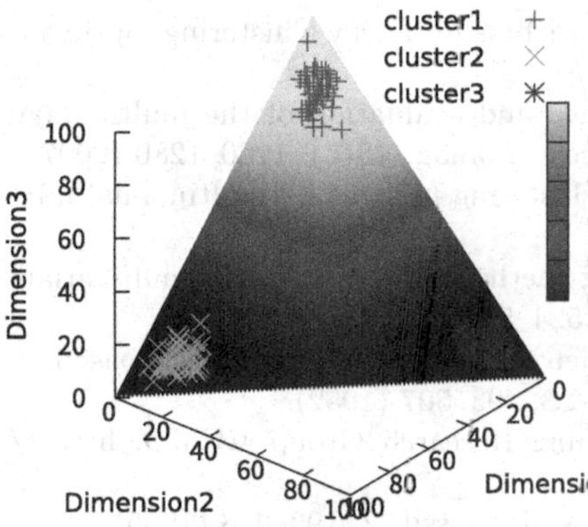

Fig. 6. FCF of the proposed method with $(m_1, m_2, \lambda, t) = (1.2, 1.2, 10, 10^{-5})$.

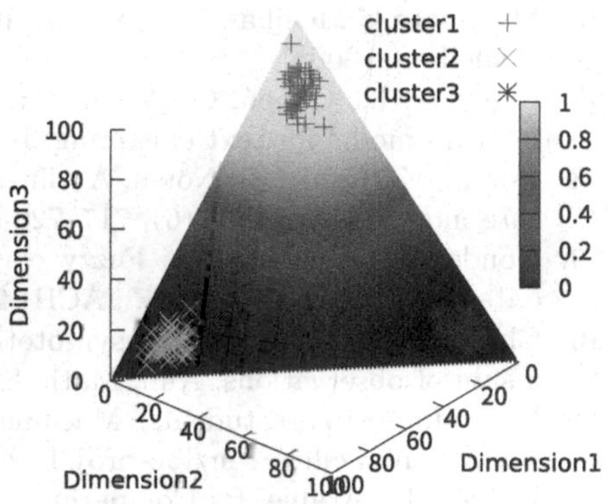

Fig. 7. FCF of the QFCCMM method with $(m, \lambda, t) = (1.2, 10, 10^{-5})$.

5 Summary

In this paper, we present the QBFCCMM clustering method. Numerical experiments were performed on an artificial dataset and the effects of the fuzzification parameters were clarified; further, it was confirmed that the proposed method reduced to the QFCCMM method via appropriate control of the fuzzification parameters. Numerical experiments were also performed on a real dataset, and the proposed method was observed to outperform the QFCCMM method in terms of clustering accuracy.

However, the experiment on real datasets is still insufficient. Then, in our future research, the proposed method will be applied to several real datasets, and the results will be compared with those obtained using conventional methods in terms of the clustering accuracies. Furthermore, the proposed fuzzification technique will applied to other types of data, such as spherical data [13–15].

References

1. MacQueen, J.B.: Some methods of classification and analysis of multivariate observations. In: Proceedings of the 5th Berkeley Symposium on Mathematical Statistics and Probability, pp. 281–297 (1967)
2. Bezdek, J.: Pattern Recognition with Fuzzy Objective Function Algorithms, Plenum Press, New York (1981)
3. Miyamoto, S., Mukaidono, M.: Fuzzy c-Means as a regularization and maximum entropy approach. In: Proceedings of the 7th International Fuzzy Systems Association World Congress (IFSA 1997), vol. 2, pp. 86–92 (1997)
4. Miyamoto, S., Kurosawa, N.: Controlling Cluster volume sizes in fuzzy c-means clustering. In: Proceedings of the SCIS&ISIS2004, pp. 1–4 (2004)
5. Ichihashi, H., Honda, K., Tani, N.: Gaussian Mixture PDF approximation and Fuzzy c-means clustering with entropy regularization. In: Proceedings 4th Asian Fuzzy System Symposium, pp. 217–221 (2000)

6. Miyamoto, S., Ichihashi, H., Honda, K.: Algorithms for Fuzzy Clustering, Springer, Heidelberg (2008)
7. Rigouste, L., Cappé, O., Yvon, F.: Inference and evaluation of the multinomial mixture model for text clustering. Inf. Process. Manag. **43**(5), 1260–1280 (2007)
8. Honda, K., Oshio, S., Notsu, A.: Fuzzy co-clustering induced by multinomial mixture models. JACIII **19**(6), 717–726 (2015)
9. Kondo, T., Kanzawa, Y.: Fuzzy clustering methods for categorical multivariate data based on q-divergence. JACIII **22**(4), 524–536 (2018)
10. Chernoff, H.: A measure of asymptotic efficiency for tests of a hypothesis based on a sum of observations. Ann. Math. Statist. **23**, 493–507 (1952)
11. Lise's Inquisitive Students, Machine Learning Research Group @UMD. http://www.cs.umd.edu/~sen/lbc-proj/LBC.html
12. Hubert, L., Arabie, P.: Comparing partitions. J. Classif. **2**, 193–218 (1985)
13. Kanzawa, Y.: On kernelization for a maximizing model of Bezdek-like spherical fuzzy *c*-means clustering. In: Torra, V., Narukawa, Y., Endo, Y. (eds.) MDAI 2014. LNCS (LNAI), vol. 8825, pp. 108–121. Springer, Cham (2014). https://doi.org/10.1007/978-3-319-12054-6_10
14. Kanzawa, Y.: A maximizing model of Bezdek-like spherical fuzzy *c*-means. J. Adv. Comput. Intell. Intell. Inform. **19**(5), 662–669 (2015)
15. Kanzawa, Y.: A maximizing model of spherical Bezdek-type fuzzy multi-medoids clustering. J. Adv. Comput. Intell. Intell. Inform. **19**(6), 738–746 (2015)

Automatic Clustering of CT Scans of COVID-19 Patients Based on Deep Learning

Pierluigi Bemportato, Gabriella Casalino, Giovanna Castellano, and Gennaro Vessio(✉)

Department of Computer Science, University of Bari "Aldo Moro", Bari, Italy
{gabriella.casalino,giovanna.castellano, gennaro.vessio}@uniba.it

Abstract. Although several vaccination campaigns have been launched to combat the ongoing COVID-19 pandemic, the primary treatment of suspected infected people is still symptomatic. In particular, the analysis of images derived from computed tomography (CT) appears to be useful for retrospectively analyzing the novel coronavirus and the chest injuries it causes. The growing body of literature on this topic shows the predominance of supervised learning methods that are typically adopted to automatically discriminate pathological patients from normal controls. However, very little work has been done from an unsupervised perspective. In this paper, we propose a new pipeline for automatic clustering of CT scans of COVID-19 patients based on deep learning. A pre-trained convolutional neural network is used for feature extraction; then, the extracted features are used as input to a deep embedding clustering model to perform the final clustering. The method was tested on the publicly available SARS-CoV-2 CT-Scan dataset that not only provides scans of COVID patients but also of patients with other lung conditions. The results obtained indicate that the radiological features of COVID patients largely overlap with those of other lung diseases. Unsupervised approaches to COVID analysis are promising, as they reduce the need for hard-to-collect human annotations and allow for deeper analysis not tied to a binary or multiclass classification task.

Keywords: Computed tomography · COVID-19 · Coronavirus · Deep learning · Clustering

1 Introduction

First appearing in Wuhan, China in December 2019, the new coronavirus (SARS-CoV-2) has spread around the world, leading to an ongoing pandemic known globally as COVID-19. Symptoms of COVID-19 are variable but often include fever, cough, fatigue, difficulty breathing, and loss of smell and taste. Symptoms begin two to fourteen days after exposure to the virus[1]. Most people who

[1] https://www.cdc.gov/coronavirus/2019-ncov/symptoms-testing/symptoms.html.

V. Torra and Y. Narukawa (Eds.): MDAI 2021, LNAI 12898, pp. 231–242, 2021.
https://doi.org/10.1007/978-3-030-85529-1_19

contract the virus develop mild to moderate symptoms and recover without needing special treatment. But some other people develop acute respiratory distress syndrome and other severe and critical symptoms, which can lead to death. SARS-CoV-2 is mainly transmitted through droplets produced by infected people when they cough, sneeze, or exhale. People can become infected by breathing in the virus if in close proximity to a person with COVID-19, or by touching a contaminated surface and then touching their eyes, nose, or mouth. For this reason, preventive measures have been taken, which mainly include physical or social distancing, quarantine, ventilation of indoor spaces, hand washing, etc. Additionally, the use of face masks in public settings has been recommended to minimize the risk of transmission.

Several vaccines have been developed and various countries have launched mass vaccination campaigns. However, although work is underway to develop drugs that inhibit the virus, primary treatment is still symptomatic. Along with laboratory tests, chest computed tomography (CT) scans can be useful in diagnosing COVID-19 in individuals with a high clinical suspicion of infection. Recent findings, in fact, have observed imaging patterns in CT scans of patients with COVID-19 [23,30]. Typical features seen on CT initially include bilateral multilobar ground glass opacities with peripheral or posterior distribution [27,44]. These patterns can help physicians not only in the early diagnosis of the disease, but also in understanding the pathogenesis of SARS-CoV-2.

A growing amount of research is aimed at developing artificial intelligence methods for identifying whether people are infected with SARS-CoV-2 through computational analysis of their CT scans. Most of this work is based on supervised approaches where the ground truth was provided directly by the doctors, e.g. [2,29,45]. However, while this approach is really useful for automatically discriminating pathological patients from normal controls, it requires tremendous effort to annotate lesions, which is not acceptable when COVID-19 is spreading rapidly. Moreover, it does not help in the differential diagnosis of different lung diseases or in the identification of subgroups within the population, as the machine has been trained to solve only a binary discrimination task.

There is currently very little work on applying unsupervised techniques to analyze COVID-19 CT scans [32]. To fill this gap, in this work we propose a method for clustering CT scans of individuals with or without COVID-19 based on unsupervised deep learning. The method is based on using a pre-trained deep convolutional neural network (CNN), i.e. VGG16 [34], as an unsupervised feature extractor, and then using a deep embedded clustering model [40] to perform the final clustering. The choice of this deep learning pipeline was motivated by the difficulty of applying traditional clustering algorithms on the raw pixel space given by rather complex CT scans. The method has been tested on the SARS-CoV-2 CT-Scan dataset recently made available by Soares et al. [35].

The rest of this paper is structured as follows. Section 2 deals with related work. Sections 3 and 4 describe materials and methods. Section 5 presents the results obtained. Section 6 concludes the paper and outlines some future developments of this research.

2 Related Work

Chest computed tomography is an important tool in the diagnosis of lung diseases including pneumonia. The CT scan procedure has faster response time than a molecular diagnostic test performed in a standard laboratory, can provide more detailed information about the disease, and is better for quantitative measurement of lesion size and extent or severity of the pulmonary involvement, which may have prognostic implications [33]. Radiological imaging is also an important diagnostic tool for COVID-19. Most COVID-19 cases have similar features on CT images, including ground-glass opacities in the early stage and lung consolidation in the late stage. Although typical CT images can help early screening of suspected cases, the images of various viral pneumonia are similar and overlap with other infectious and inflammatory lung diseases [37]. Therefore, it is difficult for radiologists to distinguish COVID-19 from other viral pneumonia. Accurate CT-based artificial intelligence systems may have the potential to aid in the early detection of COVID-19 for planning, monitoring and treatment, and to set the benchmark for longitudinal follow-ups [17,41].

Artificial intelligence methods to support the diagnosis of COVID-19 based on CT scans have proved very promising. Current investigations concern supervised learning methods based on deep neural network models. A benchmarking study for automated classification of COVID-19 has been recently reported in [46]. Pre-trained convolutional neural networks using CT data have been extensively used for COVID-19 diagnosis [6,14,24,42]. In particular, in [6] ten CNNs are compared to distinguish COVID-19 infection from non-COVID-19 groups.

However, all of the above deep learning methods for diagnosing COVID-19 rely on supervised learning, so they require annotating lesions, particularly for disease detection in CT volumes. At the moment, annotating COVID-19 lesions costs a huge amount of effort for radiologists, which is not acceptable when COVID-19 is spreading rapidly and there are major shortages for radiologists. Furthermore, supervised approaches are less suitable for the differential diagnosis of different lung diseases or for the identification of subgroups within the population, as the algorithms learn to minimize a specific previously defined loss function. Therefore, carrying out COVID-19 detection while avoiding full supervision is of great importance. Despite this, there are very few works in the literature that apply unsupervised learning to extract useful knowledge from COVID-19 CT images [32].

So far, unsupervised learning methods have been applied to COVID-19 CT images only for efficient image segmentation to find interesting ROIs that can help improve the diagnostic process. For example, in [1] the density peak clustering algorithm using generalized extreme value distribution is applied to COVID-19 CT scans collected from different datasets, while in [43] lesions in lung CT images are detected through pixel-level anomaly modeling. One of the few works that uses unsupervised learning on COVID-19 CT scan images to perform classification is [15]. Here the authors use clustering as a means to identify patterns of pulmonary tissue sequelae in a dataset consisting of X-ray and CT images in three classes: COVID-19 cases, viral pneumonia cases, and normal lungs.

The results show that clustering methods can create clusters in PCA reduced images that distinguish the three classes, thus revealing that there is latent information within COVID-19 images and there is an underlying similarity between many COVID-19 cases. Self-Organizing Feature Maps are used in [19] to group COVID chest X-ray images. Moreover, explainable results are obtained by averaging the weights of neurons in a cluster, thus producing an average image, representative of those in the cluster. Unsupervised rare pattern mining is used in [26] to discriminate patients affected by COVID-19 from CT scans of their lungs. A severity score is evaluated from the matrix profile of the image, thus suggesting the severity of the condition in the images. The results are then used as input for supervised methods, such as deep neural networks, for classification and predictive tasks. In [11] unsupervised methods, and in particular the k-means algorithm, are applied to extract groups of significant patterns in the images. Two separate groups are obtained, corresponding to healthy and sick patients, and two sub-groups can be identified based on the severity of the lesions.

Finally, several approaches have also been proposed that use partial supervision to address the lack of labels in CT scans. These include weak labeling, multiple instance learning and self supervised learning [13,16,38].

Enhanced by these results, in this work we propose a novel approach based on deep clustering to derive meaningful groups of CT scan images useful for diagnostic purposes. In particular, our work is the first attempt to apply an unsupervised learning approach to the SARS-CoV-2 CT-Scan dataset. As far as we know, only supervised approaches have been applied to this dataset, e.g. [4,5,18,28,39], to name a few.

3 Materials

The SARS-CoV-2 CT-Scan dataset [35] collects CT scan images of 2482 patients, fairly balanced between 1252 scans of patients with COVID-19 and 1230 scans of patients without COVID-19, but who had other lung diseases. The data were collected from real patients in hospitals in Sao Paulo, Brazil, and are publicly available on Kaggle[2]. The scans are all gray-scale images with varying sizes (generally a few hundred pixels per size).

The data come from 60 SARS-CoV-2 infected patients, including 32 males and 28 females, and 60 non-SARS-CoV-2 infected patients, including 30 males and 30 females. Since there are many images that come from the same patients, a random split between training set and validation set would cause data leakage in a supervised setting. However, we did not incur in this risk, because we clustered the overall data and we used the ground truth only for evaluation purposes and not for model building. All data were approved by the Ethical Committee of the Public Hospital of the Government Employees of Sao Paulo (HSPM), Brazil.

[2] https://www.kaggle.com/plameneduardo/sarscov2-ctscan-dataset.

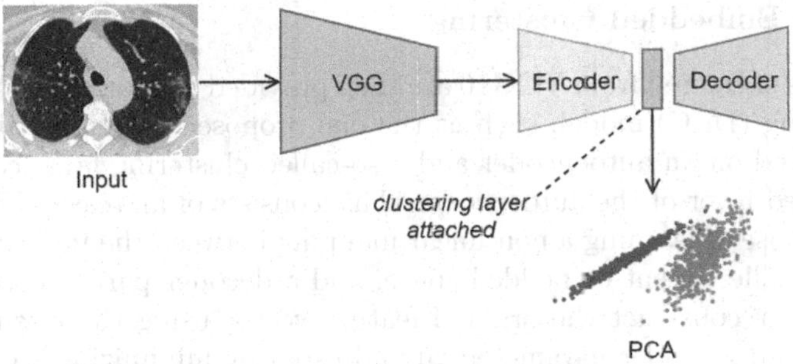

Fig. 1. Schema of the proposed deep clustering pipeline.

4 Methods

Clustering is known to be difficult due to a lack of supervision on how to guide the search for patterns in the data. Several popular algorithms, such as k-means, are effective and easy to use on structured datasets [3]. However, they usually prove ineffective when used on complex real-world image domains. On the other hand, extracting significant hand-designed features from medical images on which to apply traditional algorithms is a challenge, mainly due to the difficulty of translating domain knowledge into features and highlighting differences between images which are usually very subtle. To get around these difficulties, we resorted to a deep learning pipeline. Deep neural networks, in fact, have seen extreme popularity in recent years thanks to their ability to discover useful features for themselves directly from data [25,31]. A scheme of the proposed pipeline is shown in Fig. 1. Details are provided in the following.

4.1 Feature Extraction with VGG16

First, each image is resized to 224 × 224, which is the input typically expected by the following deep network, and normalized between 0 and 1. The deep network is a well-known VGG16 convolutional neural network [34], with weights pre-trained on ImageNet [20]. VGG16 has a classic scheme where 3 × 3 convolution and 2 × 2 max pooling are interleaved throughout the network. All hidden layers are equipped with the ReLU activation function. The network is able to build a hierarchy of visual features, starting from simple edges and shapes in the previous layers up to higher level concepts such as complex objects and shapes in the following layers. This approach is therefore suitable for obtaining high-level semantic representations from initial medical images without the need for any supervision. To obtain these features, we use the common practice of *deep transfer learning* [36] by considering the 4096 dimension feature vector from the last fully-connected layer (i.e., FC2) of the network.

4.2 Deep Embedded Clustering

The features extracted from VGG16 are then provided as input to a deep embedded clustering (DEC) model, such as the one proposed by Xie et al. [40]. This model is based on an autoencoder and a so-called clustering layer connected to the embedded layer of the autoencoder. This consists of an encoder part, which has the purpose of learning a non-linear mapping between the input feature vector and a smaller latent embedded space, and a decoder part, whose task is to learn how to reconstruct the original feature vector using the latent representation. The autoencoder parameters are adjusted by minimizing a classic mean squared reconstruction loss:

$$\mathcal{L}_r = \frac{1}{N}\sum_{i=1}^{N}(\mathbf{x}_i' - \mathbf{x}_i)^2,$$

where N is the number of scans, and $\mathbf{x}_i$ and $\mathbf{x}_i'$ are the features of the single scan and their reconstruction provided by the decoder, respectively. In addition to the input layer, which depends on the specific dataset, and which in our case is sized 4096, the encoder size have been set as in the original paper to 500–500–2000-10, with 10 as the size of the latent embedded space. The decoder mirrors this architecture. We do not directly provide input images to such framework or other similar ones based on convolutional layers, such as [12], because we want to exploit the ability of a deep network like VGG to extract useful representations and we do not have enough data to learn these representations by using only the autoencoder.

The ending part of the pipeline consists of a clustering layer attached to the embedded layer of the autoencoder. Given an initial estimate of the nonlinear mapping from scan images to embedded features and initial cluster centroids, the goal of the clustering layer is to assign the embedded features of each scan z_i to a cluster centroid μ_j using Student's t distribution:

$$q_{ij} = \frac{\left(1 + \|z_i - \mu_j\|^2\right)^{-1}}{\sum_j \left(1 + \|z_i - \mu_j\|^2\right)^{-1}},$$

where q_{ij} represents the probability of z_i of belonging to cluster j. Membership probabilities are used to calculate an auxiliary target distribution P:

$$p_{ij} = \frac{q_{ij}^2 / \sum_i q_{ij}}{\sum_j \left(q_{ij}^2 / \sum_i q_{ij}\right)},$$

where $\sum_i q_{ij}$ are soft cluster frequencies. Clustering is done by minimizing the Kullback-Leibler (KL) divergence between P and Q:

$$\mathcal{L}_c = KL(P \parallel Q) = \sum_i \sum_j p_{ij} \log\left(\frac{p_{ij}}{q_{ij}}\right).$$

In practice, the q_{ij}'s provide a measure of the similarity between each data point and the different k centroids. Higher values for q_{ij} indicate more confidence in assigning a data point to a particular cluster. The auxiliary target distribution is designed to place greater emphasis on the data points assigned with greater confidence, while normalizing the loss contribution of each centroid. Then, by minimizing the divergence between the membership probabilities and the target distribution, the network improves the initial estimate by learning from previous high-confidence predictions. Deep clustering is being used with very promising results in several complex real domains, e.g. [7,8].

The overall training of DEC is divided into two phases. In the first step, the autoencoder is trained to learn an initial set of embedded features, minimizing $\mathcal{L}_r$. After this pre-training phase, the learned features are used to initialize the cluster centroids μ_j using traditional k-means. Finally, the decoder is abandoned and embedding feature learning and clustering are jointly optimized using only $\mathcal{L}_c$. It is worth noting that, to avoid instability, P is updated by using all the data every t iterations. DEC training stops when the change in clustering assignments between consecutive updates is below a certain threshold δ.

4.3 Visualization via PCA

After training, we obtain an embedding representation of the initial CT scans based on a 10-dimensional feature space. It should be noted that this reduced feature space will differ depending on the number of clusters. In fact, since DEC simultaneously minimizes image reconstruction and clustering assignment, varying the number of clusters will lead to different feature embeddings and consequently to different arrangements of the data in the reduced space.

Starting from the learned 10-dimensional feature space, we apply a classic Principal Component Analysis (PCA) [22] to reduce dimensionality and project the feature space onto a 2-dimensional space in which clusters can be better visualized and analyzed.

5 Experiment

5.1 Setting

The experiments were performed using Google Colaboratory, which provides a cloud platform to write, execute and share Python code. As deep learning libraries, we used the well-known TensorFlow and the integrated Keras API.

As for the hyper-parameter setting, the autoencoder was first pre-trained for 50,000 iterations using stochastic gradient descent on mini-batches of size 256 using a learning rate of 0.1 with momentum of 0.9. The encoder was then fine-tuned for a maximum of 30,000 iterations using a learning rate of 0.01 and a convergence threshold δ of 0.1. To initialize the cluster centroids, k-means was run with 20 random restarts and choosing the best solution.

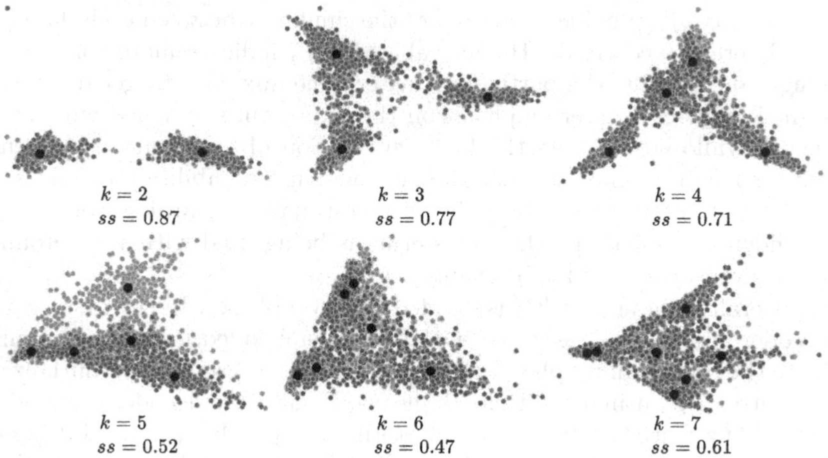

Fig. 2. PCA visualization of the clusters found with the method, as the value of k varies. Patients with COVID-19 are shown in orange; in blue those without COVID-19. The black dots are the centroids returned by the clustering method.

For the clustering evaluation, we used the well-known silhouette score [9]. The silhouette score is between -1 and 1, which represent the worst and best possible value, respectively.

Finally, to perform a qualitative assessment on the clusters found with the method we used, as mentioned earlier, PCA to visualize the embedded features in two dimensions, as well as the already known ground truth to assign labels to the data points in the visualization.

5.2 Results

Figure 2 simultaneously shows the quantitative results obtained by clustering the CT scan dataset with the proposed method and the PCA visualization. Since clustering is an unsupervised learning problem, the number of "optimal" groups is not know in advance. Moreover, the internal geometric structure of the data may not coincide with the a-priori known classes. A number of clusters ranging from 2 to 7 was then considered. As evidenced by the silhouette score, well-separated clusters are found when the granularity of clustering is coarse, with performance decreasing as the number of clusters increases.

A more precise assessment can be made integrating these results with the qualitative visualization of how data points appear separated based on their ground truth and the clustering results. As suggested by the internal clustering measures, the best separation and cohesion between clusters is obtained with $k = 2$. Indeed, we can clearly observe two geometrically distinct groups, with the centroids positioned in the middle of the clouds; however, the a-priori known labels are completely overlapping. As mentioned above, the distribution of data changes as the number of clusters varies. However, as the number of cluster increases, as suggested by the values of the internal measure, cluster separation and cohesion decrease, and the higher the number of clusters, the lower the meaningfulness of the clusters. In fact, with $k = 6$ and $k = 7$ overlapping clusters are returned, as suggested by the centroids close to each other. On the contrary, an interesting result is given by $k = 5$, where we have a group of COVID patients clearly separated from the other four clusters, which form a different group at the bottom of the two-dimensional feature space.

It is worth remembering that non-COVID patients are not healthy people, but patients hospitalized with other lung diseases. Therefore, the results shown, which were obtained without using any kind of supervision during the learning process, indicate that there are several homogeneous characteristics between chest lesions due to COVID and pre-existing and already known pathologies. Furthermore, the clusters found may reflect radiological differences between different pathologies with non-overlapping features. The disease evolves and changes in imaging characteristics have been revealed from follow-up CT scans of patients with COVID-19 [10]. In addition, some asymptomatic patients have shown changes in CT, while, conversely, no changes have been detected in CT scans of patients with COVID-19 in [21]. Therefore, the correlation between CT scans and presence/absence of the disease is not yet fully understood.

Comparative studies between the proposed method and traditional clustering methods, which yielded very poor results in our experiments, are not reported in this manuscript for space reasons.

6 Conclusion

In this paper, we have preliminary addressed the problem of automatic computation of CT scans of COVID-19 patients with deep learning methods from an unsupervised perspective instead of using supervised approaches, which are quite common in the literature. Decoupling from the need for annotation by domain experts is useful because, in the first place, annotating lesions is a laborious and error-prone activity. Moreover, it forces models to focus on specific learning tasks, which do not help the differential diagnosis of different lung diseases that share overlapping characteristics and, ultimately, elucidate novel pathogenesis mechanisms of the novel coronavirus in relation to pre-existing viral pneumonia. To this end, we proposed a deep learning pipeline for chest CT processing which, although based on well-known previous methods, integrates them into a novel flow for better clustering of complex real-world medical images, on which the application of traditional clustering algorithms would prove ineffective.

The results obtained were rather unexpected as they do not reflect the already known ground truth aimed at distinguishing between COVID and non-COVID patients, but suggest that several homogeneous radiological features are shared between COVID-19 and other lung diseases. The dataset used, in fact, does not provide normal control images but scans of patients hospitalized with other lung diseases. Unfortunately, these data do not come with additional information, so no further explanation of the results obtained without the help of medical experts is possible. Hence, the most crucial future work to do is to better interpret these preliminary results with the help of doctors.

Other future developments of this research appear to be promising. First of all, we want to collect and visually analyze the cluster medoids, i.e. data representative of each cluster but limited to being actual data, to be used as cluster prototypes for a better interpretation and explanation of the proposed model. This will help medical experts support our work. Second, we would like to extend the proposed method to other datasets, hopefully equipped with additional metadata to further investigate cluster characteristics. Finally, we would like to exploit semi-supervised strategies where knowledge of a small amount of data can drive towards better clustering or interpretability of clusters.

References

1. Abd Elaziz, M., AA Al-Qaness, M., Abo Zaid, E.O., Lu, S., Ali Ibrahim, R.A., Ewees, A.: Automatic clustering method to segment COVID-19 CT images. PloS One **16**(1), e0244416 (2021)
2. Afshar, P., et al.: COVID-CT-MD: COVID-19 computed tomography (CT) scan dataset applicable in machine learning and deep learning. arXiv preprint arXiv:2009.14623 (2020)
3. Ahmed, M., Seraj, R., Islam, S.M.S.: The k-means algorithm: a comprehensive survey and performance evaluation. Electronics **9**(8), 1295 (2020)
4. Angelov, P., Soares, E.: Explainable-by-design approach for COVID-19 classification via CT-scan. medRxiv (2020). https://doi.org/10.1101/2020.04.24.20078584
5. Angelov, P., Soares, E.: Towards explainable deep neural networks (xDNN). Neural Networks **130**, 185–194 (2020)
6. Ardakani, A.A., Kanafi, A.R., Acharya, U.R., Khadem, N., Mohammadi, A.: Application of deep learning technique to manage COVID-19 in routine clinical practice using CT images: results of 10 convolutional neural networks. Comput. Biol. Med. **121**, 103795 (2020)
7. Bhowmik, D., Gao, S., Young, M.T., Ramanathan, A.: Deep clustering of protein folding simulations. BMC Bioinform. **19**(18), 47–58 (2018)
8. Castellano, G., Vessio, G.: Deep convolutional embedding for digitized painting clustering. In: 2020 25th International Conference on Pattern Recognition (ICPR), pp. 2708–2715. IEEE (2021)
9. Dinh, D.-T., Fujinami, T., Huynh, V.-N.: Estimating the optimal number of clusters in categorical data clustering by Silhouette coefficient. In: Chen, J., Huynh, V.N., Nguyen, G.-N., Tang, X. (eds.) KSS 2019. CCIS, vol. 1103, pp. 1–17. Springer, Singapore (2019). https://doi.org/10.1007/978-981-15-1209-4_1
10. Dong, D., et al.: The role of imaging in the detection and management of COVID-19: a review. IEEE Rev. Biomed. Eng. **14**, 16 (2020)
11. Gozes, O., Frid-Adar, M., Sagie, N., Zhang, H., Ji, W., Greenspan, H.: Coronavirus detection and analysis on chest CT with deep learning. arXiv preprint arXiv:2004.02640 (2020)
12. Guo, X., Liu, X., Zhu, E., Yin, J.: Deep clustering with convolutional autoencoders. In: Liu, D., Xie, S., Li, Y., Zhao, D., El-Alfy, E.S. (eds.) ICONIP 2017. LNCS, vol. 10635, pp. 373–382. Springer, Cham (2017). https://doi.org/10.1007/978-3-319-70096-0_39

13. He, X., et al.: Sample-efficient deep learning for COVID-19 diagnosis based on CT scans. medRxiv (2020)
14. Hemdan, E.E.D., Shouman, M.A., Karar, M.E.: COVIDX-Net: a framework of deep learning classifiers to diagnose COVID-19 in X-ray images. arXiv preprint arXiv:2003.11055 (2020)
15. Householder, J., Householder, A., Gomez-Reed, J.P., Park, F., Zhang, S.: Clustering COVID-19 lung scans. arXiv preprint arXiv:2009.09899 (2020)
16. Hu, S., Gao, Y., Niu, Z., Jiang, Y., Li, L., Xiao, X., Wang, M., Fang, E.F., Menpes-Smith, W., Xia, J., et al.: Weakly supervised deep learning for COVID-19 infection detection and classification from CT images. IEEE Access **8**, 118869–118883 (2020)
17. Huang, P., et al.: Use of chest CT in combination with negative RT-PCR essay for the 2019 novel coronavirus but high clinical suspicion. Radiology **295**(1), 22–23 (2020)
18. Kechagias-Stamatis, O., Aouf, N., Koukos, J.: Deep learning fusion for COVID-19 diagnosis. medRxiv (2020)
19. King, B., Barve, S., Ford, A., Jha, R.: Unsupervised clustering of COVID-19 chest X-ray images with a self-organizing feature map. In: 2020 IEEE 63rd International Midwest Symposium on Circuits and Systems (MWSCAS), pp. 395–398 (2020). https://doi.org/10.1109/MWSCAS48704.2020.9184493
20. Krizhevsky, A., Sutskever, I., Hinton, G.E.: ImageNet classification with deep convolutional neural networks. Commun. ACM **60**(6), 84–90 (2017)
21. Lee, E.Y., Ng, M.Y., Khong, P.L.: COVID-19 pneumonia: what has CT taught us? The Lancet Infect. Diseases **20**(4), 384–385 (2020)
22. Lever, J., Krzywinski, M., Altman, N.: Points of significance: Principal component analysis (2017)
23. Li, B., et al.: Diagnostic value and key features of computed tomography in Coronavirus Disease 2019. Emerg. Microbes Infect. **9**(1), 787–793 (2020)
24. Li, L., Qin, L., Xu, Z., et al.: Using artificial intelligence to detect COVID-19 and community-acquired pneumonia based on pulmonary CT: evaluation of the diagnostic accuracy. Radiology **296**(2), E65–E71 (2020). https://doi.org/10.1148/radiol.2020200905
25. Litjens, G., et al.: A survey on deep learning in medical image analysis. Med. Image Anal. **42**, 60–88 (2017)
26. Liu, Q., Leung, C.K., Hu, P.: A two-dimensional sparse matrix profile DenseNet for COVID-19 diagnosis using chest CT images. IEEE Access **8**, 213718–213728 (2020). https://doi.org/10.1109/ACCESS.2020.3040245
27. Parekh, M., Donuru, A., Balasubramanya, R., Kapur, S.: Review of the chest CT differential diagnosis of ground-glass opacities in the COVID era. Radiology **297**(3), E289–E302 (2020)
28. Pathak, Y., Shukla, P.K., Arya, KV.: Deep bidirectional classification model for COVID-19 disease infected patients. IEEE/ACM Trans. Comput. Biol. Bioinform. (2020, Epub ahead of print). https://doi.org/10.1109/TCBB.2020.3009859. PMID: 32750891
29. Pham, T.D.: A comprehensive study on classification of COVID-19 on computed tomography with pretrained convolutional neural networks. Sci. Rep. **10**(1), 1–8 (2020)
30. Pontone, G., et al.: Role of computed tomography in COVID-19. J. Cardiovasc Comput Tomogr. **15**(1), 27–36 (2020). https://doi.org/10.1016/j.jcct.2020.08.013. Epub 4 2020 September, PMID: 32952101, PMCID: PMC7473149
31. Pouyanfar, S., et al.: A survey on deep learning: algorithms, techniques, and applications. ACM Comput. Surv. (CSUR) **51**(5), 1–36 (2018)

32. Shi, F., et al.: Review of artificial intelligence techniques in imaging data acquisition, segmentation and diagnosis for COVID-19. IEEE Rev. Biomed. Eng. **14**, 4–15 (2020)
33. Shi, H., et al.: Radiological findings from 81 patients with COVID-19 pneumonia in Wuhan. A descriptive study. The Lancet infectious diseases, China (2020)
34. Simonyan, K., Zisserman, A.: Very deep convolutional networks for large-scale image recognition. arXiv preprint arXiv:1409.1556 (2014)
35. Soares, E., Angelov, P., Biaso, S., Froes, M.H., Abe, D.K.: SARS-CoV-2 CT-scan dataset: a large dataset of real patients CT scans for SARS-CoV-2 identification. medRxiv (2020)
36. Tan, C., Sun, F., Kong, T., Zhang, W., Yang, C., Liu, C.: A survey on deep transfer learning. In: Kůrková, V., Manolopoulos, Y., Hammer, B., Iliadis, L., Maglogiannis, I. (eds.) ICANN 2018. LNCS, vol. 11141, pp. 270–279. Springer, Cham (2018). https://doi.org/10.1007/978-3-030-01424-7_27
37. Wang, S., et al.: A deep learning algorithm using CT images to screen for corona virus disease (COVID-19). Eur. Radiol. 1–9 (2021)
38. Wang, X., et al.: A weakly-supervised framework for COVID-19 classification and lesion localization from chest CT. IEEE Trans. Med. Imag. **39**(8), 2615–2625 (2020)
39. Wang, Z., Liu, Q., Dou, Q.: Contrastive cross-site learning with redesigned net for COVID-19 CT classification. IEEE J. Biomed. Health Inform. **24**(10), 2806–2813 (2020)
40. Xie, J., Girshick, R., Farhadi, A.: Unsupervised deep embedding for clustering analysis. In: International Conference on Machine Learning, pp. 478–487 (2016)
41. Xie, X., Zhong, Z., Zhao, W., Zheng, C., Wang, F., Liu, J.: Chest CT for typical 2019-nCoV pneumonia: relationship to negative RT-PCR testing. Radiology 200343 (2020)
42. Xu, X., et al.: A deep learning system to screen novel coronavirus disease 2019 pneumonia. Engineering **6**(10), 1122–1129 (2020)
43. Yao, Q., Xiao, L., Liu, P., Zhou, S.K.: Label-free segmentation of COVID-19 lesions in lung CT. arXiv preprint arXiv:2009.06456 (2020)
44. Yoon, S.H., et al.: Chest radiographic and CT findings of the 2019 novel coronavirus disease (COVID-19): analysis of nine patients treated in Korea. Korean J. Radiol. **21**(4), 494 (2020)
45. Zhang, K., et al.: Clinically applicable AI system for accurate diagnosis, quantitative measurements, and prognosis of COVID-19 pneumonia using computed tomography. Cell **181**, 1423 (2020)
46. Zhao, J., Zhang, Y., He, X., Xie, P.: COVID-CT-Dataset: a CT scan dataset about COVID-19. arXiv preprint arXiv:2003.13865 (2020)

Network Clustering with Controlled Node Size

Yukihiro Hamasuna[1,2](✉), Shusuke Nakano[3], and Yasunori Endo[4]

[1] Department of Informatics, School of Science and Engineering, Kindai University, 3-4-1 Kowakae, Higashiosaka, Osaka 577-8502, Japan

[2] Cyber Informatics Research Institute, Kindai University, 3-4-1 Kowakae, Higashiosaka, Osaka 577-8502, Japan

yhama@info.kindai.ac.jp

[3] Graduate School of Science and Engineering, Kindai University, 3-4-1 Kowakae, Higashiosaka, Osaka 577-8502, Japan

[4] Faculty of Engineering, Information and Systems, University of Tsukuba, 1-1-1, Tennodai, Tsukuba, Ibaraki 305-8573, Japan

endo@risk.tsukuba.ac.jp

Abstract. Community detection is an important problem in network clustering. This paper proposes a new network clustering method based on the control of cluster size. The word "cluster size" refers to the number of objects in a cluster. The optimization problem of the proposed method considers a constraint on the number of objects classified into a cluster. The proposed method accurately detects the community structure from the network data by adjusting the lower and upper limits of the cluster size as the parameters. Numerical experiments were conducted using two artificial and six benchmark datasets to verify the effectiveness of the proposed method. In the numerical experiments, the proposed method was compared with the k-medoids clustering, the Louvain method, and spectral clustering. The results show that the proposed method yields better results in terms of both clustering performance and community detection than the conventional methods.

Keywords: Network clustering · k-medoids · Size control · Modularity · Diffusion kernel

1 Introduction

Owing to the increasing demand for analyzing big data, data mining, which summarizes the data and extracts useful information, has attracted much research attention [1,2]. Clustering is a data analysis methods that divides a set of objects into groups called clusters. In k-means clustering, which is the most conventional clustering method, objects and cluster representatives are assumed to be vectors in Euclidean space. k-medoids clustering, which is another clustering method [3], selects cluster representatives from objects in each cluster. Fuzzy c-means clustering is one of the most active research topics, and the robustness of this

V. Torra and Y. Narukawa (Eds.): MDAI 2021, LNAI 12898, pp. 243–256, 2021.
https://doi.org/10.1007/978-3-030-85529-1_20

method and its theoretical relationship to a Gaussian mixture model have been discussed [1]. Furthermore, the kernel method is applied to conventional clustering methods to handle complex datasets and extract important properties [1,4]. Network clustering, which is referred to as community detection, is an important research topic for many real-world problems in areas such as social networking, e-commerce, and bioinformatics [5]. A community is a group of nodes that are strongly connected to each other than to the nodes in other groups. A community is considered as a cluster in the context of clustering [1].

Community detection is a task of dividing a network into edges that are dense within a community and edges that are sparse between different communities [5]. Community detection in network analysis is regarded as cluster partition in the context of clustering. The Louvain method [6] and spectral clustering [7] are typical methods for community detection from network data. It is often difficult to obtain the appropriate cluster partition because of some network-structure features such as the number and weight of the edges. In addition to these methods, network data can be divided using k-medoids clustering [8] by weighting using a Euclidean distance or diffusion kernel [9] based on the adjacency matrix. As shown in the present paper, many clustering methods for network data achieve cluster partition based on an algorithm for optimizing the formulation of community structures.

Size-control clustering is a technique that was introduced into the optimization problem as a constraint for the number of objects in a cluster [8,10,11]. The k-means clustering based methods have the advantageous in that it can be applied to nonlinearization using the kernel method [1]. Additionally, by providing an upper limit and a lower limit to the constraints on the cluster size, it has an advantage in that the number of objects included in a cluster is adjustable, which is difficult to achieve with a purely objective function. Size-control clustering is an important method to obtain an appropriate cluster structure from large-scale complex data.

The concept of size-control clustering for network data has been proposed in a previous study [8,10]. The present paper proposes a network clustering method using size control named controlled-sized clustering based on optimization for network data (COCBON). To overcome the limitations of the conventional methods, the proposed method handles weighted and unweighted network data by using a diffusion kernel to calculate the dissimilarity between nodes [9]. The proposed method also obtains cluster partitions different from those obtained using conventional k-medoids clustering by adjusting the parameters related to the size control. The effectiveness of the proposed method is analyzed using multidimensional scaling (MDS) [12] through numerical experiments, in which the clustering performance is evaluated with artificial and benchmark datasets using the adjusted rand index [13] and modularity. To demonstrate the advantages of the proposed method, MDS is used to visualize the difference in cluster partitions between the proposed method and conventional methods.

2 Preliminaries

A set of objects to be clustered is given, and it is denoted by $\boldsymbol{X} = \{\boldsymbol{x}_1, \ldots, \boldsymbol{x}_n\}$, where $\boldsymbol{x}_k$ $(k = 1, \ldots, n)$ is an object. For an unweighted graph, the elements a_{ij} of the adjacent matrix $\boldsymbol{A}$ are equal to 1 if nodes i and j are connected or equal to 0 if no edges exist between nodes i and j. For a weighted graph, the dissimilarity between two nodes i and j is denoted by $d_{ij} \geq 0$, and the dissimilarity matrix is denoted by $\boldsymbol{D} = (d_{ij})_{i=1,\ldots,n,\ j=1,\ldots,n}$. We assume D to be symmetric; thus, $d_{ij} = d_{ji}$, and $d_{ii} = 0$. Network data $(\boldsymbol{X}, \boldsymbol{D})$ are assumed to be given such that objects $\boldsymbol{X}$ and the weight d_{ij} of edges between nodes i and j are given. A cluster is denoted by $\boldsymbol{G}_i$, and a collection of clusters is denoted by $\mathcal{G} = \{\boldsymbol{G}_1, \ldots, \boldsymbol{G}_c\}$. The membership degree of $\boldsymbol{x}_k$ belonging to $\boldsymbol{G}_i$ and the partition matrix are denoted as u_{ki} and $\boldsymbol{U} = (u_{ki})_{1 \leq k \leq n,\ 1 \leq i \leq c}$, respectively.

2.1 *k*-medoids Clustering

k-medoids clustering is a variant of *k*-means clustering [2,3]. Whereas the cluster center, which is calculated as the center of gravity of the objects in the cluster, is used as a cluster representative in *k*-means clustering [2], an object in each cluster is chosen as a cluster representative in *k*-medoids clustering [3]. An objective function of *k*-medoids clustering is expressed as follows:

$$\text{min.}\ J_{\text{KMdd}}(\boldsymbol{U}, \boldsymbol{W}) = \sum_{i=1}^{c} \sum_{k=1}^{n} \sum_{l=1}^{n} u_{ki} w_{li} r_{kl}. \tag{1}$$

where r_{kl} is a measure of the relationship between objects and $\boldsymbol{W} = (w_{li})_{1 \leq l \leq n,\ 1 \leq i \leq c}$ is a variable called the prototype weight. In many cases, r_{kl} is considered as a dissimilarity between objects. An algorithm of *k*-medoids clustering is based on the alternating optimization of u_{ki} and w_{li} under the following constraints:

$$\mathcal{U}_h = \left\{ (u_{ki}) : u_{ki} \in \{0, 1\},\ \sum_{i=1}^{c} u_{ki} = 1,\ {}^{\forall}k \right\}, \tag{2}$$

$$\mathcal{W}_h = \left\{ (w_{li}) : w_{li} \in \{0, 1\},\ \sum_{l=1}^{n} w_{li} = 1,\ {}^{\forall}i \right\}. \tag{3}$$

The lth object that takes $w_{li} = 1$ is the representative of a cluster. An important feature of *k*-medoids clustering is that it handles relational data expressed as a table of distances between objects, such as network data.

Algorithm 1. KMdd

KMdd1 Set cluster number c and initial medoids $w_{li} \in \boldsymbol{W}$ by choosing objects at random.
KMdd2 Calculate $u_{ki} \in \boldsymbol{U}$ using (4).
KMdd3 Calculate $w_{li} \in \boldsymbol{W}$ using (5).
KMdd4 If the convergence criterion is satisfied, stop. Otherwise, return to **KMdd2**.

The optimal solutions for u_{ki} and w_{li} are as follows:

$$u_{ki} = \begin{cases} 1 & \left(i = \arg\min_{s} \sum_{l=1}^{n} w_{ls} r_{kl} \right) \\ 0 & (\text{ otherwise }) \end{cases}, \tag{4}$$

$$w_{li} = \begin{cases} 1 & \left(l = \arg\min_{t} \sum_{k=1}^{n} u_{ki} r_{kt} \right) \\ 0 & (\text{ otherwise }) \end{cases}. \tag{5}$$

By considering the optimization problem of J_{kd}, the optimal solution of w_{li} is expressed in (5). The KMdd algorithm is summarized as follows:

The number of repetitions, convergence of each variable, or convergence of an objective function is used as the convergence criterion in **KMdd4**.

2.2 Controlled-Sized Clustering Based on Optimization

Controlled-sized clustering based on optimization (COCBO) optimizes the objective function of k-means clustering by solving a linear programming (LP) problem under constraints on the cluster size [10]. COCBO is an extension of the approach to equalize the cluster size. The parameter K related to the cluster size satisfies certain conditions for the number of objects and clusters. More details can be found in the present paper [10].

In COCBO, the parameter K controls the cluster size in the range between α and β, and it satisfies $\alpha \leq K$ and $K + 1 \leq \beta$. Here, α and β control the lower and upper bound of cluster size, respectively. The optimization problem of COCBO is expressed as follows:

$$\text{min.} \quad J(\boldsymbol{U}, \boldsymbol{V}) = \sum_{k=1}^{n} \sum_{i=1}^{c} u_{ki} d_{ki} \tag{6}$$

$$\text{s. t.} \quad \mathcal{U}_h = \left\{ (u_{ki}) : u_{ki} \in \{0, 1\}, \ \sum_{i=1}^{c} u_{ki} = 1, \ {}^{\forall}k \right\},$$

$$\alpha \leq \sum_{k=1}^{n} u_{ki} \leq \beta, {}^{\forall}i, \tag{7}$$

$$\alpha \leq K, \tag{8}$$

$$\beta \geq K + 1. \tag{9}$$

Equation (6) is an objective function of k-means clustering, and d_{ki} is the dissimilarity between an object and a cluster representative. In most cases of k-means clustering, d_{ki} is the squared Euclidean distance. Equation (7) is a constraint on the cluster size, which is the number of objects in a cluster. Equations (8) and (9) are constraints on the lower and upper bound of the cluster size, respectively.

2.3 Kernel Method

The kernel method is an important technique to handle data with a complex structure. It is applied to both supervised learning and unsupervised learning to handle complex datasets.

First, we define symbols to introduce kernel functions. $\phi : \Re^p \to \Re^s (p \ll s)$ denotes mapping from an input space $\Re^p$ to a high-dimensional feature space $\Re^s$. An object in the feature space is denoted by $\phi(\boldsymbol{x}_k) \in \Re^s$. The kernel function $K : \Re^p \times \Re^p \to \Re$ satisfies the following relation:

$$K(\boldsymbol{x}, \boldsymbol{y}) = \langle \phi(\boldsymbol{x}), \phi(\boldsymbol{y}) \rangle.$$

A typical kernel function that handles vectorial data is The Gaussian kernel, which is expressed as follows:

$$K(\boldsymbol{x}_i, \boldsymbol{x}_j) = \exp\left(-\gamma \|\boldsymbol{x}_i - \boldsymbol{x}_j\|^2\right), \tag{10}$$

where $\boldsymbol{x}_i$ and $\boldsymbol{x}_j$ are input vectors and $\gamma > 0$ is a kernel parameter. The Gaussian kernel is a function that calculates the similarity between the input vectors.

A diffusion kernel is a typical kernel function that targets network data [9]. It can be used to calculate the similarity between nodes using an edge connection, that is, the adjacency matrix A. The diffusion kernel is expressed as follows:

$$K(t) = \exp(t\boldsymbol{L}) \tag{11}$$

In (11), $t > 0$ is also a kernel parameter, and L is a reversed-sign graph Laplacian.

The elements of $\boldsymbol{L}$ are expressed as follows:

$$l_{ij} = \begin{cases} 1 & (i \neq j \text{ and node } i \text{ is adjacent to node } j) \\ -\text{deg}_i & (i = j) \\ 0 & (\text{otherwise}) \end{cases} \tag{12}$$

Here, deg_i is a degree of node i and is expressed as $\text{deg}_i = \sum_{j=1}^{n} l_{ij}$.

3 Proposed Method

The proposed method, COCBON, is an extension of COCBO that can handle network data [8].Whereas COCBO is based on the k-means algorithm and handles vectorial data, COCBON is based on the k-medoids algorithm and handles network data. The optimization problem of COCBON is expressed as follows:

$$\begin{aligned}
\text{min.} \quad & J(\boldsymbol{U}, \boldsymbol{W}) = \sum_{i=1}^{c}\sum_{k=1}^{n}\sum_{l=1}^{n} u_{ki} w_{li} r_{kl}, \\
\text{s. t.} \quad & \mathcal{U}_h = \left\{ (u_{ki}) : u_{ki} \in \{0,1\}, \ \sum_{i=1}^{c} u_{ki} = 1, \ ^{\forall}k \right\}, \\
& \mathcal{W}_h = \left\{ (w_{li}) : w_{li} \in \{0,1\}, \ \sum_{l=1}^{n} w_{li} = 1, \ ^{\forall}i \right\}, \\
& \alpha \leq \sum_{k=1}^{n} u_{ki} \leq \beta, ^{\forall}i, \\
& \alpha \leq K, \\
& \beta \geq K + 1.
\end{aligned}$$

The objective function and the constraints on u_{ki} and w_{li} of COCBON are the same as those of the k-medoids algorithm, which are expressed in (1), (2), and (3). The constraints on the cluster size and the parameter for the lower and upper bound are the same as those of COCBO, which are expressed in (7), (8), and (9).

If $\alpha = 1$ and $\beta \to n$, COCBON is equivalent to k-medoids clustering. In many clustering methods, the cluster partition is obtained by optimizing evaluation indexes such as the objective function without including a specific number of objects in the cluster. It may be difficult to obtain a good cluster partition using conventional methods when the distribution of data is biased. COCBON is expected to yield a better cluster partition for biased data by introducing the constraint on the cluster size.

The COCBON algorithm is summarized in Algorithm 2:

The number of repetitions, convergence of each variable, or convergence of an objective function is used as the convergence criterion in **COCBON4** as well as **KMdd**. The solver for solve LP problems in **COCBON2** is PuLP 1.6.9. The PuLP is an LP modeler written in Python and is open source (https://pypi.org/project/PuLP/).

Algorithm 2. COCBON

COCBON1 Set cluster size K, cluster number c, parameters α and β, and initial medoids $w_{li} \in \boldsymbol{W}$ by choosing objects at random.

COCBON2 Calculate $u_{ki} \in \boldsymbol{U}$ by solving the LP problem.

COCBON3 Calculate $w_{li} \in \boldsymbol{W}$ using (5).

COCBON4 If the convergence criterion is satisfied, stop. Otherwise, return to **COCBON2**.

---

4 Numerical Experiments

We conducted numerical experiments with two artificial datasets and six benchmark datasets to verify the effectiveness of COCBON. First, we describe the calculation conditions of the numerical experiments. Second, we compare the results among COCBON and three conventional methods: k-medoids clustering [3], the Louvain method [6], and spectral clustering [7]. Third, we summarize the results and the features of the proposed method.

4.1 Experimental Setup

The above-mentioned methods are compared using two artificial datasets and six benchmark datasets in terms of the evaluated value of the adjusted rand index (ARI) [13]. ARI is a measure of similarity between two cluster partitions. The value of ARI is 1 when the two cluster partitions completely match. Of the six benchmark datasets, four are network data created from vectorial data. For k-medoids and COCBON, the maximum number of iterations of the algorithm was set to 100.

Table 1 lists the numbers of nodes, edges, and clusters of the two artificial datasets, which are visualized in Figs. 1 and 2. In these figures, each node is color-coded by the cluster it belongs to. Table 2 lists the numbers of objects, attributes, and clusters of the four vectorial benchmark datasets, while Table 3 lists the numbers of nodes, edges, and clusters of the two network benchmark datasets.

For the experiments, the two artificial and two benchmark network datasets are converted to a weighted graph by using a diffusion kernel [9]. First, a similarity $K(\boldsymbol{x}_i, \boldsymbol{x}_j)$ is obtained between the nodes by using the diffusion kernel given in (11). Next, the similarities are converted to dissimilarities by using the following equation:

$$K_{\text{dissim}}(\boldsymbol{x}_i, \boldsymbol{x}_j) = 1 - \frac{K(\boldsymbol{x}_i, \boldsymbol{x}_j)}{\max_{i,\ j} K} \tag{13}$$

Because the dissimilarity is based on the diffusion kernel calculated using (13), it is possible to obtain the cluster partition from the network data in the k-medoids framework.

The four vectorial benchmark datasets are also converted to a weighted graph by using a Gaussian kernel [4]. First, a similarity between objects $\boldsymbol{x}_i$ and $\boldsymbol{x}_j$, denoted as $K(\boldsymbol{x}_i, \boldsymbol{x}_j)$, is obtained using the Gaussian kernel given in (10). Because the value of the Gaussian kernel ranges from 0 to 1, the dissimilarity between all pairs of nodes can be calculated through the following transformation:

$$K_{\text{dissim}}(\boldsymbol{x}_i, \boldsymbol{x}_j) = 1 - K(\boldsymbol{x}_i, \boldsymbol{x}_j) \tag{14}$$

The vectorial data are converted into network data by calculating the dissimilarity based on the Gaussian kernel. Through the above procedure (14), as in the case of the diffusion kernel, it is possible to obtain the cluster partition from the network data in the k-medoids framework.

In the proposed method, there are several parameters for adjusting the cluster size and kernel function. The diffusion kernel is applied to the artificial and benchmark datasets for weighting, as summarized in Tables 1 and 3, while the Gaussian kernel is applied to vectorial data, as summarized in Table 2. Because parameter selection is required for each dataset, numerical experiments were conducted under the conditions listed in Table 4.

The cluster partitions of all datasets used in the numerical experiments are known. ARI, as mentioned above, is used as an index to evaluate the accuracy of the clustering result [13]. Moreover, the modularity used in the Louvain method is used as an evaluation index for network partition [6]. The modularity is expressed as follows:

$$Q = \frac{1}{2M} \sum_{i=1}^{c} \sum_{k=1}^{n} \sum_{l=1}^{n} \left[a_{kl} - \frac{\mathrm{deg_k} \mathrm{deg_l}}{2M} \right] u_{ki} u_{li} \tag{15}$$

where M is the sum of all edges, a_{kl} is the weight between nodes k and l, and $\mathrm{deg_k}$ is the sum of all weights connecting node k. The modularity ranges from −1 to 1. The closer its value is to 1, the denser are the edges within a cluster and the sparser are the edges between clusters. In addition to the quantitative evaluation, MDS [12].

4.2 Experimental Results

Table 5 lists the ARI values for k-medoids clustering [3], the Louvain method [6], spectral clustering [7], and the proposed method. The ARI values in Table 5 are calculated from the cluster partition obtained using each method and the cluster labels. The value in bold in each row is the highest ARI value for each dataset. The results with COCBON are obtained by setting the parameters α and β according to the conditions listed in Table 4. For example, the ARI with COCBON for Artificial dataset 1 is 0.856 (5, $\{6, 20, 35\}$), which indicates that ARI is 0.856 when the lower limit of the cluster size $\alpha = 5$ and the upper limit of the cluster size $\beta = \{6, 20, 35\}$.

In six out of the eight datasets, the proposed method yields the highest ARI value. In Digits and email-Eu-core network, for which spectral clustering yields the highest ARI, there is little difference in the ARI values between the proposed method and spectral clustering. Further, the proposed method yields better results than the Louvain method in all cases. For Glass, xAPI-Edu-Data, and email-Eu-core network, which are cases in which the proposed method behaves similarly to k-medoids clustering, the proposed method yields better results than k-medoids clustering.

Table 6 lists the modularity values for the conventional and proposed methods. The modularity values in Table 6 are calculated from the cluster partition obtained using each method. The value in bold in each row is the highest modularity value for each dataset. The COCBON results are obtained in the same

Table 1. Details of the artificial datasets.

	Num. of nodes (n)	Num. of edges	Num. of clusters (c)
Artificial dataset 1	35	177	7
Artificial dataset 2	120	2200	12

Table 2. Details of the four vectorial benchmark datasets.

	Num. of objects (n)	Num. of attributes (p)	Num. of clusters (c)
Iris [14]	150	4	3
Digits [14]	1797	64	10
Glass [14]	214	9	6
xAPI-Edu-Data [15, 16]	480	4	3

manner as in Table 5. Because the Louvain method is based on the local maximization of modularity, it shows the highest modularity values for all datasets. The modularity is 0 or less in all of the results for Glass, and better results are not obtained in terms of modularity evaluation.

Figures 3, 4, 5, 6, 7, 8, 9, and 10 show the results of MDS with respect to the similarity (ARI) between cluster partitions obtained using each method for each dataset. These figures illustrate the similarity between cluster partitions obtained using each method. The symbols displayed on the right in the figures correspond to each method. The COCBON results were obtained using the parameters listed in Table 4. For example, in Fig. 3, spectral clustering and the Louvain method yield different cluster partitions to k-medoids clustering and COCBON. Furthermore, in Fig. 3, the results of k-medoids, COCBON (1, 20), and COCBON (1, 35) are below the middle of the graph, while the results of COCBON (5, 6), COCBON (5,20), and COCBON (5, 35) are towards the left. In addition, k-medoids and COCBON (1, n) yield the same cluster partition. A comparison of the COCBON results obtained with different parameters indicates that the cluster partition changes when changing the lower bound β.

4.3 Discussions

The results of the numerical experiments show that the proposed method yields better cluster partitions than conventional network clustering methods by adjusting the cluster-size parameters α and β.

Table 3. Details of the two network benchmark datasets.

	Num. of nodes (n)	Num. of edges	Num. of clusters (c)
American College football [17]	115	613	12
email-Eu-core network [18,19]	986	25552	42

Table 4. Details of the parameters for each dataset.

Data	α	β	t in (11)	γ in (10)
Artificial dataset 1	$\alpha \in \{1, 3, 5\}$	$\beta \in \{6, 20, 35\}$	$t = 0.4$	–
Artificial dataset 2	$\alpha \in \{1, 6, 10\}$	$\beta \in \{11, 66, 120\}$	$t = 0.1$	–
Iris	$\alpha \in \{1, 40, 50\}$	$\beta \in \{51, 60, 150\}$	–	$\gamma = 0.1$
Digits	$\alpha \in \{1, 143, 179\}$	$\beta \in \{180, 215, 1797\}$	–	$\gamma = 0.1 \times 10^{-3}$
Glass	$\alpha \in \{1, 28, 35\}$	$\beta \in \{36, 42, 214\}$	–	$\gamma = 0.1 \times 10^{-4}$
xAPI-Edu-Data	$\alpha \in \{1, 128, 160\}$	$\beta \in \{161, 192, 480\}$	–	$\gamma = 0.1$
American College football	$\alpha \in \{1, 5, 9\}$	$\beta \in \{10, 63, 115\}$	$t = 0.3$	–
email-Eu-core network	$\alpha \in \{1, 12, 23\}$	$\beta \in \{24, 515, 986\}$	$t = 0.05$	–

The features of the proposed method COCBON are summarized as follows:

- Table 5 shows that the proposed method obtained better classification performance than the Louvain method and spectral clustering. Further, as summarized in Table 6, the proposed method showed better performance than the Louvain method in terms of modularity. These results are considered to be affected by the cluster-size parameters α and β, in addition to the weighting by the diffusion kernel and Gaussian kernel.
- The proposed method obtained both k-medoids-like cluster partitions and different ones by adjusting the parameters α and β. Figures 3, 4, 5, 6, 7, 8, 9, and 10 show the relation between the cluster partition and parameter selection. In particular, Figs. 3, 4, 6, 8, and 10 show that the cluster partitions obtained using k-medoids clustering and the proposed method are different from the ones obtained using the Louvain method and spectral clustering. Furthermore, Figs. 5, 7, and 9 show that the cluster partition obtained using the Louvain method is different from those obtained using the other methods.

Table 5. ARI results for the conventional and the proposed methods.

	k-medoids	Louvain	Spectral clustering	COCBON(α, β)
Artificial dataset 1	0.854	0.307	0.481	**0.856** (5, $\{6, 20, 35\}$)
Artificial dataset 2	0.679	0.088	0.172	**0.716** ($\{1, 6\}$, 11)
Iris	0.715	0.528	0.716	**0.786** (50, $\{51, 60, 150\}$)
Digits	0.446	0.337	**0.641**	0.601 (179, 180)
Glass	**0.252**	0.018	0.243	**0.252** (1, 214)
xAPI-Edu-Data	**0.242**	0.231	0.185	**0.242** ($\{1, 128\}$, 161)
American College football	**0.012**	0.001	**0.012**	**0.012** (1, $\{63, 115\}$)
email-Eu-core network	0.004	0.003	**0.007**	0.005 (23, $\{505, 986\}$)

Table 6. Modularity results for the conventional and the proposed methods.

	k-medoids	Louvain	Spectral clustering	COCBON(α, β)
Artificial dataset 1	0.005	**0.011**	0.005	0.005 (1, 6)
Artificial dataset 2	0.001	**0.006**	0.003	0.001 ($\{1, 6\}$, 11)
Iris	0.192	**0.221**	0.191	0.195 (50, $\{51, 60, 150\}$)
Digits	0.010	**0.014**	0.011	0.010 (179, $\{180, 215, 1797\}$)
Glass	−0.003	**0.000**	−0.003	−0.004
xAPI-Edu-Data	0.015	**0.016**	0.013	0.015 (1, 480)
American College football	0.304	**0.308**	0.304	0.304 (1, $\{63, 115\}$)
email-Eu-core network	0.184	**0.285**	0.188	0.184 (1, $\{505, 986\}$)

- The proposed method can yield various cluster partitions with the control of the parameters α and β. Additionally, in the proposed method, all the clusters have the same cluster-size constraints. It is possible to extend the proposed method by applying cluster-wise constraints on the cluster size. The proposed method based on cluster-wise constraints may have the potential to handle imbalanced data.

The above numerical experiments and discussion suggest that the proposed method is a useful technique in network clustering. With the control of the parameters α and β, it can yield flexible cluster partitions. Additionally, the proposed method can be a useful tool for network data with different types of communities if cluster-wise constraints on the cluster size are adopted.

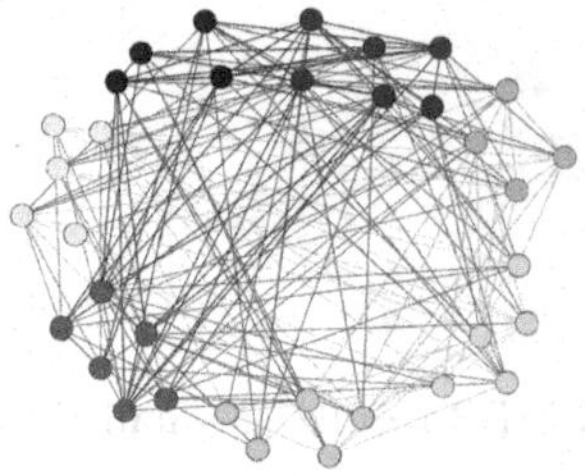

Fig. 1. Artificial dataset 1.

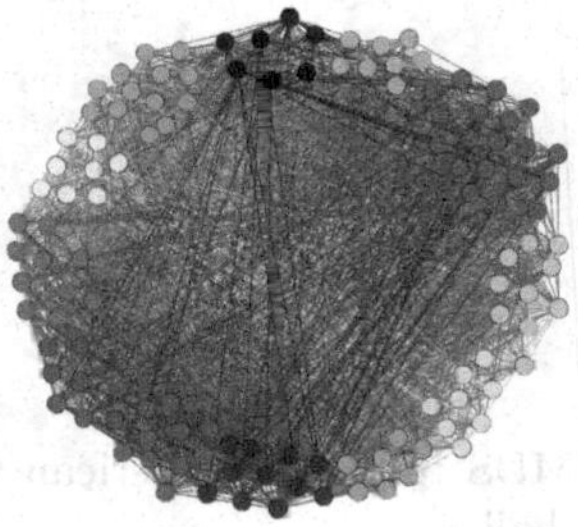

Fig. 2. Artificial dataset 2.

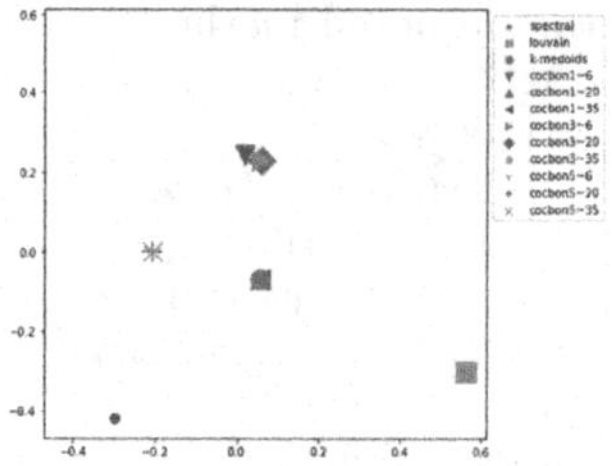

Fig. 3. MDS results for Artificial dataset 1.

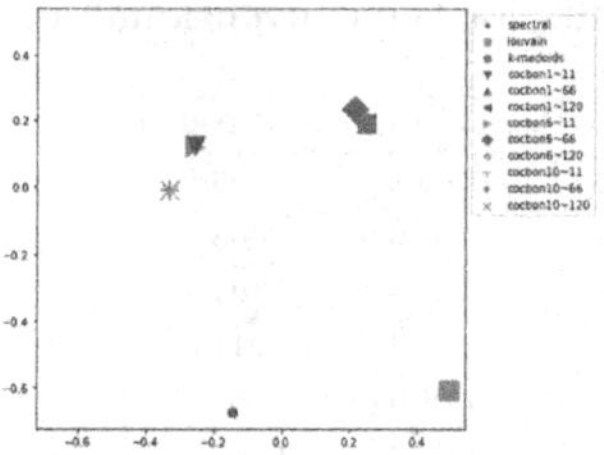

Fig. 4. MDS results for Artificial dataset 2.

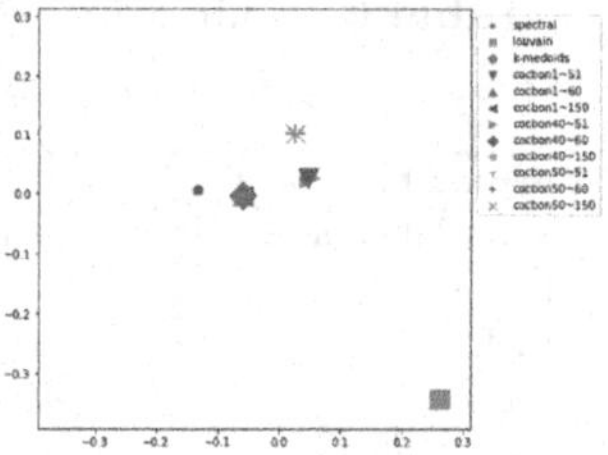

Fig. 5. MDS results for Iris.

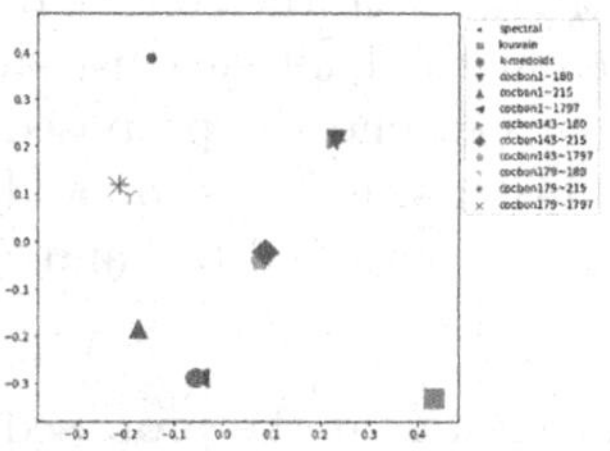

Fig. 6. MDS results for Digits.

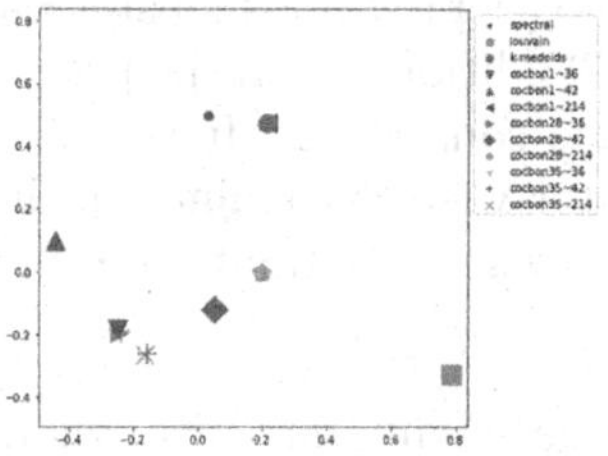

Fig. 7. MDS result for Glass.

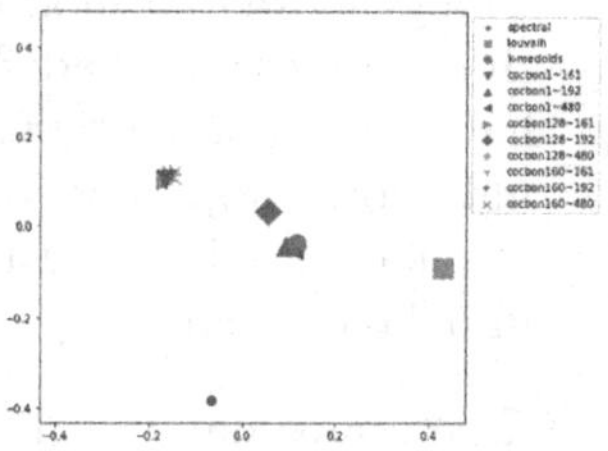

Fig. 8. MDS result for xAPI-Edu-Data.

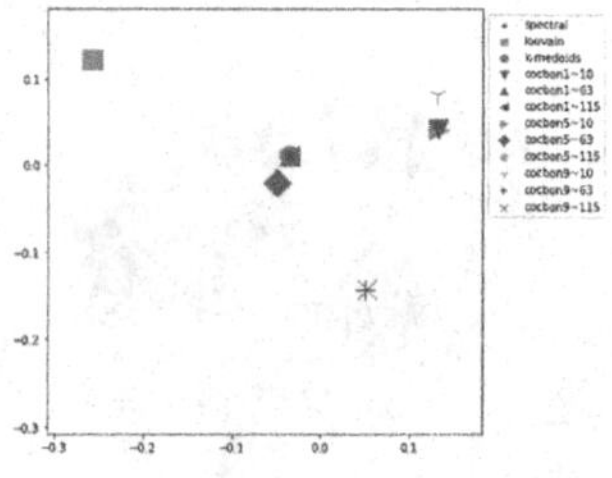

Fig. 9. MDS result for American College football.

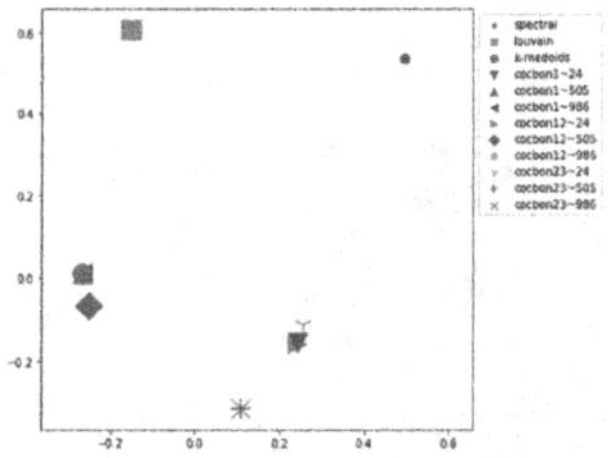

Fig. 10. MDS result for email-Eu-core network.

5 Conclusions

In this paper, a new network clustering method was proposed based on the concept of cluster-size control. The proposed method is an extension of k-medoids clustering and can yield flexible cluster partitions by adjusting the cluster-size parameters α and β. The effectiveness of the proposed method was verified through numerical experiments using artificial and benchmark datasets. The experimental results showed that the proposed method has better clustering performance than the Louvain method and spectral clustering, which is a typical network clustering method.

In future work, we will extend the proposed method by adopting cluster-wise constraints on the cluster size to handle complex network data with different types of communities. Furthermore, we will investigate a way to adjust suitable cluster-size parameters by parallelization, which is considered to be a useful approach to fuzzify the proposed method to demonstrate its robustness according to a previous study [11]. Additional numerical experiments with large-scale data are necessary to comprehensively establish the effectiveness of the proposed method for network clustering.

Acknowledgements. This work was partly supported by JSPS KAKENHI Grant Number JP19K12146.

References

1. Miyamoto, S., Ichihashi, H., Honda, K.: Algorithms for Fuzzy Clustering. Springer, Heidelberg (2008)
2. Jain, A.K.: Data clustering: 50 years beyond K-means. Patt. Recogn. Lett. **31**(8), 651–666 (2010). https://doi.org/10.1016/j.patrec.2009.09.011
3. Kaufman, L., Rousseeuw, P.J.: Finding Groups in Data: An Introduction to Cluster Analysis, Wiley, New York (1990)
4. Girolami, M.: Mercer kernel-based clustering in feature space. In: IEEE Trans. Neural Netw. **13**(3), 780–784 (2002). https://doi.org/10.1109/TNN.2002.1000150
5. Newman, M.: Networks: An Introduction, Oxford University Press, New York (2010)
6. Blondel, V.D., Guillaume, J.-L., Lambiotte, R., Lefebvre, E.: Fast unfolding of communities in large networks. J. Statist. Mech. Theory Exp. **2008**, P10008 (2008). https://doi.org/10.1088/1742-5468/2008/10/P10008
7. von Luxburg, U.: A tutorial on spectral clustering. Stat. Comput. **17**, 395–416 (2007). https://doi.org/10.1007/s11222-007-9033-z
8. Nakano, S., Hamasuna, Y., Endo, Y.: A study on controlled node sized network clustering for unweighted network data. In: Proceedings of SCIS & ISIS 2018, pp. 826–831 (2018). https://doi.org/10.1109/SCIS-ISIS.2018.00136
9. Kondor, R.I., Lafferty, J.D.: Diffusion kernels on graphs and other discrete input spaces. In: Proceedings of the Nineteenth International Conference on Machine Learning (ICML2002), pp. 315–322 (2002)
10. Endo, Y., Ishida, S., Kinoshita, N., Hamasuna, Y.: On various types of controlled-sized clustering based on optimization. In: IEEE International Conference on Fuzzy Systems (FUZZIEEE2017), pp. 1–6 (2017). https://doi.org/10.1109/FUZZ-IEEE.2017.8015556
11. Kitajima, K., Endo, Y., Hamasuna, Y.: Fuzzified even-sized clustering based on optimization. J. Adv. Comput. Intell. Intell. Inform. **22**(4), 537–543 (2018). https://doi.org/10.20965/jaciii.2018.p0537
12. Hastie, T., Tibshirani, R., Friedman, J.: The Elements of Statistical Learning: Data Mining, Inference, and Prediction, 2nd edn, Springer Series in Statistics, Springer, New York (2009)
13. Hubert, L., Arabie, P.: Comparing partitions. J. Classifi. **2**(1), 193–218 (1985)
14. Dheeru, D., Casey, D.: UCI Machine Learning Repository. University of California, School of Information and Computer Science, Irvine (2017). http://archive.ics.uci.edu/ml

15. Amerieh, E.A., Hamtini, T., Aljarah, I.: Mining educational data to predict student's academic performance using ensemble methods. Int. J. Database Theor. Appl. 119–136 (2016). http://dx.doi.org/10.14257/ijdta.2016.9.8.13
16. Amerieh, E.A., Hamtini, T., Aljarah, I.: Preprocessing and analyzing educational data set using X-API for improving student's performance. In: IEEE Jordan Conference on Applied Electrical Engineering and Computing Technologies, pp. 1–5 (2015). https://doi.org/10.1109/AEECT.2015.7360581
17. Girvan, M., Newman, M.E.J.: Community structure in social and biological networks. Proc. Natl. Acad. Sci. USA **99**(12), 7821–7826 (2002)
18. Leskovec, J., Kleinberg, J., Faloutsos, C.: Graph evolution: densification and shrinking diameters. ACM Trans. Knowl. Discov. Data **1**, 2–es (2007). https://doi.org/10.1145/1217299.1217301
19. Yin, H., Benson, A.R., Leskovec, J., Gleich, D.F.: Local higher-order graph clustering. In: Proceedings of the 23rd ACM SIGKDD International Conference on Knowledge Discovery and Data Mining (2017). https://doi.org/10.1145/3097983.3098069

Data Science and Data Privacy

Fairly Private Through Group Tagging and Relation Impact

Poushali Sengupta[1(✉)] and Subhankar Mishra[2,3]

[1] University of Kalyani, Nadia, India
[2] School of Computer Sciences, NISER, Bhubaneswar 752050, India
smishra@niser.ac.in
[3] Homi Bhabha National Institute, Anushaktinagar, Mumbai 400094, India

Abstract. Privacy and Fairness both are very important nowadays. For most of the cases in the online service providing system, users have to share their personal information with the organizations. In return, the clients not only demand a high privacy guarantee to their sensitive data but also expected to be treated fairly irrespective of their age, gender, religion, race, skin color, or other sensitive protected attributes. Our work introduces a novel architecture that is balanced among the privacy-utility-fairness trade-off. The proposed mechanism applies *Group Tagging Method* and *Fairly Iterative Shuffling (FIS)* that amplifies the privacy through random shuffling and prevents linkage attack. The algorithm introduces a fair classification problem by *Relation Impact* based on *Equalized Minimal FPR-FNR* among the protected tagged group. For the count report generation, the aggregator uses TF-IDF to add noise for providing longitudinal Differential Privacy guarantee. Lastly, the mechanism boosts the utility through risk minimization function and obtain the optimal privacy-utility budget of the system. In our work, we have done a case study on gender equality in the admission system and helps to obtain a satisfying result which implies that the proposed architecture achieves the group fairness and optimal privacy-utility trade-off for both the numerical and decision making Queries.

Keywords: Group tagging · Relation impact · Privacy · Risk minimizer · Group fairness

1 Introduction

To deal with information leakage [1,2], nowadays the organizations and industries have all concentration on the "privacy" of the dataset which is the most important and necessary thing to protect the dataset from some unexpected attacks by the attackers that causes data breach. The main aim of a statistical disclosure control is to protect the "privacy" of "individuals" in a database from the sudden attack. To solve this problem, in 2006, Dwork introduces the notion of "Differential Privacy" (DP) [3–5]. Some of the important works on DP have been summarized in Table 1.

P. Sengupta—This work was done while the author was visiting NISER.

V. Torra and Y. Narukawa (Eds.): MDAI 2021, LNAI 12898, pp. 259–272, 2021.
https://doi.org/10.1007/978-3-030-85529-1_21

Table 1. Related works on Differential Privacy

Paper	Year	Brief Summary
Differential Privacy [5]	2006	Initial work on DP
RAPPOR [7]	2014	Fastest implementation of LDP
Deep Learning with Differential Privacy [8]	2016	Introduces a refined analysis of privacy costs within DP
Communication-Efficient Learning of Deep Networks from Decentralized Data [9]	2016	Introduces practical method for the federated learning of deep networks based on iterative model averaging
PROCHLO [10]	2017	Encode Shuffle and Analyse (ESA) for strong CDP
Adding DP to Iterative Training Procedures [11]	2018	Introduces an algorithm with the modular approach to minimize the changes on training algorithm, provides a variety of configuration strategies for the privacy mechanism
Amplification by Shuffling [12]	2018	$\mathcal{O}(\epsilon\sqrt{\log\frac{1}{\delta}/n}, \delta)$ centralized differential privacy (CDP) guarantee
Controlled Noise [13]	2019	Duel algorithm based on average consensus (AC)
Capacity Bounded DP [14]	2019	Location Privacy with matrix factorization
ARA [15]	2020	Aggregation of privacy reports & fastest CDP
Privacy at Scale [6]	2020	Privatization, ingestion and aggregation
ESA Revisited [34]	2020	Improved [12] with longitudinal privacy
BUDS [16]	2020	Optimize the privacy-utility trade-off by Iterative Shuffling and Risk Minimizer

Adding to the importance of trade-off between privacy-Utility guarantee to the sensitive data, the demand for fairness also belongs to the clients' top priority list. Nowadays, in most online service systems, users not only ask for high privacy but also expect to be treated fairly irrespective of religion, skin color, native language, gender, cast, and other sensitive classes. Table 2 shows some existing algorithms on fairness mechanism.

In their 2012 work [17] Dwork et al. have explicitly shown that individual fairness is a generalization of $\epsilon-$differential privacy. There is a distinction in fair ML research between 'group' & 'individual' fairness measures. Much existing research assumes that both have importance, but conflicting. [33] argues that this seeming conflict is based on a misconception and shows that individual and group fairness are not fundamentally in conflict. [27–32] are some state-of-the-art algorithms on Fairness with Differential Privacy. However, none of them provide either the tight bounds on privacy & loss function or an optimal trade-off between privacy-utility-fairness. Our proposed mechanism aims to achieve both.

Table 2. Related works on Fairness

Paper	Year	Brief summary
Fairness Beyond Disparate Treatment Disparate Impact [18]	2017	Introduces Disparate Mistreatment for group fairness
Algorithmic decision making and the cost of fairness [19]	2017	Deals with a tension between constraint and unconstrained algorithms
On Fairness and Calibration [20]	2017	Explore the stress between minimizing error discrepancy across different population groups while maintaining calibrated probability estimates
An Information-Theoretic Perspective Fairness and Accuracy [21]	2019	Provide trade-off between fairness and accuracy
Fairness Through the Lens of Proportional Equality [22]	2019	Measure the fairness of a classifier and provide group fairness through proportional equality
FACT: A Diagnostic for Group Fairness Trade-offs [23]	2020	Enable systematic characterization of performance-fairness trade-offs among the group
Fair Learning with Private Demographic Data [24]	2020	Allows individuals to release their sensitive information privately while allowing the downstream entity to learn predictors which are non-discriminatory giving theoretical guarantees on fairness
Fairness, Equality, and Power in Algorithmic Decision-Making [25]	2021	Focuses on inequality and the causal impact of algorithms and the distribution of power between and within groups

In this work, the idea is to introduce a group fairness mechanism along with a differential privacy mechanism with the help of the Fairly Iterative Shuffling technique which provides optimal privacy-utility trade-off. The main aim of this work is to make the balance among privacy-utility-fairness during the process. The contributions of this work are:

1. *Group Tagging* method tags every individual by their protected attribute class so that the FIS can not change their protected data during the shuffling. If it will change, the classifier's decision will not be fair as the system wants to generate group fairness among the protected attribute class in the decision-making problem.
2. FIS itself posses a strong privacy guarantee to the data that prevents all types of linkage attack
3. The proposed architecture applies the notion of *Relation Impact* for maintaining group fairness to decide the allowance or giving the chance of participation of the individuals in any particular event without the leakage of their protective attributes.

4. Here, every individual is treated according to their due rather than only "merit". So, the mechanism not only focuses on the directly related attribute but also considers the highly related background factor which affects the results indirectly.
5. *Risk Minimzer* amplifies the utility by minimizing expected loss of the model and helps to achieve the optimal privacy-utility trade-off through regularization parameter.

2 Methodology

The goal of work is to develop an unbiased secured mechanism bringing fairness which is also differentially private. To achieve, before fairness is implemented, the collected data must differentially private without hampering the number of persons corresponding to the respected protected groups, i.e. the data will be divided into separate groups according to the variety of protected class.

Let assume we have a data set with the records of n individuals and the number of dependent attributes is m. Among the n individuals, let there be n_i number of individual having $i^{th}; i = 1(1)n_p$ protected attribute. The whole mechanism is divided into seven parts, they are One-hot Encoding of collected data, Query Function, Group tagging, Correlation Matrix, Fairly Iterative Shuffling (*FIS*), 2-Layers Fair Classification through *Relation Impact* and Boosting Utility through *Risk Minimizer* & Count Report Generation.

2.1 One-Hot Encoding

One hot encoding [4,16] is a method of converting categorical variables into a structured form that could be used in ML algorithms to do better performance in prediction. Now, the correct choice of encoding has a great impression on the utility guarantee. In this architecture, the preference is one-hot encoding instead of any other process of encoding to generate a fair & privatized report with the least noise insertion and provides maximum utility.

2.2 Query Function

In the first phase of this architecture, a query function [16] is used on the dataset to get the important attributes related to a particular query as shown in Table 3. Though our proposed DP mechanism works for generalized group fairness problems, we have taken **Example 2** as our case study and analyzed the case of gender unbiasedness more elaborately in upcoming sections. Also, it is to be noted, This mechanism performs for both (**Example 1 & 2**) type of Queries; related to numerical and decision-making problems and always provide the privatized DP report against user's protected attributes.

Table 3. Query Function: assume the database is structured with records of individuals where *name, age, sex, marks of maths, marks of physics, marks of computer science, annual family income, living area, religion* are ten attributes.

Query	*Analysis*
Example 1: How many Hindus live in the urban area with the score in maths greater than ninety percent?	The attributes 'Religion', 'living area', and 'marks in maths' will be delivered as answers from the query function. This implies that only these three attributes are important for generating the final report to that particular numerical (count-related) query. This type of Question does not need any fair decision but should be privatized against the protected attribute "Religion"
Example 2: Which students are eligible the admission based on the marks of maths?	Now for this type of question, not only the attribute 'marks of maths' is important, but also the gender of the students matter as we want to develop a fair mechanism. For this, it is necessary to give special importance to the attribute 'sex' in every case which is achievable by the attribute tagging method that is discussed later. For this mechanism, the query function considers the attribute 'sex' as the important attribute and returns it no matter what the question is just to maintain a fair balance between male and female candidates

2.3 Correlation Matrix

After the query function [16], the mechanism calculates a correlation matrix that measures the relation distance of each attribute with the required events. The previously given **Examples 2** concentrate on the decision that should fair related to gender in the admission system based on the maths score. But in reality, other sensitive background factors are indirectly related to the examination results that are obtained by the students. For example, the economical status, electricity facility, tuition opportunity, etc. It is obvious that the students came from the strong economical background or high-class society usually take more than one tuition or extra classes to improve scores. On the other hand, the poor students can't afford such facilities. Not only that, in some cases, they don't have the proper environment to study like electricity, proper living area, etc. Now, these factors have a great effect on their results and should not be avoided. So, here, the correlation matrix is used to get the highly related background factors that have a great impact on the required event.

For the general case in, having a dataset with n rows and k attributes, the applied query function returns c the number of attributes which are required to generate the answer along with protected attribute (such as skin color, religion, gender, etc.) and the correlation matrix returns d number of related background factors that affects the results. If $m = c + d$ (excluding protected attribute) number of total attributes are important for generating the final report, then these m attributes will tie up together to represent a single attribute and the reduced number of attributes will be $g = k - m$. Let assume, there are $2S(S > 1)$ number of shufflers and if g is divisible by S, these attributes are divided into S group with g/S attributes in each group. Now, If g is not divisible by S, then extra elements will choose a group randomly without replacement. After that, the group tagging method is applied which helps to preserve the individuals' protected attributes after fairly iterative shuffling [16] method. As the tied-up attributes include the highly related background attribute to the given event, the following group tagging method assists the fair classification in the future.

2.4 Group Tagging Method

Group tagging is a method to classify the individuals based on their protected attribute where each tagged group contains S group of attributes each. This process can be happened by tagging each row with their corresponding protected class. If we consider the given **Example 2** which concentrates on fairness between gender groups (we have taken male and female groups for case study) for admission, the method will work as follow. If the first row contains a record of a female candidate, then this row is tagged by its gender group 'female'. For the male candidates, the rows are tagged by 'male'. Now rows having the same tagging element will shuffle among each other i.e. In the case of female candidates, the iterative shuffling will occur only with the rows tagged as 'female' and on the other hand, rows tagged with 'male' goes for iterative shuffling separately. In this way, the records of male and female candidates never exchange with each other and help to maintain their gender in the generated report. This process never hampers the accuracy of the classifiers for taking a fair decision based on their gender group in the given problem. Also, the iterative shuffle technique establishes a strong privacy guarantee which shuffles all the records of female and male candidates separately among their tagged group. In this way, the probability of belonging the record of a particular individual to their unique ID becomes very low, and the unbiased randomization shuffles technique keeps the data secured. This technique prevents all types of linkage attacks, similarity attacks, background knowledge attacks and also reduces the exposure of user data attributes.

2.5 Fairly Iterative Shuffling (FIS)

A randomized mechanism is applied with Fairly Iterative Shuffling which occurs repeatedly to the given data for producing the randomized unbiased report to a particular query. Let's have a dataset with n rows containing the records of n individuals among which n_1 persons are female and $n_2 = n - n_1$ persons are male. That means after the group tagging method, there are exactly n_1 rows tagged as 'female' whereas n_2 rows are tagged as 'male'. In this stage of our proposed architecture, each row of 'male' and 'female' groups is shuffled iteratively by the $2S, (S > 1)$ number of shufflers. Before shuffling each gender group is divided into some batches containing S attribute groups and every attribute group of each batch chooses a shuffler from the S number of shufflers randomly without replacement for independent shuffling. Now the female tagged group contain n_1 individuals and assume that it is divided into t_F batches where $1, 2, ..., t_F$ batches have $n_{11}, n_{12}, ..., n_{1t_F}; (n_{11} \simeq n_{12} \simeq .... \simeq n_{1t_F})$ number of rows respectively. Similarly, let assume the male tagged group have n_2 individuals' records and it is divided into $1, 2, ..., t_M$ batches having $n_{21}, n_{22}, ..., n_{2t_M}; (n_{21} \simeq n_{22} \simeq .... \simeq n_{2t_M})$ rows respectively. Now for one gender tagged group, each attribute group of every batch chooses a shuffler randomly without replacement from any S number of shufflers. On the other hand, For another gender-tagged group, each group of attributes chooses a shuffler similarly from the remaining S number of shufflers. This shuffling technique occurs repeatedly or iteratively until the last batches of both the gender tagged group goes for independent shuffling.

This architecture is applicable in the generalized case also and not only limited to gender unbiasedness problem If there is i number of variety for a protected attribute (skin color, religion, etc.), the individuals tagged with their considerable protected class and then divided into i number of groups. Each group divided into batches (t_i number of batches for ith group) with approximately same batch size ($n_i1 \simeq n_i2 \simeq ... \simeq n_{it_i}$) and go for shuffling [16, 34].

2.6 2-Layers Fair Classification Through Relation Impact

This mechanism uses 2 layer classifications. After FIS, firstly the classifiers use the group classifications that classify the individuals into gender groups (male and female) from the shuffled data sets. After that, the classifiers make the subgroup-classifications between the groups depends on the Relation Impact (details in Sect. 3.2).

Group Classification: After FIS the classifiers divided the shuffled data into the number of classes according to their gender. In our problem setting, the mechanism focuses on the gender unbiasedness between males and females. This step is the 1st layer of classification where the classifiers use the tagged gender label to classify the data. As in the group tagging method each individual is tagged with his gender category, the classifiers concentrate on this tagged label for classification. This classification is made based on user-given gender information which cannot be affected or changed by the shuffling as it is tied up before going for IS. So, the chance of miss classification error is very less (in fact negligible) here.

Subgroup-Classification. This is the second layer of classification where the classifier classifies the groups into positive and negative subgroups by the notion of Relation Impact. The classifiers decide the eligibility, i.e. if an individual should be allowed to be admitted to the course based on their related background information. For example 2, the classifiers make the decision based on the student's math score, their economical background, living area (rural/urban), electricity facility, etc. The positive class of each gender group contains the eligible individuals and the negative class contains the rejected individuals for the admission.

2.7 Risk Minimization and Count Report Generation

This is an important stage of this architecture where the report is generated from a shuffled and unbiased dataset for a given q query related to sum or count aggregation i.e. numerical answer (consider the **Example 1**). The report is generated by average count calculation by an aggregator function that uses TF-IDF [15,16] calculation to add minimum noise to the report. But, before the final report generation, the loss function [16] is calculated, and risk minimizing assessment [16] is done to obtain the optimal solution. This optimal solution refers to the situation where the architecture achieves a strong privacy guarantee, as well as maximum utility [26] measure while providing an unbiased report against a protected class. The details of risk minimization technique is discussed in Sect. 3.3.

3 Analysis

Considering a data set containing n rows and g attributes and there exists a iS number of shufflers. The attributes are divided into S groups as described previously. Now there is n_i number of individuals tagged as i^{it} protected attribute; $i = 1, 2, ..n_p$ where each row of the dataset has the records of individuals corresponding to their crowd ID. The tagged attributes divide the data sets into groups, For example, if there is i number of variety in a protected attribute, tagged individuals will be divided into i groups. Now, Each tagged group are divided into some batches containing approximately the same rows i.e. $1, 2, ..., t_i$ number of batches for group tagged with i^{th}; $i = 1(1)n_p$ attribute. After that, each group of attribute from every batch chooses a random shuffler and go for shuffling.

3.1 FIS Randomised Response Ratio and Privacy Budget

To proceed further with the proof, we will consider the following theorem from BUDS [16]:

Theorem 1 *(Iterative Shuffling: IS [16]). A randomisation function $\mathcal{R}_S^*$ applied by S $(S > 1)$ number of shuffler providing iterative shuffling to a data set X with n rows and g attributes, where the data base is divided into $1, 2, ..., t$ batches containing $n_1, n_2, ..., n_t$ number of rows respectively, will provide ϵ-differential privacy to the data with privacy budget-*

$$\epsilon = \ln\left(RR_\infty\right) = \ln\left[\frac{t}{(n_1 - 1)^S}\right] \qquad (1)$$

only when, $n_1 \simeq n_2 \simeq ..... \simeq n_t$.
Here RR_∞ = *Randomised Response Ratio, and*

$$RR = \frac{P(Response = YES \mid Truth = YES)}{P(Response = YES \mid Truth = NO)} \tag{2}$$

Lemma 1 *The Randomised Response Ration of Iterative Shuffling* (RR_{IS}) *[4, 5]=* $\frac{P(Row\ belongs\ to\ its\ own\ unique\ ID)}{P(Row\ does\ not\ belongs\ to\ its\ own\ unique\ ID)}$ *satisfy the condition of Differential Privacy.*

Now, the proposed architecture provides a fair mechanism along with a strong privacy budget to secure the sensitive information of individuals. This algorithm is the non-discriminant of biases regarding different groups based on protected attributes (for example gender, skin color, etc.). Now the help of Theorem 1, the privacy budget of *FIS* is developed next.

Theorem 2 *(Fairly Iterative Shuffling- FIS). A randomisation function* $\mathcal{R}^*{}_{iS}$ *applied by* iS ($S > 1$) *number of shuffler providing iterative shuffling to a data set* X *with* n *rows and* g *attributes among which* n_i *number of individuals are tagged with ith protected class* ($i = 1, 2, ...n_p$)*; i.e. there are* n_p *number of user groups according to their protected class where tagged groups are divided into* $1, 2, ..., t_i$ *batches containing* $n_{i1}, n_{i2}, ..., n_{it_i}$ *number of rows for ith group, will provide* ϵ*-differential privacy to the data with privacy budget-*

$$\epsilon_{Fair} = \ln\left[\frac{1}{n_p!}\Sigma_{i=1}^{n_p}\frac{t_i}{(n_{it_i} - 1)^S}\right] \tag{3}$$

only when, $n_{i1} \simeq n_{i2} \simeq ..... \simeq n_{it_i}; i = 1, 2, ..., n_p$

Proof. The records in the group of attributes for the different tagged groups go for independent shuffling separately. According to the Theorem 1 [16] we can say that, For the records tagged with ith group,

$$RR_i = \frac{P(Row\ belongs\ to\ its\ own\ unique\ ID\ tagged\ with\ ith\ group)}{P(Row\ does\ not\ belongs\ to\ its\ own\ unique\ ID\ tagged\ with\ ith\ group)} \tag{4}$$

$$= \frac{t_i}{(n_{it_i} - 1)^S} \tag{5}$$

as, $n_{i1} \simeq n_{i2} \simeq ..... \simeq n_{it_i}; i = 1, 2, ..., n_p$. So, For the whole architecture the fair randomized response will be:

$$RR_\infty = \frac{1}{n_p!}\left[\frac{t_1}{(n_{1t_1} - 1)^S} + \frac{t_2}{(n_{2t_2} - 1)^S} + ... + \frac{t_{n_p}}{(n_{n_p t_{n_p}} - 1)^S}\right] \tag{6}$$

Now the privacy budget for this proposed architecture is:

$$\epsilon_{Fair} = \ln\frac{1}{n_p!}\left[\frac{t_1}{(n_{1t_1} - 1)^S} + \frac{t_2}{(n_{2t_2} - 1)^S} + ... + \frac{t_{n_p}}{(n_{n_p t_{n_p}} - 1)^S}\right] \tag{7}$$

That means,

$$\epsilon_{Fair} = \ln \left[\frac{1}{n_p!} \Sigma_{i=1}^{n_p} \frac{t_i}{(n_{it_i} - 1)^S} \right] \tag{8}$$

In our case study, we focuses on the decision problem of gender unbiasedness between male and female groups for allowing them to take admission in a particular course (considering example 2). For this case, the FIS posses a privacy budget of :

$$\epsilon = \ln \frac{1}{2} \left[\frac{t_F}{(n_{1t_F} - 1)^S} + \frac{t_M}{(n_{2t_M} - 1)^S} \right] \tag{9}$$

Where, $S > 1$, n_1 is the number of female in database, $n_2 = (n - n_1)$ is number of male in database, t_F is the number of batches in female group, t_M is the number of batches in male group, n_{1t_F} is the average female batch size, n_{2t_M} is average male batch size and ϵ_{Fair} is the privacy parameter of this mechanism. The small value of ϵ_{Fair} refers to a better privacy guarantee.

3.2 Fair Classifiers and Relation Impact

After FIS, we introduce the notion of *Relation Impact* that helps the classifiers to take a fair decision. This mechanism applies fair classifiers that use the principle of Relation Impact to avoid the miss classification error and able to create unbiased classes or groups to consider for giving chance in an event.

Definition 1. *Relation Impact: The aim of a set of classifiers is to study a classification function $\hat{h}_c : \mathcal{X} \longrightarrow \mathcal{Y}$, defined in a hypothesis space $\mathcal{H}$ where $\hat{h}_c$ minimizes some aimed loss function to reduce miss-classification error based on the related background attribute class $\mathcal{A} = (a_1, a_2, ..., a_m)$:*

$$\hat{h}_c := \underset{h_c \in \mathcal{H}}{\operatorname{argmin}}\, E_{(\mathcal{X},\mathcal{Y})\sim P} L[h_c(\mathcal{X}|\mathcal{A}), \mathcal{Y}] \tag{10}$$

The Efficiency of the classifier will be estimated using testing dataset $\mathcal{D}_{tst} = (x_{j_t}, y_{j_t})^{n_t}{}_{j=1(1)n_t}$, based on the observation that how accurate the predicted labels $\hat{h}_c(x_{j_t}|\mathcal{A})$'s, are corresponds to the true labels y_j's.

Therefore, it means the decision regarding the relevant persons or groups is taken according to their due based on all the highly related factors to that particular event. This implies that distribution will not possess numerical equality whereas it will be influenced by the relation distance of important background factors and consider the individuals according to their rightful needs. For example, in the classroom, this might mean the teacher spending more time with male students at night classes rather than female students because most of the time female students do not choose to take the night classes for safety issues. Not only that, the students (irrespective of gender) who came from the weaker economical background cannot always participate in the extra classes or cannot afford any extra books or study material. So, it is clear that, in the admission system, if someone wants to choose the students fairly, not only the sex group matters, but also the economical background, and other factors like appropriate subject scores, living locations, electricity facility, etc. also matters. So, these types of factors which are directly and indirectly highly related to the final exam score of the students should be taken to account. The notion of *Relation Impact* exactly does that, where it trains the model by considering all the highly related attributes for taking the decision and minimizes the expected error of the classification by *Equalizing Minimal FPR-FNR* of different groups.

In our work, the idea is to provide a fair decision with the help of an unbiased classification based on the correlation of the attributes to the required event. The mechanism first calculates the correlation matrix and any attributes which have the higher (usually greater than 0.5 or less than −0.5) positive and negative correlation to the required events are only taken to be accountable for the final decision-making procedure. Here the classifiers use the Relation Impact where the main aim is to build a model that can minimize the expected decision loss to reach the maximum accuracy. The classifiers attend that minimum loss, i.e. minimum miss classification error, by achieving the Minimal Equalised FPR-FNR of the different groups; FPR: False Positive Rate, FNR: False Negative Rate.

Definition 2. *Equalized Minimal FPR-FNR: Let's have a set of classifiers* $\hat{h_c}\{h_{c1}, h_{c2}, ...h_{ci}\} : \mathcal{X} \rightarrow \mathcal{Y}$ *based on the hypothesis space* $\mathcal{H}$ *where* $i = 1, 2..n_p$ *denotes the number of groups, and* $\mathcal{A} = \{A_1, A_2, ...A_m\}$ *is the set of related background attributes, then the classifier is said to satisfy the condition of Equalised Minimal FPR-FNR ratio if:*

$$\underset{h_c \epsilon \mathcal{H}}{\operatorname{argmin}} FPR_1(\mathcal{A}) = \underset{h_c \epsilon \mathcal{H}}{\operatorname{argmin}} FPR_2(\mathcal{A}) = ... = \underset{h_c \epsilon \mathcal{H}}{\operatorname{argmin}} FPR_{n_p}(\mathcal{A}) \tag{11}$$

and,

$$\underset{h_c \epsilon \mathcal{H}}{\operatorname{argmin}} FNR_1(\mathcal{A}) = \underset{h_c \epsilon \mathcal{H}}{\operatorname{argmin}} FNR_2(\mathcal{A}) = ... = \underset{h_c \epsilon \mathcal{H}}{\operatorname{argmin}} FNR_{n_p}(\mathcal{A}) \tag{12}$$

Here, $\hat{h_c}(\mathcal{X})$ *is the predicted label by the classifier,* $\mathcal{Y}$ *is the true label and both of them take the label value either* y_1 *or* Y_0 *where* y_1 *denotes the positive label,* y_0 *denotes the negative label and*

$$FPR_i(\mathcal{A}) = P_i(\hat{h_c}(\mathcal{X}) = y_1 | \mathcal{Y} = y_0, \mathcal{A}) \tag{13}$$

and

$$FNR_i(\mathcal{A}) = P_i(\hat{h_c}(\mathcal{X}) = y_0 | \mathcal{Y} = y_1, \mathcal{A}) \tag{14}$$

For our mentioned problem in example 2, the classifier will achieve the higher accuracy for predicting the positive and negative subgroups for both the gender groups (Male and Female), by minimizing the target loss when,

$$\underset{h_c \epsilon \mathcal{H}}{\operatorname{argmin}} FPR_F(\mathcal{A}) = \underset{h_c \epsilon \mathcal{H}}{\operatorname{argmin}} FPR_M(\mathcal{A}) \tag{15}$$

and

$$\underset{h_c \epsilon \mathcal{H}}{\operatorname{argmin}} FNR_F(\mathcal{A}) = \underset{h_c \epsilon \mathcal{H}}{\operatorname{argmin}} FNR_M(\mathcal{A}) \tag{16}$$

Remark 1. *FIS does not hamper the Equalised Minimal FPR-FNR (Detailed discussion is in appendix).*

3.3 Boosting Utility Through Risk Minimization

The real information which can be gained from the data by a particular query or set of queries is defined as the utility [26] of the system. The goal of this section is to discuss the gained utility from the data by applying Fair-BUDS and provide a tight bound for loss function between input and output average-count. This bound has a great impact on the utility of the system and this depends on the previously obtained privacy budget. Before obtaining the final result, an optimization function with a risk assessment technique [16] is applied to get the maximum utility from the data. Let's assume, before FIS shuffling, the true data set will give a average count report for a time horizon $[d] = \{1, ..d\}$ which is denoted as $\mathcal{C_T} = \sum_{i=1}^{n} \sum_{j \epsilon \mathcal{Q}} \sum_{T \epsilon [d]} x_{ij}[T]$; where, x_{ij} is the jth record of the ith individual from the true data set for the time horizon $[d]$ and $\mathcal{Q} = \{a_1, a_2, ...\}$ is the set of attributes given by the query function for a particular set query or set of queries. Now, $\mathcal{C_{FS}} = \sum_{i=1}^{n} \sum_{j \epsilon \mathcal{Q}} \sum_{T \epsilon [d]} x_{ij}^{FS}[d]$; where x_{ij}^{FS} is the jth record of the ith individual from the FIS data set for the time horizon $[d]$. The calculated loss function [16] regarding the input and output count will be:

$$\mathcal{L}(\mathcal{C_T}, \mathcal{C_{FS}}) \leq \mathcal{C_{FS}} \times \left| e^{\ln \left[\frac{1}{n_p!} \Sigma_{i=1}^{n_p} \frac{t_i}{(n_{it_i} - 1)^S} \right]} - 1 \right| \tag{17}$$

When, $e^{\epsilon} \to 0$, the Utility $\mathcal{U}(\mathcal{C_T}, \mathcal{C_{FS}} | \mathcal{X}, \mathcal{Y}) \to 1$ as $\mathcal{L}(\mathcal{C_T}, \mathcal{C_{FS}}) \to 0$. Now the aim of this architecture is not only to be fair and secured but also it should pose a high utility guarantee which can be obtained by risk-minimizing assessment [16]. Proceeding with risk minimization technique, a mechanism $\mathcal{R}^*(iS)$ can be found (when there is i number of tagged groups) which minimizes the risk [16], where

$$\mathcal{R}^*(iS) = \underset{\mathcal{R}^*(iS) \epsilon \mathcal{H}}{\operatorname{argmin}} \; Risk_{(\mathrm{Ep})}(\mathcal{R}^*(iS)) \tag{18}$$

4 Conclusion and Future Scope

The Proposed architecture provides group fairness to the user based on their protected attribute class while giving a strong privacy guarantee to their sensitive attribute. The Risk Minimizer amplifies the utility Guarantee to the system which makes the algorithm possess an optimal balance between privacy-utility-fairness. This balanced architecture full-filled user's top priorities related to fairness and Privacy in the Online Service system. Though the mechanism shows good performance in privacy-utility trade-off for all kinds of Queries and generates unbiased classification for decision-making problems, the use of One-Hot encoding is one kind of constraint for big data analysis. Trying other encoding option (which does not depend on data dimension) is our future target. On the other hand, the work performance and both the privacy-utility upper bounds are given on theoretical aspects only. So, various experiments with different benchmark datasets for both online and offline settings are also in our plan.

References

1. Chatzikokolakis, K., Chothia, T., Guha, A.: Statistical measurement of information leakage. In: Esparza, J., Majumdar, R. (eds.) TACAS 2010. LNCS, vol. 6015, pp. 390–404. Springer, Heidelberg (2010). https://doi.org/10.1007/978-3-642-12002-2_33
2. Alvim, M.S., et al.: Measuring information leakage using generalized gain functions. In: 2012 IEEE 25th Computer Security Foundations Symposium. IEEE (2012)
3. Dwork, C.: Differential privacy: a survey of results. In: Agrawal, M., Du, D., Duan, Z., Li, A. (eds.) TAMC 2008. LNCS, vol. 4978, pp. 1–19. Springer, Heidelberg (2008). https://doi.org/10.1007/978-3-540-79228-4_1
4. Sengupta, P., Paul, S., Mishra, S.: Learning with differential privacy. In: Handbook of Research on Cyber Crime and Information Privacy, pp. 372–395. IGI Global (2021)
5. Dwork, C., Roth, A.: The algorithmic foundations of differential privacy. Found. Trends Theoret. Comput. Sci. **9**(3–4), 211–407 (2014)
6. Cormode, G., et al.: Privacy at scale: local differential privacy in practice. In: Proceedings of the 2018 International Conference on Management of Data (2018)
7. Erlingsson, Ú., Pihur, V., Korolova, A.: RAPPOR: randomized aggregatable privacy-preserving ordinal response. In: Proceedings of the 2014 ACM SIGSAC Conference on Computer and Communications Security (2014)
8. Abadi, M., et al.: Deep learning with differential privacy. In: Proceedings of the 2016 ACM SIGSAC Conference on Computer and Communications Security (2016)
9. McMahan, B., et al.: Communication-efficient learning of deep networks from decentralized data. In: Artificial Intelligence and Statistics. PMLR (2017)
10. Bittau, A., et al.: PROCHLO: strong privacy for analytics in the crowd. In: Proceedings of the 26th Symposium on Operating Systems Principles (2017)
11. McMahan, H.B., et al.: A general approach to adding differential privacy to iterative training procedures. arXiv preprint arXiv:1812.06210 (2018)
12. Erlingsson, Ú., et al.: Amplification by shuffling: from local to central differential privacy via anonymity. In: Proceedings of the Thirtieth Annual ACM-SIAM Symposium on Discrete Algorithms. Society for Industrial and Applied Mathematics (2019)
13. Hanzely, F., et al.: A privacy preserving randomized gossip algorithm via controlled noise insertion. arXiv preprint arXiv:1901.09367 (2019)
14. Chaudhuri, K., Imola, J., Machanavajjhala, A.: Capacity bounded differential privacy. arXiv preprint arXiv:1907.02159 (2019)
15. Paul, S., Mishra, S.: ARA: aggregated RAPPOR and analysis for centralized differential privacy. SN Comput. Sci. **1**(1), 1–10 (2020)

16. Sengupta, P., Paul, S., Mishra, S.: BUDS: balancing utility and differential privacy by shuffling. In: 2020 11th International Conference on Computing, Communication and Networking Technologies (ICCCNT). IEEE (2020)
17. Dwork, C., et al.: Fairness through awareness. In: Proceedings of the 3rd Innovations in Theoretical Computer Science Conference (2012)
18. Zafar, M.B., et al.: Fairness beyond disparate treatment & disparate impact: learning classification without disparate mistreatment. In: Proceedings of the 26th International Conference on World Wide Web (2017)
19. Corbett-Davies, S., et al.: Algorithmic decision making and the cost of fairness. In: Proceedings of the 23rd ACM SIGKDD International Conference on Knowledge Discovery and Data Mining (2017)
20. Pleiss, G., et al.: On fairness and calibration. arXiv preprint arXiv:1709.02012 (2017)
21. Dutta, S., et al.: An information-theoretic perspective on the relationship between fairness and accuracy. arXiv preprint arXiv:1910.07870 (2019)
22. Biswas, A., Mukherjee, S.: Fairness through the lens of proportional equality. In: AAMAS (2019)
23. Kim, J.S., Chen, J., Talwalkar, A.: FACT: a diagnostic for group fairness trade-offs. In: International Conference on Machine Learning. PMLR (2020)
24. Mozannar, H., Ohannessian, M., Srebro, N.: Fair learning with private demographic data. In: International Conference on Machine Learning. PMLR (2020)
25. Kasy, M., Abebe, R.: Fairness, equality, and power in algorithmic decision-making. In: Proceedings of the 2021 ACM Conference on Fairness, Accountability, and Transparency (2021)
26. Li, T., Li, N.: On the tradeoff between privacy and utility in data publishing. In: Proceedings of the 15th ACM SIGKDD International Conference on Knowledge Discovery and Data Mining (2009)
27. Hardt, M.: A study of privacy and fairness in sensitive data analysis (2011)
28. Cummings, R., et al.: On the compatibility of privacy and fairness. In: Adjunct Publication of the 27th Conference on User Modeling, Adaptation and Personalization (2019)
29. Xu, D., Yuan, S., Wu, X.: Achieving differential privacy and fairness in logistic regression. In: Companion Proceedings of the 2019 World Wide Web Conference (2019)
30. Bagdasaryan, E., Shmatikov, V.: Differential privacy has disparate impact on model accuracy. arXiv preprint arXiv:1905.12101 (2019)
31. Farrand, T., et al.: Neither private nor fair: impact of data imbalance on utility and fairness in differential privacy. In: Proceedings of the 2020 Workshop on Privacy-Preserving Machine Learning in Practice (2020)
32. Ekstrand, M.D., Joshaghani, R., Mehrpouyan, H.: Privacy for all: ensuring fair and equitable privacy protections. In: Conference on Fairness, Accountability and Transparency. PMLR (2018)
33. Binns, R.: On the apparent conflict between individual and group fairness. In: Proceedings of the 2020 Conference on Fairness, Accountability, and Transparency (2020)
34. Erlingsson, Ú., et al.: Encode, shuffle, analyze privacy revisited: formalizations and empirical evaluation. arXiv preprint arXiv:2001.03618 (2020)

MEDICI: A Simple to Use Synthetic Social Network Data Generator

David F. Nettleton[1(✉)], Sergio Nettleton[2], and Marc Canal i Farriol[1]

[1] Universitat Pompeu Fabra, Barcelona, Catalunya, Spain
david.nettleton@upf.edu

[2] Universitat Politécnica de Catalunya, Barcelona, Catalunya, Spain

Abstract. The motivation of the work in this paper is due to the need in research and applied fields for synthetic social network data due to (i) difficulties to obtain real data and (ii) data privacy issues of the real data. The issues to address are first to obtain a graph with a social network type structure, label it with communities. The main focus is the generation of realistic data, its assignment to and propagation within the graph. The main aim in this work is to implement an easy to use standalone end-user application which addresses the aforementioned issues. The methods used are the R-MAT and Louvain algorithms, with some modifications, for graph generation and community labeling respectively, and the development of a Java based system for the data generation using an original seed assignment algorithm followed by a second algorithm for weighted and probabilistic data propagation to neighbors and other nodes. The results show that a close fit can be achieved between the initial user specification and the generated data, and that the algorithms have potential for scale up. The system is made publicly available in a Github Java project.

Keywords: Online social networks · Synthetic data · Graphs

1 Introduction

The use of online social networks has been steadily increasing and evolving over the last decade, since they first became available to the general public in the 2000s. This has created a great interest for commercial and academic reasons in studying human behavior in the online social network environment in particular and Internet in general. Behavior rules have emerged on how users tend to group by affinities, how are they interconnected, which are the key demographics, how to capture and evaluate activity, information propagation, and the general dynamics of social networks, all of which has made this a popular field of study [1]. On the other hand, privacy issues have raised concerns that major corporations use our data in a market where the control is lost to third parties [2, 3]. However, after the initial mercantilist focus, efforts to redirect research and applications for social good and taking into account ethical considerations are now becoming main-stream, especially with think-tank and government backing [4].

As those who are familiar to the field of social network data analysis will know, two major issues are (i) difficulties to obtain real data and (ii) data privacy issues of the real

V. Torra and Y. Narukawa (Eds.): MDAI 2021, LNAI 12898, pp. 273–285, 2021.
https://doi.org/10.1007/978-3-030-85529-1_22

data. Although the focus of the current work is not graph generation or scale-up, it is useful to mention related work in these fields [5–7]. With respect to the generation of synthetic knowledge and population of the graph with attributes and attribute-values a recent work is found in [8]. Other relevant works on populating topologies with realistic data are [9–12] however many are specific to a given domain/data type, and/or require complex frameworks in order to work.

The name "Medici" refers to the 15th century family in Florence who developed a sophisticated social network [1, 13]. Robins [13] cites Medici as a specific analogy of local behavior and global structure in a real social network. Medici addresses the need for an application which is accessible for general users with medium level skills, whereas it also can be used as a research support tool for more advanced users. The graphical user interface is designed to lower the entry barrier for the former type of users, and whose focus is on the data analysis per se rather than the process to obtain the data.

The Medici application described in this paper responds to some of the aforementioned issues, by offering a synthetic data generator for online social network graphs, which mitigates the need for accumulating real personal data of users, and can be useful for applied research and development in fields such as population studies for public resource assignment and medical care, pandemic data analysis, among others.

The algorithm for synthetic data generation was first developed by Nettleton in [14, 15], and in [16] Nettleton and Salas applied a preliminary version of the seed assignment approach to a data privacy application, in which the seeds were assigned to k-anonymous groups in order to anonymize them. The authors state that the current paper is an extension of the arXiv pre-print [17].

The paper is organized as follows: in Sect. 2 some theoretical background is given regarding the algorithms: R-MAT ("recursive matrix", graph structure generator), Louvain (community labeling). This is described in relation to the functionality of the application and the sequence of steps for data processing. Then, in Sect. 3 a detailed description is given of the Medici algorithm and how it assigns and propagates user data through the graph structure. This is followed in Sect. 4 with the empirical testing of the system, first a description of the experimental setup, then examples of the data generated and benchmarking of deviations between different executions. Finally, Sect. 5 gives the conclusions. The full system source code and runtime is available at our public Github project [18].

2 Background to Third Party Algorithms

The following describes the two "third party" algorithms: R-MAT and Louvain, which generate the graph and label the communities, respectively. The third (proprietary) algorithm, Medici, which is the key focus of the work, assigns and propagates the data and is described later in Sect. 3.

With respect to the choice of algorithms, the focus of the paper is not the graph generation step, it is the data assignment and propagation step. R-MAT and Louvain and two widely used and reliable methods in the literature, hence these were chosen. And then focus was made on the data generation per se. However, it is useful to mention alternatives, such as those presented in [7, 19] which address issues such as scalability and degree sequence and distance between nodes.

2.1 R-MAT

Chakrabarti et al. [20] present R-MAT, a recursive model for graph mining. The objective of R-MAT is to model an existing graph of real data, thus deriving its parameterization in terms of given descriptor variables. A typical adjacency matrix of {0, 1} values is used to represent the graph (nodes, edges). The authors state that one of the challenges in modeling real graphs, such as social networks, is replicating the power law distributions, skew distributions, and other reported structures, such as the "bow-tie" and the "jellyfish", while maintaining a small diameter for the graph. In order to represent this, a recursive partitioning is carried out, which can be considered as a binomial cascade in two dimensions. The expected number of nodes c_k with out-degree k is given by:

$$C_k = \binom{E}{k} \sum_{i=0}^{n} \binom{n}{i} \left[\rho^{n-i}(1-\rho)^{i}\right]^{k} \left[1-\rho^{n-i}(1-\rho)^{i}\right]^{E-k} \quad (1)$$

where 2^n is the number of nodes in the R-MAT graph (typically $n = \log_2 N$), N is the number of nodes in the real graph, $\rho =$ probability of an edge falling into partition $a +$ probability of an edge falling into partition b, and E is the number of edges in the real graph. The authors tested the method on two real datasets, "epinions" and "clickstream". Descriptive parameters are used such as degree distributions, number of reachable pairs, number of hops, effective diameter and stress distribution.

2.2 Louvain

The 'Louvain' method [21] can be considered an optimization of Newman and Girvan's method [22], in terms of computational cost. Firstly, it looks for smaller communities by optimizing modularity locally. As a second step, it aggregates nodes of the same community and builds a new network whose nodes are the communities. These two steps are repeated iteratively until the modularity value is maximized. The optimization is based on evaluating the modularity gain, which is done by performing a local calculation of the change in modularity for a given community, caused by moving each node from it to an adjacent community.

Modularity [22]: During processing, the graph is successively divided in components, and the correctness of the community partitions is measured. The quality metric used for a given community is called the modularity. For a graph divided into k communities, a symmetrical matrix e is defined of order k^2 whose elements e_{ij} are the subset of edges from the total graph which connect the nodes of communities i and j.

The trace of matrix e, denoted as $Tre = \sum_i e_{ii}$ gives the fraction of edges in the graph which connect nodes of the same community. Hence, a good division in communities should obtain a high value for the trace of matrix e. As a quality indicator, the sum of the rows $a_i = \sum_j e_{ij}$ is defined to represent the fraction of edges which connect nodes of community i. Following on from this, the modularity metric was defined as:

$$Q = \sum_i (e_{ii} - a_i^2) = Tre - ||e^2|| \quad (2)$$

Where $||x||$ indicates the sum of the elements of matrix x. This parameter measures the fraction of edges in the graph which connect vertices in the same community, minus the expected value of the same number of edges in the graph with the same community partitions but with random connections between their respective nodes. If the number of intra-community edges shows no improvement on the expected value, then the modularity would be $Q = 0$. On the other hand, Q approaches a maximum value of 1 when the community structure is strong. According to [22], the usual empirical range for Q is between 0.3 and 0.7.

For the current release of Medici, we have adapted Louvain to produce exactly 10 communities (0 to 9), using the "resolution" as an optimization parameter (approximating to 1). Note that the "resolution" parameter was incorporated in the Gephi implementation of Louvain from the idea of Lambiotte et al. [23]. A resolution closer to 1 implies a better "quality" of the result, which is related to the "modularity" quality measure and the optimum partitioning. In our implementation, if the algorithm does not converge to 10 communities in the predefined timeout (30 s), we aggregate communities 9 to N in one community (labeled as 9). The community size after resolution optimization tends to fragment quickly for communities 9 to N, and the aggregate community (9) is generally one of the smallest with respect to communities 0 to 8 (less than 10% of the sum).

3 Medici

In the following, first an overall vision is provided in Sect. 3.1, then details of the data assignment and propagation algorithm are given in Sect. 3.2.

3.1 Overall Vision

The following describes the functionality of the synthetic data generator for online social networks. It follows an easy to use "workflow" sequential structure guiding the user from the initial data entry through the data definition and generation, followed by analysis of the generated data and export (Fig. 1). Improvements in this version include the integration of the R-MAT [20] and Louvain [21] algorithms to create the graph and assign the community labels, respectively. However, the user can provide their own graph and community assignment files if s/he wishes. Apart from the "static" information about each user, the system now also allows the assignment of dynamic data, such as simple "likes". This can serve as the basis so that the user can customize and develop more sophisticated simulations. For this, we make available the Java source code [18, 24] in our Github repository.

Hence, the file inputs to the program are the social network skeleton structure (nodes and links) where each node has been assigned a "community label". If the user wants to do this, it can be done manually or it can be done with an automated algorithm such as Louvain [21] (also available in Gephi [23] and included in Medici "as is"). One limitation is that there must be 10 communities and 10 profiles. This limit was a simplification for implementation purposes, and as a first step a fixed number of communities/profiles worked and gave reliable results. As future work a general N communities/profiles version could be developed. Though this limitation seems a bind,

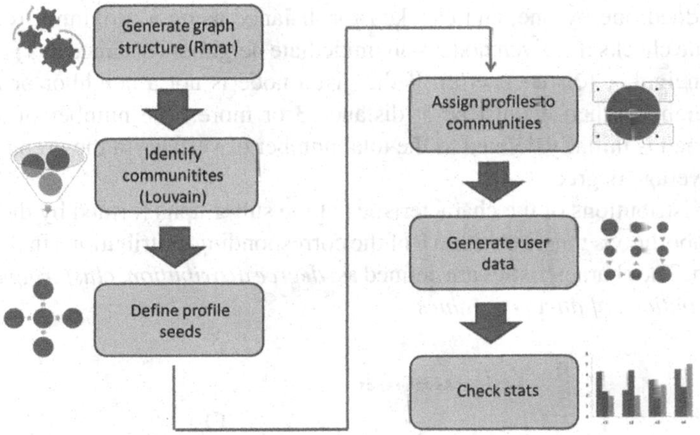

Fig. 1. Processing sequence

if you want less communities you can define some with very few members (by controlling the profile percentage definitions) so they are residual and have a minimum effect on the overall network. If you want fewer profiles you can define some which are identical.

For the graph definition, we have included the default R-MAT algorithm. If you want to supply a graph from outside the program you can do that but then it's possible that the Louvain algorithm cannot establish 10 communities (0 to 9), depending on the size and complexity of the graph. Note the Louvain algorithm has been modified to aggregate communities 9 to N as community 9, with a timeout of 30 s on the process.

3.2 Data Assignment and Propagation

The following gives a detailed description of the data assignment and propagation algorithm, the assignment of seeds, neighbors and any other nodes.

The **data generator** has the following four main steps:

(i) Choose which nodes will be seeds in each community
(ii) Assign prototype profile to seeds in each community (profile x → community y)
(iii) Assign data to neighbors of each seed in function of seed profile (i.e. neighbors tend to be similar to seeds). This will have a similarity component (to neighbors with data assigned) and a random component (to promote a degree of diversity).
(iv) Assign data to remaining nodes (those which still have no data assigned from steps (ii) and (iii). This will have a similarity component (to neighbors with data assigned) and a random component (to promote a degree of diversity).

Seed Assignment: The seed assignment has to comply with two criteria:

(v) Each seed node must be at least at distance 3 from any other seed node (in a given community), so that each sub-graph (with the seed as center) can be modified without perturbing the adjacent sub-graphs. In order to achieve this, seed nodes are added one by one, and checked for distance using a proximity routine. This routine checks if a given node is an immediate neighbor (distance = 1) or neighbor of a neighbor (distance = 2). If the given node is not a neighbor or a neighbor of neighbor, then it must be at distance 3 or more. The number of seed nodes assigned is initially defined as the total number of vertices in the graph divided by the average degree.

(vi) The distributions of the characteristics of the sub-graphs formed by the seed node neighborhoods must be within δ of the corresponding distributions in the complete graph. The characteristics are defined as: *degree distribution, clustering coefficient, distributions of attribute-values.*

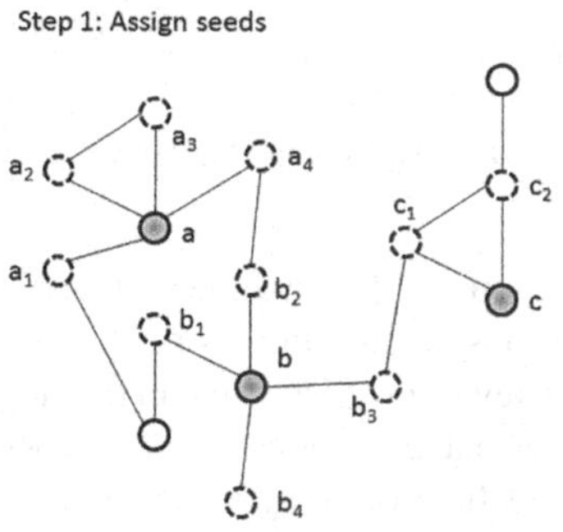

Fig. 2. Seed assignment.

Figure 2 shows a generic example of the seed assignment process, for which the seeds are labeled as a, b and c. We note that each seed is at a shortest distance of three from any other seed, hence the immediate neighborhoods do not overlap.

The assignment of the seed nodes is actually an optimization process. It is possible that the random assignment of seeds, especially the first seed, can result in a sub-optimal assignment. For example, if the first seed is assigned to a major hub node, the average path length to it of many nodes in the graph will be short.

Recall that we denote the number of seeds as σ and we take as an upper-bound for the number of seeds to be assigned, the number |V|/dAVG.

The seed assignment process has T tries at randomly assigning y see nodes. If, after I iterations, σ seeds have not been assigned, then σ is reduced by one and another try is made. σ is progressively reduced by one until σ seeds are assigned. Finally, the process tries to add more seeds to the current configuration (to avoid the assignment of a sub-optimal number of seeds).

Coverage: the assignment of the seed nodes in the manner described has a coverage of between 20% and 50% of the nodes in the graph. This is because isolated nodes tend to remain between sub-graphs which cannot be assigned due to the minimum distance requirement between seeds (≥3) in the same community, and because the seed sub-graphs cannot overlap.

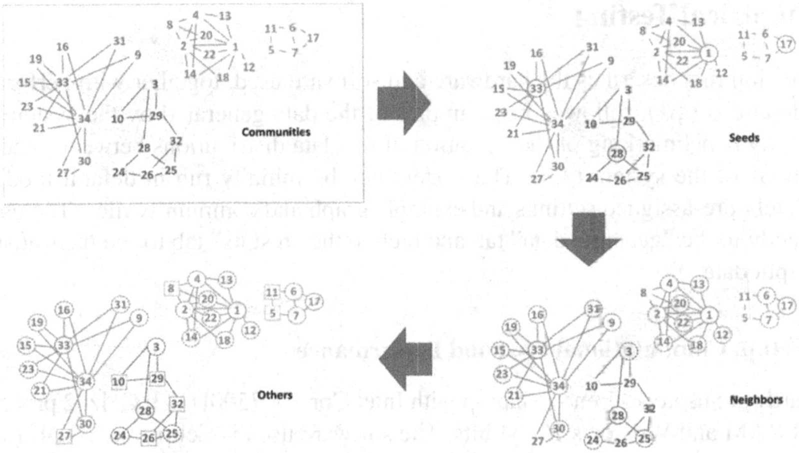

Fig. 3. Example of data propagation steps (Karate graph dataset): (Communities) four communities are identified; (Seeds) seeds assigned to each community; (Neighbors) neighbors of seeds are assigned; (Others) any remaining nodes are assigned.

Figure 3 illustrates the data propagation mechanism. Figure 3 "Communities" shows four communities identified in the Karate graph which will have been assigned by the Louvain algorithm, as indicated by the green, red, blue and grey colors. Next, in Fig. 3 "Seeds" the seeds are assigned to each community, as indicated by the circled nodes 33, 28, 1 and 17, respectively. In this example, due to the small size of the graph, only four seeds are used, one per community. At this point the profiles corresponding to each community are assigned to the seeds. This is followed by the assignment of the data to the neighbors of each seed (Fig. 3 "Neighbors") as indicated by the dashed circles. The neighbors are assigned values which "similar" to their seed profiles, depending on the "similarity threshold" defined. Finally, as shown in Fig. 3 "Others", any remaining nodes are assigned values which are either similar to the neighbors or randomly assigned, depending on the control parameter values defined. These are indicated as dashed squares.

The number of seeds assigned guarantees a good "coverage" of the data, and tries to minimize the number of unassigned nodes (which are neither seeds nor neighbors of seeds). So the user can see the coverage and try different number of seeds to see which gives the best. By default, the number of seeds is equal to about 11% of the number of nodes in the graph. Also, seed nodes are chosen based on the HITS statistic, which indicates the nodes which are best connected in a community (i.e. those which have the most neighbors).

Thus, two aspects are optimized during the process: (i) the maximum number of possible seeds is assigned, resulting in an optimal or close to optimal coverage of the complete graph, given the restriction that seeds should be at a distance of three or more from each other; (ii) the distributions of the characteristics of the seed sub-graphs (seeds and their neighbors) has a given similarity to the corresponding distributions of the characteristics in the complete graph (all the nodes of the graph).

4 Empirical Testing

This section first describes the hardware and software used, together with performance considerations (4.1), followed by examples of the data generated by the system (4.2), and finally benchmarking of the deviation of the data distributions between successive executions of the system (4.3). The system can be initially run in default mode with completely pre-assigned settings and example graph and community files. The user can go directly to the "generate data" tab and then to the "results" tab to see the statistics of the output data.

4.1 Setup, Computational Cost and Performance

The hardware used is a Lenovo laptop with Intel Core i7-7500U, 2.9 GHz, 2 processors, 12 GB RAM and Windows 10, 64 bits. The software used is Eclipse Java IDE for Web Developers Version 4.9.0 and Java Fx. The R-MAT algorithm is highly efficient and generates the 1k, 10k and 50k graphs almost instantly on the hardware and software indicated. The Louvain algorithm is less efficient - during N-fold testing all processing times for community labeling were less than 30 s for the 1k and 10k graphs. For the 50k graph a 30 s time limit was assigned, which labels 10 communities with a high quality modularity. The seed assignment is a computationally expensive process - during N-fold testing all processing times for seed assignment were less than 60 s for the 1k and 10k graphs (11% seeds) and less than 300 s for the 50k graph (15% seeds). For graphs up to 1M nodes it would be recommended to run the system on higher range hardware and for bigger graphs some of the system would need to be reprogrammed for parallel computation in a cloud environment. The attributes (each with their respective categories) which describe users (nodes) of the graph are: age, gender, residence, religion, marital status, profession, political orientation, sexuality, likes.

4.2 Example Data Generated

In Tables 1 and 2 are shown the data generated from one of the executions of the data generation for the 50k by 250k graph. The three rightmost columns give the number of users (or nodes) assigned to each attribute-value for (i) the whole graph, (ii) community 4 and (iii) community 1.

In order to interpret the results, we refer to the overall percentages assigned for each attribute value, and the profiles assigned to each community. For example, in the overall graph the users are weighted towards young people (age 18–25 and 26–35), with somewhat more females than males, residence is fairly equitative, and so on.

In Community 4, there is a weighting towards older people (age 66–75 and 76–85), mainly female, who live in San Jose, mainly Jewish, and so on. This is because the seed Profile 5 assigned to Community 4 has these characteristics. Likewise, in Community 1, there is a weighting towards younger people (as in the whole graph), slightly more males, who live in Palo Alto, mainly Christian, and so on. This is because the seed Profile 2 assigned to Community 1 has these characteristics.

By comparing the results with the overall attribute-value proportions for the whole graph and the seed profiles assigned to each community, we have validated that the output

Table 1. Attribute-value assignments (age to marital) - 50k nodes by 250k edges.

		All	Community 4	Community 1
Attribute	Value	Frequency	Frequency	Frequency
Age	18–25	10065	43	1935
	26–35	10247	48	706
	36–45	7517	19	173
	46–55	2620	17	87
	56–65	8726	92	71
	66–75	2922	841	60
	76–85	1136	107	88
Gender	Male	17041	101	1904
	Female	26457	1067	1220
Residence	Palo Alto	5040	105	1989
	Santa Barbara	9541	109	350
	Winthrop	8123	50	141
	Boston	8371	31	162
	Cambridge	8032	41	176
	San Jose	4097	831	302
Religion	Buddhist	8617	16	115
	Christian	9056	280	2242
	Hindu	10962	13	137
	Jewish	4007	794	222
	Muslim	3861	46	266
	Sikh	749	0	0
	Traditional Spirituality	0	0	0
	Other Religion	0	0	0
	No Religious Affiliation	4743	2	32
Marital	Single	14720	190	1839
	Married	14596	815	742
	Divorced	6043	78	285
	Widowed	7860	84	255

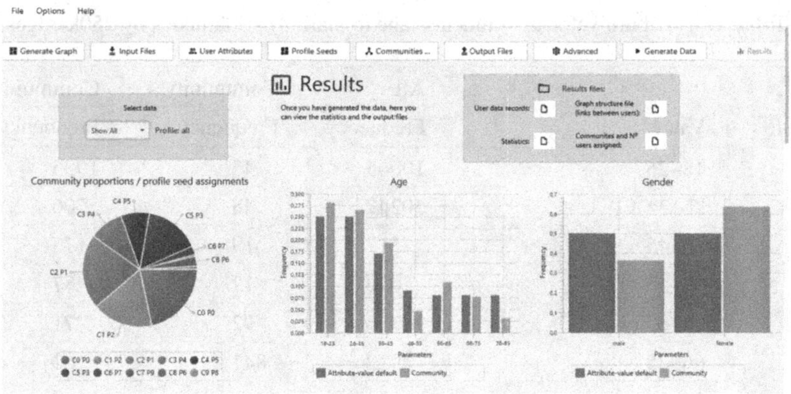

Fig. 4. Medici statistics page (all communities)

produced is as expected, taking into account the stochastic factor in the assignment which will cause some variations. Figure 4 shows part of the statistics page of the Medici application, showing the distributions for the age and gender attributes for the whole graph. The target distribution is shown in orange and the real one in yellow, which takes into account a grade of random variability in data assignment.

4.3 Stability of the Data Generator: Deviation Between Successive Executions

The system has been initially tested by 3 fold runs (for each parameter setting) for different graph dimensions (nodes, edges): 1k by 10k (Table 2), 10k by 100k and 50k by 250k (Table 3). The deviation for successive executions is detailed for each overall graph and for two selected community profile assignments. The results show a good stability, taking into account the stochastic nature of the data generator. In Table 2, for example, shows the deviations between three successive runs for populating the same graph and community structure with 1k nodes and 10k edges. It can be seen that for the whole graph, the average and standard deviation is in general below 4%, with the exception of gender which is slightly higher. For two example communities, 1 and 4, the average and standard deviation does not pass 10%, with the exception of Gender (Community 1) and Like1 (Community 4). It is noted that, due to the stochastic nature of the see profile assignment and propagation, variations are expected, but the general distributions should be fairly consistent for the same graph, communities and seed profile assignments.

Table 3 shows similar statistics for scale up to progressively bigger graphs, and it can be seen that the deviations become progressively smaller for scale-up: the maximum average deviation for the 50k graph is 1.5%.

Table 2. 1k nodes by 10k edges.

	All - deviation		Community 1 – deviation		Community 4 – deviation	
	Avg.	Stdev.	Avg.	Stdev.	Avg.	Stdev.
Age	1.8	1.2	6.4	4.1	5.3	3.2
Gender	6.7	4.5	18.4	10.2	9.2	7.2
Residence	1.5	0.9	4.4	3.1	5.1	3.7
Religion	1.8	1.1	2.7	2.0	2.3	1.3
Marital	3.9	2.2	5.6	3.1	9.2	7.1
Profession	0.8	0.6	6.7	4.4	2.0	1.7
Political	1.5	0.9	4.9	3.2	3.5	2.0
Sexuality	1.7	0.9	4.6	3.0	6.6	3.6
Like1	2.8	1.6	9.2	6.1	10.2	8.2
Like2	2.1	1.6	9.7	6.1	9.2	5.1
Like3	3.3	1.8	6.6	3.6	7.1	4.1

Table 3. 50k nodes by 250k edges.

	All - deviation		Community 1 – deviation		Community 4 – deviation	
	Avg.	Stdev.	Avg.	Stdev.	Avg.	Stdev.
Age	0.3	0.1	0.7	0.4	0.7	0.4
Gender	0.3	0.2	0.9	0.6	0.7	0.5
Residence	0.3	0.2	0.9	0.5	0.7	0.5
Religion	0.2	0.1	0.4	0.2	0.5	0.3
Marital	0.4	0.2	1.1	0.8	0.6	0.4
Profession	0.2	0.1	0.4	0.3	0.6	0.3
Political	0.2	0.1	0.6	0.4	0.9	0.5
Sexuality	0.3	0.1	1.1	0.6	0.7	0.5
Like1	0.5	0.3	1.3	0.6	1.1	0.7
Like2	0.4	0.3	1.5	0.9	0.6	0.4
Like3	0.7	0.4	1.3	0.7	1.1	0.5

5 Conclusions

The Medici application embodies a self-contained tool for generating synthetic data for small to medium size social network graphs. Using the R-MAT and Louvain algorithms, the user can create a network from scratch and label the communities. Then, the user can define the seed profiles based on user demographics and likes, assign the seeds to the communities, and finally generate the data using the Medici algorithm.

Although the system currently has the stated limitations, the data diversity and volume offered can serve for many useful studies and applications. It will also serve for software developers to scale up the data processing and add more flexible functionalities, such as a variable number of communities and adding new attributes and their values. As part of future work of scale-up, R3Mat [7] can be evaluated. As future work, dynamic user activity data is planned to be incorporated into the user affinities with time and place information.

References

1. Nettleton, D.F.: Data mining of social networks represented as graphs. Comput. Sci. Rev. **7**, 1–34 (2013)
2. The ethics of Big Data: Balancing economic benefits and ethical questions of Big Data in the EU policy context", Study of the European Economic and Social Committee (2017), Published by: "Visits and Publications" Unit EESC-2017–41-EN (2017)
3. Newman, N.: The costs of lost privacy: consumer harm and rising economic inequality in the age of Google. Wm. Mitchell L. Rev. **40**, 849 (2013)
4. Tomašev, N., et al.: AI for social good: unlocking the opportunity for positive impact. Nat. Commun. **11**(1), 1–6 (2020)
5. Park, H., Kim, M.S.: TrillionG: a trillion-scale synthetic graph generator using a recursive vector model. In: Proceedings of the 2017 ACM International Conference on Management of Data, pp. 913–928, May 2017
6. Samsi, S., et al.: Static graph challenge: Subgraph isomorphism. In: 2017 IEEE High Performance Extreme Computing Conference (HPEC), pp. 1–6. IEEE, September 2017
7. Angles, R., Paredes, R., & García, R. (2020). R3MAT: A Rapid and Robust Graph Generator. IEEE Access, 8, 130048–130065
8. Feng, Z., et al.: A schema-driven synthetic knowledge graph generation approach with extended graph differential dependencies (GDDxs). IEEE Access. **30**, 5609 (2020)
9. Pérez-Rosés, H., Sebé, F.: Synthetic generation of social network data with endorsements. J. Simul. **9**, 279 (2014). https://doi.org/10.1057/jos.2014.29
10. Ali, A.M., Alvari, H., Hajibagheri, A., Lakkaraj, K., Sukthankar, G.: Synthetic generators for cloning social network data. In: Proceedings SocInfo 2014 (2014)
11. Barrett, C.L., et al.: Generation and analysis of large synthetic social contact networks. In: Proceedings of the 2009 Winter Simulation Conference, 13–16 December 2009, pp.1003–1014 (2009)
12. Boncz, P., et al.: Benchmark Design for Navigational Pattern Matching Benchmarking. LDBC Cooperative Project FP7 – 317548. Coordinators: Arnau Prat, Alex Averbuch. Issue 3 28/09/2014 (2014)
13. Robins, G., Pattison, P., Woolcock, J.: Small and other worlds: global network structures from local processes. Am. J. Sociol. (AJS) **110**(4), 894–936 (2005)
14. Nettleton, D.F.: Generating synthetic online social network graph data and topologies. In: 3rd Workshop on Graph-based Tech. & Apps, UPC, Barcelona, Spain, March 2015
15. Nettleton, D.F.: A synthetic data generator for online social network graphs. Soc. Netw. Anal. Min. **6**(1), 1–33 (2016). https://doi.org/10.1007/s13278-016-0352-y
16. Nettleton, D.F., Salas, J.: A data driven anonymization system for information rich online social network graphs. Expert Syst. Appl. **55**, 87–105 (2016)
17. Nettleton, D.F., Nettleton, S., Canal i Farriol,, M. (2021). MEDICI: A simple to use synthetic social network data generator. arXiv preprint arXiv:2101.01956
18. Nettleton, D.F.: Social Network Synthetic Data Generator [Source code, Git repository] (2021). https://github.com/dnettlet/MEDICI

19. Torra, V., Jonsson, A., Navarro-Arribas, G., Salas, J.: Synthetic generation of spatial graphs. Int. J. Intell. Syst. **33**(12), 2364–2378 (2018)
20. Chakrabarti, D., Zhan, Y., Faloutsos, C., R-MAT: a recursive model for graph mining. In: Proceedings of the 2004 SIAM International Conference on Data Mining. Society for Industrial and Applied Mathematics, pp. 442–446 (2004)
21. Blondel, V.D., Guillaume, J.L., Lambiotte, R., Lefebure, E.: Fast unfolding of communities in large networks. J. Stat. Mech. Theory Experiment **10**, 1000 (2008)
22. Newman, M.E.J., Girvan, M.: Finding and evaluating community structure in networks, Phys. Rev. E **69**, 026113 (2004)
23. Lambiotte, R., Delvenne, J.-C., Barahona, M.: Laplacian dynamics and multiscale modular structure in networks. IEEE Trans. Network Sci. Eng. **1**(2), 76–90 (2015)
24. Canal i Farriol, M.: Interfície d'usuari per a una aplicació de generació de dades sintètiques per a xarxes socials, Final year undergraduate project, DTIC, Universitat Pompeu Fabra (2019)
25. Bastian, M., Heymann, S., Jacomy, M.: Gephi: an open source software for exploring and manipulating networks. In: Proceedings 3rd International AAAI Conference on Weblogs and Social Media, pp. 361–362 (2009)

Answer Passage Ranking Enhancement Using Shallow Linguistic Features

Bahadorreza Ofoghi[1(✉)] and Armita Zarnegar[2]

[1] School of Information Technology, Deakin University, Burwood Victoria 3125, Australia
b.ofoghi@deakin.edu.au

[2] Faculty of Science, Engineering, and Technology, Swinburne University of Technology, Melbourne, Australia
azarnegar@swin.edu.au

Abstract. Question Answering (QA) systems play an important role in decision support systems. Deep neural network-based passage rankers have recently been developed to more effectively rank likely answer-containing passages for QA purposes. These rankers utilize distributed word or sentence embeddings. Such distributed representations mostly carry semantic relatedness of text units in which explicit linguistic features are under-represented. In this paper, we take novel approaches to combine linguistic features (such as different part-of-speech measures) with distributed sentence representations of questions and passages. The QUASAR-T fact-seeking questions and short text passages were used in our experiments to show that while ensembling of deep relevance measures based on pure sentence embedding with linguistic features using several machine learning techniques fails to improve upon the passage ranking performance of our baseline neural network ranker, the concatenation of the same features within the network structure significantly improves the overall performance of passage ranking for QA.

Keywords: Question answering · Passage ranking · Deep learning · Shallow linguistic features

1 Introduction

Natural language Question Answering (QA) systems have recently been utilized to support decision analysis and modeling (e.g., [7,19]). Previous studies in the QA domain show that answer extraction is more effective from passage-level information compared with the analysis of full-text documents [12]. There is evidence of positive correlation between the effectiveness of QA and answer passage ranking [9]. For QA, both the general semantic relevance of a passage to the question and answer recall are of importance. For instance, given the question *"When did Google start?"*, the passage *"Google was launched by Larry Page and Sergey Brin, students at Stanford University"* will not be counted as an effective, answer-containing passage since it does not include the actual answer *1998*.

V. Torra and Y. Narukawa (Eds.): MDAI 2021, LNAI 12898, pp. 286–298, 2021.
https://doi.org/10.1007/978-3-030-85529-1_23

Specificity of passages (i.e., containing answer candidates) in the QA domain necessitates the utilization of explicit and shallow *linguistic* features, especially from within the passages to be considered in the process of passage ranking, those that are under-represented in the general semantics of sentences (see Fig. 1).

	Semantically relevant passage	Specific answer-containing passage
Q1	The.det most.adv frequent.adj symptom.noun is.verb a.det stiff.adj jaw.noun, caused.v by.adp spasm.noun of.adp the.det muscle.noun that.adj closes.v the.det mouth.noun – accounting.v for.adp the.det disease.noun 's.part familiar.adj name.noun lockjaw.noun.	The.det first.adj sign.noun of.adp tetanus.noun is.v a.det tightening.noun of.adp the.det jaw.noun muscles.noun that.adj gives.v the.det disease.noun its.adj common.adj name.noun, lockjaw.noun.
Q2	He.prp composed.v many.adj operas.noun, but.conj his.adj greatest.adj triumph.noun was.v "I.noun Pagliacci.noun" for.adp which.adj he.prp wrote.v both.adj the.det libretto.noun and.conj the.det music.noun.	It.prp 's.v a.det setting.noun that.adj would.v have.v brought.v tears.noun of.adp joy.noun to.adp Ruggero.noun Leoncavallo.noun, the.det composer.noun and.conj librettist.noun who.noun gave.v Pagliacci.v life.noun in.adp 1892.num.
Q3	At.adp that.det time.noun, it.prp was.v called.v Edo.noun.	Tokyo.noun was.v formerly.adv called.v Edo.noun.

Q1: Lockjaw is another name for which disease? (tetanus)

Q2: Who wrote the opera Pagliacci? (leoncavallo)

Q3: What city was originally called edo? (tokyo)

Fig. 1. Example question and passage cases where part-of-speech of tokens in passages, named entities, and query term coverage are shown. Answer-containing passages demonstrate specific linguistic characteristics, e.g., more nominal terms, less pronouns, and sufficient query term coverage. Note: The questions and passages are taken from the QUASAR-T development set (see Sect. 2.1 for more details).

Recent advances in fact-seeking QA and passage retrieval have been based on the utilization of distributed word representations as well as deep learning structures. The works in [9,15] are based on the utilization of distributed word embeddings learned using word2vec [10] and GloVe [13] to represent the text of questions and passages with word-level embeddings and to find the most relevant passages. The work in [15] relies mainly on Convolutional Neural Networks and word embeddings as the discriminant analyzer to score and rank passages. The several rankers developed in [9] with LSTMs have been reported to outperform another deep learning-based passage retrieval system in [18] that developed a Reinforced Ranker-Reader QA system. The LSTMs were also employed in [2] in combination with word embeddings and character embeddings, and resulted in improvements over several baseline traditional and deep learning-based answer passage retrieval systems. The work in [11] made use of Bidirectional Encoder Representations from Transformers (BERT) [5]. The evaluation results of this technique on the TREC CAR and MS MARCO data sets show significant improvements over some of the state-of-the-art passage ranking techniques.

Beyond the distributed representation of characters and terms, there have been efforts to capture semantics of the larger portions of text, such as sentences [3,8] . InferSent [3], for instance, developed sentence embeddings to encode the overall meaning of sentences. The InferSent encodings have been shown to generalize well to several natural language processing (transfer) tasks, such as multi-genre natural language inference and semantic textual similarity analysis. InferSent embeddings were used in [9] with feed-forward neural networks to train a passage ranking system which performed well on the QUASAR data set [6].

While previous works in the domain of answer passage retrieval and ranking have made significant progress in retrieving and ranking answer candidate passages at top ranks through the use of deep textual features (e.g., word and sentence embeddings), the possible contribution of more explicit utilization of linguistic features has not been studied.

Our approach to fill this gap is based on the utilization of both sentence-level semantics and explicit representation of several linguistic passage features. Focusing on passage ranking only and leaving aside answer extraction to further machine comprehension stages of QA that are not part of this work, we represent each sentence with its sentence embedding using InferSent [3]. The sentence embeddings are fed into a deep feed-forward neural network to predict whether or not a passage is likely to contain a candidate answer to a question. We then make use of the linguistic features of passages including token count, noun count, verb count, adverb count, pronoun count, query coverage, and named entity count to analyze the contribution of these features in passage ranking and to improve the final passage ranking effectiveness in terms of mean rank (MR) and answer recall of passages.

2 Methods

2.1 Data Set

The QUASAR-T QA data set [6] was used in our experiments which includes training, development, and test subsets, each with short and long passages retrieved per question (100 short and 20 long passages). We focused on the short passages in the development and test subsets, each of which containing 3,000 questions. While there are other data sets for QA, e.g., SQUAD [14], the 1-to-many question-to-passage requirements are not met by such data sets to facilitate passage ranking experiments.

2.2 Deep Neural Network Ranker

The baseline ranker in our analyses was a feed-forward deep neural network model. We constructed the input feature vector X_i to this ranker similar to the work in [9] and by concatenating question embedding (qe_i) and passage embedding (pe_i) that go through the network structure to find the answer-containing probability for passage i, as shown in the following equations.

Table 1. MR and recall (r@top) analysis of our baseline model (BL-NN) and relevant methods on the test set. The results that are not available (not reported in referenced works) are shown with n/a.

Method	mr	r@1	r@2	r@3	r@4	r@5
BL-NN	9.58	0.25	0.39	0.47	0.54	**0.58**
R^3	n/a	**0.40**	n/a	0.51	n/a	0.54
InferSent	n/a	0.36	n/a	**0.52**	n/a	0.56

$$X_i = [qe \oplus pe_i \oplus (qe - pe_i) \oplus (qe \odot pe_i)] \tag{1}$$

$$X_i^{flat} = \text{flatten}(X_i) \tag{2}$$

$$D^{(1)} = \text{ReLU}(W^{(1)} X_i^{flat}) \tag{3}$$

$$O_i = \text{softmax}(W^{(2)} D^{(1)}) \tag{4}$$

The embeddings for both questions and passages were constructed using InferSent where the output embedding per sentence has 4096 features. The input vector X, therefore, included 16,384 features. In cases where the text included more than one sentence, the embeddings of sentences were vector summed to create the representative sentence embedding of the entire text. To train this baseline model, the QUASAR-T development set of questions and short contexts were utilized. For each question, the contexts were first pre-processed and pseudo-labeled according to whether they contained the actual answer to the question. Then, a subset of 1 positive context and 5 negative contexts were extracted per question to train the baseline model in 500 epochs without any early stopping criteria or any regularization method. The pseudo-labeling of contexts resulted in a 2-feature output vector per context; hence, the output layer of the model is a dense layer including two neurons with Softmax activation. We modeled the passage ranking task as a classification problem similar to [11]; thus, the binary cross-entropy loss function was used, i.e., $L = -\sum_{i \in P_{pos}} log(p_i) - \sum_{j \in P_{neg}} log(p_j = 1 - p_i)$ where the P_{pos} and P_{neg} index sets represent the positive and negative pseudo-labeled contexts, p_i is the answer-containing probability (class = 1), and p_j is the probability of class = 0.

The trained model would then generate a probability per class, i.e., answer-containing versus answer-free. The answer-containing probability of each passage was used to rank passages. The loss and accuracy of the model in the training phase is shown in Fig. 2. Table 1 shows the detailed results of this model when applied on the test set as compared with two existing, relevant (neural network-based) systems R^3 [18] and InferSent ranker [9] without the utilization of other lexical semantic features.

2.3 Explicit Shallow Linguistic Features

Distributed sentence representations capture several surface, syntactic, and semantic characteristics of text [4]. However, in the context of QA, there are

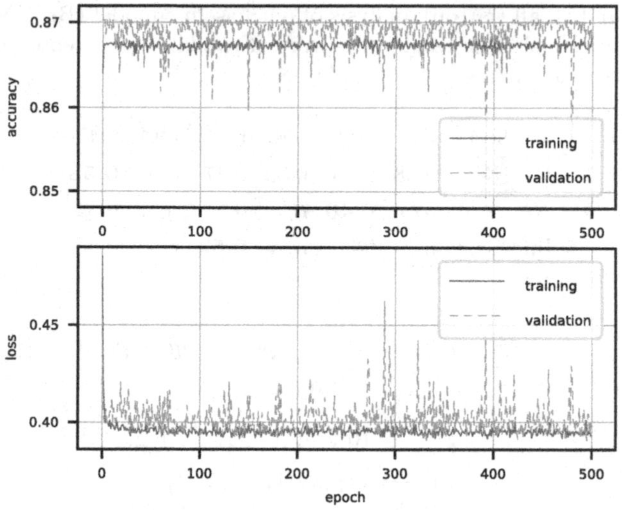

Fig. 2. Loss and accuracy analysis of the baseline neural network ranker on the QUASAR-T development set. Note: Higher validation accuracy values are due to the validation accuracy being calculated at the end of each epoch versus training accuracy being calculated batch-wise.

other explicit linguistic features that can potentially enhance answer passage retrieval and ranking and yet they are under-represented in distributed embeddings. The explicit linguistic characteristics that we focused on included the following categories.

First, terms of specific part-of-speech can distinguish between an answer-containing passage and the one that is less likely to include an explicit answer to a question. These features include the number of nouns, verbs, adverbs, pronouns, as well as the general count of tokens within the text of a passage. It can be argued that for fact-seeking questions, it is less likely that the answer will be in the form of an adverb (or a verb) while it is more likely to be a noun or a nominal predicate. In addition, the larger number of tokens can be argued to have a positive impact on the chances of a passage containing the actual or candidate answer. Pronouns, on the other hand, can mask the actual answer within a passage and as such, the smaller number of pronouns may result in higher quality passages. As one example from the QUASAR-T data set, the question *"Which is considered the most powerful piece on the chess board?"* has contexts such as *"The queen is the most powerful piece in the board"* and *"She is the most powerful piece on the board"*. The correct answer is masked by the pronoun *"She"* in the second passage, which makes the passage less effective for QA purposes.

Second, in fact-seeking QA, the answer is most likely a text snippet that refers to the name of a location, organization, or a person. In some other cases, the date, time, or a monetary reference is sought. While named entities have been used as a category of features for matching answer candidates and questions in [16], they have not been used for featurizing passages for their likelihood of answer-containing. We will show that a larger number of named entities are found in answer-containing passages.

Third, query term coverage within a passage was selected as another signal to correctly identify answer-containing passages. The argument is on the basis that correct actual answers to a given question may mostly be positioned in the close proximity of the same query terms that are mentioned in the text of the question. This is besides the fact that a larger proportion of query term coverage also contributes to the semantic relatedness of passages and questions. These two concepts (proximity of answers to query terms and coverage of terms) can be found in a more traditional passage retrieval and ranking system called MultiText [1].

We conducted an exploratory analysis of the above features in the QUASAR-T development set. The contexts were first pseudo-labeled; then, features were extracted for every context. There were 35,162 positive and 263,804 negative (answer-free) contexts. Separated by pseudo-labels (class = 1 indicating answer-containing passages), Table 2 summarizes the descriptive statistics of the two cohorts of passages. The chart demonstrates that the medians and distributions of feature counts have meaningful differences between the two classes of passages in most cases. A two-tailed statistical t-test was then conducted on the distribution of each feature in the two passage classes and it was found that, except in the case of verb counts ($p = 0.31$), the means of all the other linguistic features were significantly different from each other at the 95% confidence level (with $p = 0.00$). The distributions show that answer-containing passages, on average, have larger token counts, noun counts, named entity counts, and query coverage while they also include smaller adverb and pronoun counts. These results were contradictory to one previous work in [17] which found that verbs can substantially contribute to the task of QA passage ranking. Our findings are, however, in agreement with the same work in terms of noun counts as [17] reported that nominal predicates can positively impact on answer passage ranking. As a result, we preserved all the explicit linguistic features in our experiments.

2.4 Fusion of Linguistic Features and Deep Semantics

2.4.1 Traditional Machine Learning Fusion

In the first attempt to enhance our baseline deep neural network ranker using explicit linguistic features, we used the answer-containing probability generated by the baseline ranker in combination with the explicit features extracted for each passage as the predictor set to re-classify the passages into positive versus negative classes. A number of traditional machine learning algorithms, including logistic regression, Gaussian naive Bayes, decision tree, random forest, linear support vector machines, and Sigmoid support vector machines, were utilized. To train each classifier, we applied our baseline neural network ranker on the QUASAR-T development data set to obtain the answer-containing probabilities (1 positive and 5 negative contexts per question), where a Gaussian noise

Table 2. Descriptive analysis of linguistic features between answer-free (class = 0) and answer-containing (class = 1) contexts in the development data set.

Feature	class = 0		class = 1		*p*-value
	mean	stdv	mean	stdv	
noun count	8.44	6.27	10.94	6.36	0.00
verb count	2.33	1.86	2.32	1.77	0.31
token count	21.83	10.72	25.49	9.79	0.00
adverb count	0.55	0.87	0.53	0.84	0.00
named entity count	2.00	2.01	2.94	2.21	0.00
query coverage	1.86	1.18	2.36	1.25	0.00
pronoun count	0.56	0.97	0.42	0.81	0.00

Table 3. The AUC analysis of the second-level classification of passages using answer-containing probabilities of the BL-NN ranker and the explicit linguistic features.

Measure	LR	RF-bCV	SVM-sigmoid	SVM-linear	GNB	DT-bCV
AUC	0.59	0.81	0.53	0.48	0.56	0.58

(mean = 0.0 and standard deviation = 0.1) was added to the probabilities for the baseline ranker was first trained using the same data set. Then, the linguistic features of the same development passages were extracted. These features and the answer-containing probabilities were then normalized using the $L2$ normalization technique and were fed into the machine learning techniques for training. Table 3 summarizes the AUC results of the different techniques, where the random forest and decision tree models went through a 5-fold cross-validation process to find the best maximum depth of the trees.

From the above machine learning techniques, the best random forest model found using cross validation (RF-bCV) had the best AUC; thus, it was selected for ranking of passages in the QUASAR-T test set. This model did not perform well as shown in Table 4.

2.4.2 Deep Learning-Based Score and Linguistic Feature Fusion

A similar procedure to the traditional machine learning fusion approach (detailed in the previous section) was taken to train a deep feed-forward neural network model (with the same structure as in the baseline neural network ranker) in 50 epochs this time. The input to the second-level network (2nd-NN) was low-dimensional and included the same answer-containing probabilities of the first baseline model (plus the Gaussian noise for training) as well as the linguistic features of passages; hence, the number of epochs was set to a much smaller number in this experiment (50 epochs). This model, when tested on the QUASAR-T test set, resulted in a better set of performance measures compared with the traditional random forest model in the previous experiment; however, the baseline neural network model was not improved upon as detailed in Table 4.

2.4.3 Deep Learning-Based Augmentation

In another experiment, the baseline deep neural network model was augmented with the explicit linguistic features extracted for passages. The augmentation of these features was done in the middle layer of the network by concatenating the outputs of the first dense layer (including 10 nodes) with the 7 linguistic features. The overall process, involving the augmented neural network ranker, is shown in Fig. 3. The linguistic feature augmentation process is especially structured in the middle layer instead of the input layer with a large number of nodes (16,384) to more directly and strongly infuse the effect of the linguistic characteristics of passages into the neural model. The input vectors to this model include the same X_i in Eq. 1 as well as LFs_i for linguistic features of passage i which go through the network structure to find the answer-containing probability of the passage as shown in Eqs. 5–7.

$$X_i^{flat} = \text{flatten}(X_i) \tag{5}$$

$$C^{(1)} = \text{ReLU}(W^{(1)} X_i^{flat}) \oplus LFs_i \tag{6}$$

$$O_i = \text{softmax}(W^{(2)} C^{(1)}) \tag{7}$$

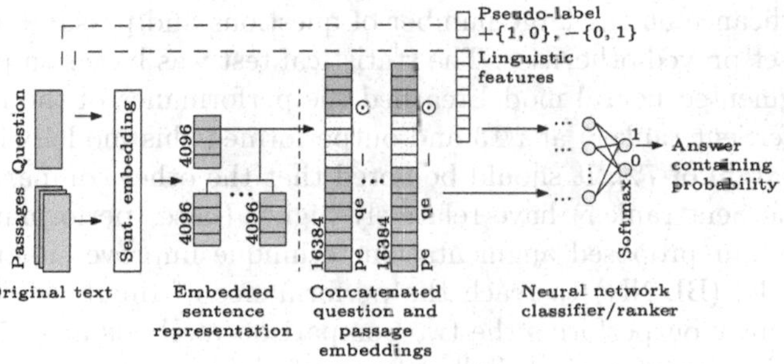

Fig. 3. The schematic view of the linguistically augmented passage ranking process for QA using a feed-forward deep neural network.

The augmented neural model was trained using the QUASAR-T development set with the same settings as in the baseline deep neural network model; a cross-entropy loss function, 500 training epochs, 1 positive passage, 5 negative passages, no early stopping criteria, and without drop-out or any other type of regularization. The loss and accuracy of the model in the training cycles were similar to those of the baseline ranker as shown in Fig. 2.

Table 4. MR and recall analysis of the rankers developed. aug-NN is the neural model augmented with linguistic features. The †s indicate statistically significant differences compared with BL-NN at the 95% confidence level.

Method	mr	r@1	r@2	r@3	r@4	r@5
BL-NN	9.58	0.25	0.39	0.47	0.54	0.58
RF-bCV	21.44	0.07	0.13	0.18	0.22	0.26
2nd-NN	13.61	0.24	0.36	0.45	0.52	0.56
aug-NN	**8.79†**	**0.30†**	**0.44†**	**0.52†**	**0.59†**	**0.63†**

Table 5. Average loss and average accuracy analysis of the neural rankers developed on the test questions/passages.

Method	avg. loss	avg. accuracy
BL-NN	0.3624	0.8869
2nd-NN	0.3580	0.8875
aug-NN	0.3610	0.8876

More importantly, this model outperformed the baseline deep neural network ranker (BL-NN) with respect to all of the QA-based evaluation metrics in our experiments, i.e., MR and recall at different levels. The detailed results of this model along with the other experimental models are summarized in Table 4. Although the improvements may seem marginal on the surface, the statistical test of significance on the large number of questions and passages in the benchmark data set proved otherwise. The statistical test was based on paired t-tests.

Our augmented neural model reached the performance of the best model in Table 1 (InferSent ranker) at r@3 and outperformed this model with respect to r@5 by a margin of 7%. It should be noted that the other comparison methods (R^3 and InferSent ranker) have relatively higher (base) performance values at r@$\{1, 3\}$ yet our proposed augmentation technique improves upon our weaker baseline model (BL-NN) to reach the performance of InferSent ranker at r@3 and significantly outperforms the two comparison methods at r@5. Also, while the proposed augmentation of shallow linguistic features was only applied on our BL-NN model and resulted in statistically significant improvements, a similar positive effect can be expected on the other comparison rankers too.

In terms of the classification performance of the several neural network models developed, the resulted of a detailed analysis of the average cross-entropy loss and average accuracy of the models are summarized in Table 5. These results are on the 3,000 QUASAR-T test questions and passages. As shown in Table 5, the average loss and accuracy of the models do not differ significantly (all within 1% variance); however, the QA-based metrics of final passage ranking have been shown to significantly improve using the augmented model.

3 Discussion

Passages that are more likely to contain specific answers to fact-seeking questions were shown to present with several linguistic features, mostly at the syntactic and lexical levels, that can further separate them from those that are less likely to recall any candidate answers. Even in presence of deep semantic relatedness between questions and passages, surface and explicit features can eventually be assisting in distinguishing between positive answer-containing and negative passages and thus in better ranking of answer passages with a vision of improvements in overall QA effectiveness. The explicit lexical and syntactic characteristics of passages intrinsically increase chances of the text of a passage to contain a candidate or a correct answer to the question. The descriptive statistical analysis that we conducted on the two cohorts of pseudo-labeled passages (positive versus negative) along with statistical tests demonstrated significant differences between the distributions of the textual features in the two passage classes with the exception on verb counts. The latter finding regarding verbs is contradictory to the previous studies that showed verbs play a substantial role in answer passage retrieval [17].

In addition to the exploratory and descriptive statistical analysis of the surface linguistic features within passages, that suggest there are lexical differences between likely answer-bearing passages and those that are less likely to recall a candidate answer, the procedure taken to utilize these features was demonstrated to play an important role. The mere fusion of surface passage characteristics with answer-containing probability calculated through a more sophisticated deep semantic-oriented neural model was shown not to reach high levels of eventual answer passage ranking effectiveness measures. This failed experiment with both traditional machine learning and deep neural network models indicates that using the explicit linguistic features at a late stage of passage (re)classification and ranking is not effective.

To understand the relationship between the semantic relatedness measure of question-passage pairs and the explicit linguistic characteristics of passages, we used the answer-containing probabilities calculated for the contexts in the development set as a proxy for semantic relatedness and found the correlation between this measure and each of the linguistic features. We used the same set of 1 positive and 5 negative passages per question, the same data set that was used to train the second-level classifiers. As shown in Table 6, verb, adverb, and pronoun counts have the lowest correlations with the answer-containing probability of a passage, the latter two are negative. Noun and named entity counts have the largest (fair) correlations with the probability measure. None of the features were overly correlated with the probability of answer-containing, which removes the possibility of multicollinearity on answer-containing probability, and yet the method fails in better positioning answer passages.

The set of the same surface textual features combined internally within the structure of the deep neural network model (concatenated with the middle layer) fulfill the expectation of improvement over the effectiveness of the linguistic-feature-free baseline neural model. The statistically significant improvements

Table 6. Correlation analysis between BL-NN answer-containing probabilities and the linguistic features of development passages. Note: a.pr = answer-containing probability.

Feature	#tokens	#nouns	#verbs	#adverbs	#pronouns	query coverage	#named entities
a.pr	0.48	0.57	0.01	−0.05	−0.14	0.22	0.55

over the performances of the baseline neural ranker support our hypothesis that the combination of the semantic relatedness of question-passage pairs (the output of the middle dense layer of the neural network model) and the surface passage features can improve answer passage ranking for fact-seeking QA.

The three neural network rankers we developed have very similar average cross-entropy loss and average accuracy values over the test questions and passages; however, in terms of the passage retrieval-based evaluation metrics (i.e., MR and recall), the ability of the rankers in positioning answer-containing passages at better ranks significantly differ from each other when linguistic features of passages are augmented within the network structure.

4 Conclusions

We analyzed the effect of several explicit, shallow linguistic features of textual passages that can enhance the overall effectiveness of answer passage ranking for fact-seeking QA. Several experiments were carried out to improve upon a baseline neural network ranker that makes use of deep semantics in sentence embeddings. The fusion of token count, noun count, verb count, adverb count, pronoun count, query coverage, and named entity count within passages with the answer-containing probabilities obtained through the application of the baseline neural model using traditional machine learning as well as a second-level neural network did not result in improved passage ranking effectiveness. However, when the same features were internally augmented with the middle layer of the baseline neural network ranker, the augmented model significantly outperformed the baseline ranker with respect to MR and recall at different levels. Our next steps will focus on more complex neural models and the effect of the infusion of a more comprehensive set of linguistic features, such as scenario-based and chunk-based textual relations as well as dependency trees/relationships.

References

1. Clarke, C.L.A., Cormack, G., Lynam, T., Li, C., McLearn, G.: Web reinforced question answering (MultiText experiments for TREC 2001) (2001)
2. Cohen, D., Croft, W.B.: A hybrid embedding approach to noisy answer passage retrieval. In: Pasi, G., Piwowarski, B., Azzopardi, L., Hanbury, A. (eds.) ECIR 2018. LNCS, vol. 10772, pp. 127–140. Springer, Cham (2018). https://doi.org/10.1007/978-3-319-76941-7_10

3. Conneau, A., Kiela, D., Schwenk, H., Barrault, L., Bordes, A.: Supervised learning of universal sentence representations from natural language inference data. In: Palmer, M., Hwa, R., Riedel, S. (eds.) EMNLP. pp. 670–680. Association for Computational Linguistics (2017)
4. Conneau, A., Kruszewski, G., Lample, G., Barrault, L., Baroni, M.: What you can cram into a single $&!#* vector: probing sentence embeddings for linguistic properties. In: Proceedings of the 56th Annual Meeting of the Association for Computational Linguistics (Volume 1: Long Papers), pp. 2126–2136. Association for Computational Linguistics, Melbourne, Australia, July 2018. https://doi.org/10.18653/v1/P18-1198
5. Devlin, J., Chang, M.W., Lee, K., Toutanova, K.: BERT: pre-training of deep bidirectional transformers for language understanding. In: Burstein, J., Doran, C., Solorio, T. (eds.) NAACL-HLT (1), pp. 4171–4186. Association for Computational Linguistics (2019)
6. Dhingra, B., Mazaitis, K., Cohen, W.W.: Quasar: datasets for question answering by search and reading. CoRR abs/1707.03904 (2017)
7. Goodwin, T.R., Harabagiu, S.M.: Medical question answering for clinical decision support. In: Proceedings of the 25th ACM International on Conference on Information and Knowledge Management, CIKM 2016, pp. 297–306. Association for Computing Machinery, New York, NY, USA (2016). https://doi.org/10.1145/2983323.2983819
8. Hill, F., Cho, K., Korhonen, A.: Learning distributed representations of sentences from unlabelled data. In: Knight, K., Nenkova, A., Rambow, O. (eds.) HLT-NAACL, pp. 1367–1377. The Association for Computational Linguistics (2016)
9. Htut, P.M., Bowman, S., Cho, K.: Training a ranking function for open-domain question answering. In: Proceedings of the 2018 Conference of the North American Chapter of the Association for Computational Linguistics: Student Research Workshop, pp. 120–127. Association for Computational Linguistics, New Orleans, Louisiana, USA, June 2018. https://doi.org/10.18653/v1/N18-4017
10. Mikolov, T., Chen, K., Corrado, G., Dean, J.: Efficient estimation of word representations in vector space. In: Bengio, Y., LeCun, Y. (eds.) ICLR (Workshop Poster) (2013)
11. Nogueira, R., Cho, K.: Passage re-ranking with BERT. CoRR abs/1901.04085 (2019)
12. Oh, H.J., Myaeng, S.H., Jang, M.G.: Semantic passage segmentation based on sentence topics for question answering. Inf. Sci. **177**(18), 3696–3717 (2007). https://doi.org/10.1016/j.ins.2007.02.038
13. Pennington, J., Socher, R., Manning, C.D.: Glove: Global vectors for word representation. EMNLP **14**, 1532–1543 (2014)
14. Rajpurkar, P., Jia, R., Liang, P.: Know what you don't know: unanswerable questions for squad. CoRR abs/1806.03822 (2018). http://arxiv.org/abs/1806.03822
15. Rosso-Mateus, A., González, F.A., Montes-y-Gómez, M.: A two-step neural network approach to passage retrieval for open domain question answering. In: Mendoza, M., Velastín, S. (eds.) CIARP 2017. LNCS, vol. 10657, pp. 566–574. Springer, Cham (2018). https://doi.org/10.1007/978-3-319-75193-1_68
16. Suzuki, J., Sasaki, Y., Maeda, E.: SVM answer selection for open-domain question answering. In: COLING 2002: The 19th International Conference on Computational Linguistics (2002)
17. Verberne, S., Boves, L., Oostdijk, N., Coppen, P.A.: Using syntactic information for improving why-question answering. In: Scott, D., Uszkoreit, H. (eds.) COLING, pp. 953–960 (2008)

18. Wang, S., et al.: R3: reinforced ranker-reader for open-domain question answering. In: AAAI (2018)
19. Wen, A., Elwazir, M.Y., Moon, S., Fan, J.: Adapting and evaluating a deep learning language model for clinical why-question answering. JAMIA Open. **3**(1), 16–20 (2020). https://doi.org/10.1093/jamiaopen/ooz072

Neural Embedded Dirichlet Processes for Topic Modeling

Miguel Palencia-Olivar[1,2](✉), Stéphane Bonnevay[1,2], Alexandre Aussem[3], and Bruno Canitia[1]

[1] Lizeo IT, 42 quai Rambaud, 69002 Lyon, France
{miguel.palencia-olivar,stephane.bonnevay,bruno.canitia}@lizeo-group.com

[2] Laboratoire ERIC, Université de Lyon, 5 Avenue Pierre Mendès France, 69500 Bron, France

[3] Laboratoire LIRIS, Université de Lyon, 25 avenue Pierre de Coubertin, 69622 Villeurbanne Cedex, France
alexandre.aussem@univ-lyon1.fr

Abstract. This paper presents two novel models: the neural Embedded Dirichlet Process and its hierarchical version, the neural Embedded Hierarchical Dirichlet Process. Both methods extend the Embedded Topic Model (ETM) to nonparametric settings, thus simultaneously learning the number of topics, latent representations of documents, and topic and word embeddings from data. To achieve this, we replace ETM's logistic normal prior over a Gaussian with a Dirichlet Process and a Hierarchical Dirichlet Process in a variational autoencoding inference setting. We test our models on the 20 Newsgroups and on the Humanitarian Assistance and Disaster Relief datasets. Our models present the advantage of maintaining low perplexity while providing analysts with meaningful document, topic and word representations that outperform other state of the art methods, while avoiding costly reruns on large datasets, even in a multilingual context.

Keywords: Topic modeling · Text mining · Natural language processing · Deep learning

1 Introduction

Widely used in both industry and academia [6], topic models are among the go-to set of tools when it comes to unsupervised text exploration. Since its introduction, *the Latent Dirichlet Allocation (LDA)* [2] has been used as a basic canvas for a variety of topic models with different hypothesis sets and use-cases [4,7,30]. LDA is a two-level model that hypothesizes that in a corpus, pairwise-independent documents are mixtures of topics that themselves are mixtures of multinomials applied at word-level. The inference processes [5,13] aim at determining the mixture proportions for each level. These inference processes have different inspirations that best fit in different settings [5]. In this paper, we focus

V. Torra and Y. Narukawa (Eds.): MDAI 2021, LNAI 12898, pp. 299–310, 2021.
https://doi.org/10.1007/978-3-030-85529-1_24

on neural variational inference [18], as it enables benefitting from both neural networks and LDA-like topic model properties and advances, while maintaining good interpretability of the results. The *Embedded Topic Model* (ETM) [11] is a particularly interesting example, as it extracts document, topic and word representations while handling topic and word similarities. Despite these features, and as most neural-enhanced topic models, the topic number is considered a hyperparameter. As a consequence, the user is forced to perform multiple runs for successful model selection, as it is impossible to guess the number of topics in a massive amount of text whose topics are unknown by definition. Reruns are relatively easily feasible for small corpora analysis, however, they come with important cost concerns when the experiments are run at large scale, such as in Big Data contexts.

To circumvent the issue, we extend the ETM [11] to nonparametric settings by devising Embedded Dirichlet Processes and Embedded Hierarchical Dirichlet Processes. These two *Variational autoencoder-based (VAE)* topic models automatically infer the number of topics from data thanks to Beta and Gamma distributions and a stick-breaking construction. Our models present the advantage of maintaining low perplexity while providing analysts with meaningful document, topic and word meaning representations that outperform other state of the art methods, while avoiding costly reruns on large datasets. Our work is an additional step towards building extensive text summarization algorithms that can handle massive streams of text. We primarily focus on maintaining a trade-off between predictive power and interpretability, as literature shows that the sole goodness of fit does not correlate with human judgement [8].

The paper is organized as follows: Sect. 2 introduces general background on neural topic models and their capabilities to scale to massive amounts of data. Section 3 also provides background on our work, with an emphasis on the ETM and the Dirichlet Process. We present our models in Sect. 4, along with their optimization objectives. Section 5 is dedicated to empirical studies. Finally, Sect. 6 presents our both our conclusions and future work directions.

2 Related Work

Neural topic models emerged following works on black-box variational inference [18, 24]. These works aim at providing a framework for graphical model inference without having to entirely devise a new algorithm at each modification. The framework makes working with processes comparable to G. Box's loop [3] easier, as it alleviates the bottleneck of inference by using neural networks to learn both model and variational parameters from data in directed probabilistic models. VAEs [18] are the most famous realization of the framework. They consist of two neural networks working together: a neural network called encoder first learns the parameters for a given family of variational distributions and is trained to approximate a probabilistic prior. A common probabilistic setting for VAEs is a Gaussian prior on a Gaussian variational distribution. The second neural network called decoder uses a sample from the variational distribution (the "code", or latent variable) to reconstruct the data. The latent variable has inferior dimensionality with respect to the original data, thus making VAEs a technique for both latent variable extraction and dimensionality reduction. The way VAEs work reflects on the training objective, where the reconstruction loss, - the log-likelihood of the reconstructed data with respect to the original data - is regularized with the Kullback-Leibler divergence between the family of variational distributions and the prior. The setting applies to topic modeling, where the latent variables to infer are usually document-topic and word-topic matrices. Last but not least, model optimization is done using *stochastic variational inference*, thus enabling the whole setting to scale to massive streams of data, text included.

One of the earliest neural topic extractors that follow the VAE logic is the *Neural Variational Document Model* [19]. The setting, however, suffers from posterior collapse due to its Gaussian prior, which led to the development of the *ProdLDA* [25]. The latter model tries to approximate a Dirichlet prior thanks to its proximity to the logistic normal distribution. ProdLDA makes topic modeling with variational autoencoders stabler, and has served as a basis for further developments. Notable works include *TopicRNN* [10], *Gaussian softmax* and *steak-breaking constructions* [20], and ETM [11]. Dieng et al.'s work both include unsupervised word information, under both sequential form [10] and embedding form [11]. Diverging with ours, these models are fully parametric with respect to the number of topics. Contrasting with these works, Miao et al. [20] and Ning et al.[23] propose unsupervised settings, thanks to RNNs and to VAE for Dirichlet Processes [22], respectively. Still, there is no notion of word linking within these works. Literature presents an extensive research that aim at including word similarity in topic models. Most of these works treat semantic units as supporting information useful for topic rendering, thus making them auxiliary additions. Other methods switch priors to include some linking between words, tweak word assignment to the topics [9,16,29], or use pre-trained embeddings. The process even works in nonparametric settings, as reported in the *spherical Hierarchical Dirichlet Process* (sHDP) [1]; sHDP, though, does not conjointly learn word embeddings with topics, and needs analysts to provide previously trained word vectors. Contrasting with these approaches, our models' learning mode is fully unsupervised with respect to both the number of topics and latent representations (word, documents and topics), and do not need pre-training of any kind.

3 Background

Embedded Topic Model. ETM [11] extends the Latent Dirichlet Allocation [2] to include CBOW-like embeddings [21] in its generative process. The algorithm learns word embeddings from data by using entire document contexts instead of surrounding words. The contexts themselves are topic embeddings that are visualizable in the same space as the word embeddings. It is also possible - yet optional - to initialize the embedding layers with pre-fitted and more complex word representations, including words that do not appear in the dataset at hand. In this context, the model will still fit representations for the additional words according to their neighbourhood. ETM shares LDA's mixture assumptions, except that in contrast with the latter, words and topics can show similarities in their embedding space. ETM scaling is achieved by means of a VAE setting [18], which comes to the cost of a logistic normal prior instead of a Dirichlet to work with the explicit reparameterization trick.

Dirichlet Process and Stick-Breaking Construction. Stick-breaking construction for mixture model priors is often involved in *Dirichlet Process*-based models, topic models included [14,26]. Ishwaran & James [17] define a *stick-breaking prior* (SBP) as a random measure of the form $G(\cdot) = \sum_{k=1}^{\infty} \pi_k \delta_{\zeta_k}$, where δ_{ζ_k} is a discrete measure concentrated at $\zeta_k \sim G_0$, a draw from a base distribution G_0. The π_k terms are random weights that do not depend on G_0, chosen such $0 \leq \pi_k \leq 1$, and $\sum_k \pi_k = 1$ almost surely. The following procedure enable weight sampling:

$$\pi_k = \begin{cases} v_k \ if \ k = 1 \\ v_k \Pi_{j<k} (1 - v_j) \ for \ k > 1 \end{cases} \tag{1}$$

with $v_k \sim Beta(\alpha, \beta)$. When $\alpha = 1$, the *Beta* distribution is the stick-breaking construction for the Dirichlet Process. This setting is also referred to as *Griffiths, Engen and McCloskey* (GEM) distribution. The GEM distribution takes a single concentration parameter α_0 that equals to the β parameter in the Beta distribution. Nalisnick et al. [22] have successfully adapted stick-breaking for Dirichlet Processes to *VAE* settings. They use a GEM prior over a *Kumaraswamy* distribution instead of a Beta due to its challenging sampling with the explicit reparameterization trick.

Implicit Reparameterization Gradients. *Explicit reparameterization* [18,24] is key to computing low-variance gradients of continuous random variables. This technique, however, works better with location-scale distributions, and exclude a number of continuous ones. The Beta family of distributions is among them, Dirichlet included. Figurnov et al. [12] proposed *implicit reparameterization*, that not only permits using Gamma-based distributions[1], but also faster sampling and computing more accurate gradients with unbiased estimators. Comparably to the explicit reparameterization, implicit reparameterization builds on a standardization function $S_\phi(z)$ that removes the dependence on the parameters of the variational distribution when applied to a sample from the very same distribution. In explicit reparameterization, this function needs to be continuously differentiable with respect to its argument and parameters, and invertible. In implicit reparameterization, however, computing the gradients for the latent variables with respect to the variational distribution parameter only requires differentiating the standardization function, thus making the inference process resource savvy.

[1] Beta samples are computable from Gamma samples, see [12] for more details.

4 Our Models

4.1 The Embedded Dirichlet Process

The *Embedded Dirichlet Process* (EDP) is a nonparametric, VAE-based [18] topic model that infers the number of topics from the dataset at hand. Depending of the reparameterization trick, the model places a GEM prior either on a Kumaraswamy distribution (explicit) or a Beta distribution (implicit) to achieve this adjustment. Contrasting with ETM, EDP uses a GEM prior instead of a logistic normal. GEM not only enables nonparametric topic inference; it is also more expressive as it is strongly related to the Dirichlet distribution. The Dirichlet Process itself is indeed considered an extension of the Dirichlet distribution. Also, there is no need to constrain the prior to the simplex as shown in Dieng et al. [11]. Similarly to ETM, the EDP decomposes the word-level in a dot product between the (transposed) context embeddings ϕ and the word embeddings ρ. As such, it benefits from the same abilities to find topics and embedding spaces, to handle unseen words and to regroup stopwords.

Let $\{\mathbf{w}_1, \ldots, \mathbf{w}_D\}$ be a corpus of D documents, where $\mathbf{w}_d$ is a collection of N_d words. Each document is represented as a bag of words $\mathbf{x}_d$. To complete the generative process, we need to compute the joint distribution:

$$Pr\left(\mathbf{w}_{1:N}, \pi, \hat{\theta}_{1:N} \mid \alpha_0, \Theta, \beta\right) = Pr\left(\pi \mid \alpha_0\right) \Pi_{i=1}^{N} Pr\left(w_i \mid \hat{\theta}_{\mathrm{i}}, \beta\right) Pr\left(\hat{\theta}_{\mathrm{i}} \mid \pi, \Theta\right) \tag{2}$$

where $Pr\left(\pi \mid \alpha_0\right) = GEM\left(\alpha_0\right)$, $Pr\left(\theta \mid \pi, \Theta\right) = G\left(\theta; \pi, \Theta\right)$, and $Pr\left(w \mid \theta, \beta\right) = \sigma\left(\theta\beta\right)$. For simplicity, we denote the dot product between embedding matrices with $\beta = \sigma\left(\rho^T\phi\right)$ where $\sigma\left(\cdot\right)$ is the softmax function. We use a family of variational distributions whose parameters are inferred with multilayer perceptrons to bound the log-marginal likelihood, as described in [18]. We refer to this bounding as the *evidence lower bound* (ELBO), which is a function of both the model parameters and the variational parameters, and that we seek to maximize:

$$\begin{aligned}\mathcal{L}\left(\mathbf{w}_{1:N} \mid \Theta, \psi, \beta\right) = \mathbb{E}_{q_\psi\left(\nu \mid \mathbf{w}_{1:N}\right)}\left[\log Pr\left(\mathbf{w}_{1:N} \mid \pi, \Theta, \beta\right)\right] \\ - KL\left(q_\psi\left(\nu \mid \mathbf{w}_{1:N}\right) \parallel Pr\left(\nu \mid \alpha_0\right)\right)\end{aligned} \tag{3}$$

where $q_\psi\left(\cdot\right)$ denotes the family of variational distributions, ψ denotes the neural network parameters, and v denotes the weights for the stick-breaking step. We denote the neural networks with NN, and feed them bag of words $\mathbf{x}$. We use the Adam optimizer to fit the model for both the variational parameters and the distribution parameters.

Algorithm 1: Inference process for the EDP model

```
Initialize model and variational parameters;
for i ← 1 to N_batches do
    Compute β = σ(ρ^T φ);
    Choose a minibatch B of documents;
    foreach document d in B do
        Get a bag of words representation x_d;
        Compute a = NN(x_d; ψ_a);
        Compute b = NN(x_d; ψ_b);
        Sample v ~ Beta(a, b) or v ~ Kumaraswamy(a, b);
        Compute π = (π_1, π_2, ..., π_{K-1}, π_K) =
        (v_1, v_2(1 - v_1), ..., v_{k-1} Π_{l=1}^{K-2}(1 - v_l), Π_{l=1}^{K-1}(1 - v_l));
        foreach word w in the document do
            Compute Pr(w_dn | π) = σ(π β_{.,w_dn})
        end
    end
    Estimate the ELBO (Eq. 3);
    Update model and variational parameters through backpropagation;
end
```

4.2 The Embedded Hierarchical Dirichlet Process

In the precedent setting, the concentration parameter α_0 is treated as a hyperparameter. However, this hyperparameter greatly influences the number of breaks and, consequently, the number of topics. To constrain topic growth and avoid topic redundancy, we set a corpus-level prior on the base distribution. As the GEM distribution is equal to a Beta distribution whose first shape parameter is equal to 1, we can use the Gamma distribution parameterized with δ_1 and δ_2 as a conjugate prior for the GEM distribution. The ELBO then becomes the following:

$$\begin{aligned}\mathcal{L}(\mathbf{w}_{1:N} \mid \Theta, \psi, \beta) = \mathbb{E}_{q_\psi(\nu \mid \mathbf{w}_{1:N})}\left[\log Pr(\mathbf{w}_{1:N} \mid \pi, \Theta, \beta)\right] \\ + \mathbb{E}_{q_\psi(\nu \mid \mathbf{w}_{1:N}) q(\alpha \mid \gamma_1, \gamma_2)}\left[\log Pr(\nu \mid \alpha_0)\right] \\ - \mathbb{E}_{q_\psi(\nu \mid \mathbf{w}_{1:N})}\left[\log q_\psi(\nu \mid \mathbf{w}_{1:N})\right] \\ - KL\left(q(\alpha_0 \mid \gamma_1, \gamma_2) \parallel Pr(\alpha_0 \mid \delta_1, \delta_2)\right)\end{aligned} \tag{4}$$

5 Experiments

5.1 Datasets

The experiments feature the 20 Newsgroups (20NG) and the Humanitarian Assistance and Disaster Relief articles (HADR) [15] annotated datasets. Note that HADR comes with a lexicon we'll use for qualitative estimation of results.

Both datasets consist of collections of articles about several topics: 20 topics in the case of 20 Newsgroups, and 25 topics for HADR. The 20 Newsgroups contain 18846 articles, while HADR contains approx. 504000 ones in different languages. Due to technical limitation for this work, we retained a random subset of 20000 HADR articles for our experimentations. In each case, we used 85% of the entire dataset for the training sets, 10% for the validation sets and 5% for the test sets. We filtered out words that do not appear in at least 4 documents and removed stopwords to accommodate for our computational capabilities, thus yielding V-vocabularies of 28307 words from 20 Newsgroups and 32794 words from HADR.

5.2 Training Settings

We compare our results with Ning and et. al.'s iTM-VAE-Prod, iTM-VAE-G [23] and ETM [11]. These three models all include black box neural inference with explicit reparameterization tricks, although with different probabilistic settings. ETM uses a logistic normal prior on a Gaussian, and is a parametric topic model. iTM-VAE-Prod, is a nonparametric topic model that also places a GEM prior on a Kumaraswamy distribution, but the model does not include any kind of word similarity. For further study, we also adapted iTM-VAE-Prod to include implicit reparameterization, thus placing a GEM prior on a Beta distribution. To avoid posterior collapse and stabilize VAE trainings, we used batch normalization with a batch size of 1000 documents, and chose Adam with a learning rate of 0.002. For each model, we optimized the ELBOs for both the model and the variational parameters simultaneously. We performed exponential decay on both first (0.95) and second moment (0.99) estimates. We also used weight decay (1.2×10^{-6}), but we did not use KL-annealing. Last but not least, and following [11], we normalized bag of word representations of documents by dividing them by the number of words for document length accommodation. For each model, we used multilayer perceptrons with two hidden layers of 100 neurons. As for prior parameters, we set $\alpha = 1$ and $\beta = 5$ for both iTM-VAE-Prod and EDP, $\delta_1 = 1$ and $\delta_2 = 20$ for both iTM-VAE-G and EHDP, and a standard Gaussian for ETM. We kept the exact same settings for both datasets. We give parametric models capacities for 50 and 200 topics, and nonparametric models capacities for up to 200 topics, with a truncation level of 50 for hierarchical versions of the Dirichlet Process. All the aforementioned parameters, including distribution parameters and encoder sizes were selected with cross-validation using the metrics described below.

5.3 Metrics

In practice, analysts require a topic model to provide both good insights about the topics and good predictability of unseen documents. Most topic models, however, are only trained and selected from a statistical point of view, with topic coherence computed periodically due to its expensiveness. In this configuration, coherence comes as an additional indicator that is almost set apart from the training process. As our work focuses on maintaining a fair trade-off between goodness of fit and interpretability, we select our models based on a topic quality - perplexity ratio during the validation step.

Measuring Goodness of Fit. Perplexity is a common metric in both topic and language modeling. Its formula is the following: $\exp\Big(-\frac{1}{D}\sum_{d=1}^{D}\frac{1}{|\mathbf{w}^d|}\log Pr\left(\mathbf{w}^d\right)\Big)$. D stands for the number of documents, and $\mathbf{w}^d$ is the number of words in the d-th document. As the ELBO is an upper bound on perplexity, we use it to compute the indicator. Perplexity's main interest is for assessing the model's predictive power; the lower its value, the better. We compute perplexity following Wallach et al. [28], i.e. during a document completion task.

Measuring Topic Quality. Similarly to [11], we consider topic quality to depend on both redundancy and meaningful contents. As a consequence, we compute topic quality as the product of topic diversity and topic coherence. Topic diversity is the ratio between the number of unique tokens among the top 10 words in the topic list and the total number of words in this top 10. The higher, the better. As for coherence, we use the normalized pointwise mutual information (NPMI) to measure term co-occurrence in the corpora. Topic coherence (TC) is computed with the following equations:

$$TC_{NPMI} = \frac{1}{K}\sum_{k=1}^{K}\frac{1}{45}\sum_{i=1}^{10}\sum_{j=i+1}^{10} f\left(w_i^{(k)}, w_j^{(k)}\right) \tag{5}$$

where $f\left(w_i^{(k)}, w_j^{(k)}\right) = \frac{\log\frac{Pr(w_i,w_j)}{Pr(w_i)Pr(w_j)}}{-\log Pr(w_i,w_j)}$.

5.4 Results

Nonparametric topic models that use the explicit reparameterization trick all suffered posterior collapse during the experiments. They started producing NaNs as soon as the second epoch; as a consequence, we exclude them from our analysis. The phenomenon, however, confirms that implicit reparameterization yields value as it enables building more robust probabilistic settings. As for the other algorithms, we normalized perplexity following Dieng et al. [11], and found that they all have similar predictive powers, i.e., normalized perplexities. Thus, and according to our selection criteria, topic quality is predominant to determine the best models. Table 1 displays topic quality for every model that did not experience posterior collapse. Our Embedded Hierarchical Dirichlet Process significantly outperforms the other techniques in terms of topic quality, even ETM and EDP with implicit reparameterization, despite these algorithms sharing the same decoder as EHDP. In addition, note that the NPMI only takes into account the co-occurrence of words in a single document, when word embeddings is cross-corpora. Imagine two documents (1 and 2) and three words A, B, and C. Let A and B appear together in the first document, and B and C appear together in the second document. Similarly to a transitive relation, word embeddings will find that if A and B are close, and if B and C are close, then A and C also exhibit some similarity. The NPMI will not take this aspect into account. As a consequence, and for these very reasons, we believe that the NPMI underestimates coherence in document models with word embeddings. Besides, model optimization is only performed with respect to the goodness of fit. Model coherence is treated as an additional indicator for model selection, i.e. that we use coherence to choose amongst models with similar goodness of fit. We hypothesize that involving model quality in the optimization process may improve the results. This inclusion could be some form of regularization. As for topic coverage (Table 2), EHDP falls second to iTM-VAE-G, but does not collapse to a single topic as its nonparametric pairs. These results clearly show that EHDP is a robust technique with strong ability to adapt to datasets with large vocabularies, even when augmenting the number of words by nearly 16% when switching from 20 Newsgroups to HADR.

Table 1. Topic quality per dataset and per number of topics. Best results are in bold.

Models	Diversity				Coherence				Quality			
	20NG		HADR		20NG		HADR		20NG		HADR	
# Topics	50	200	50	200	50	200	50	200	50	200	50	200
ETM	0.47	0.32	0.45	0.28	0.15	−0.06	0.15	0.07	0.07	−0.02	0.07	0.02
iTM-VAE-G	0.91		**1.0**		0.10		0.16		0.09		0.16	
EHDP	0.52		**1.0**		**0.38**		**0.32**		**0.20**		**0.32**	
iTM-VAE-Prod - implicit	**1.0**		**1.0**		0.04		0.12		0.04		0.12	
EDP - implicit	**1.0**		**1.0**		0.04		0.08		0.04		0.08	

Table 2. Topic coverage with respect to human judgement.

Datasets	20NG (%)	HADR (%)
iTM-VAE-G	**10 (50%)**	1 (4%)
EHDP	9 (45%)	**8 (32%)**
iTM-VAE-Prod	1 (5%)	1 (4%)
EDP	1 (5%)	1 (4%)

Table 3 shows the topics extracted from HADR with EHDP[2]. While our model did not detect as much topics as human annotation, it fused some of them.

[2] In topic 1, rly refers to the railway. In topic 2, gva refers to Geneva, Switzerland, while dhagva means UN Department of Humanitarian Affairs in Geneva, and Spaak is a Belgian politician involved with humanitarian relief. In topic 3, tpc refers to the Tropical Prediction Center, and nwc refers to the National Weather Service. Finally, in topic 5, drc means Democratic Republic of the Congo.

For instance, in HADR, landslides, rains and floods are three distinct topics, while EHDP's fourth topic mixes them. While this feature is useful for getting more summarized results, it can hinder document classification due to intertwined variables. The last topic, i.e. the French stopwords, is pure serendipity. In [11], the authors show that ETM can handle stopwords and separate them in a distinct topic, but ETM is only tested in monolingual settings. HADR, however, is a multilingual dataset. As it seems, some documents in French remained after we filtered the dataset. EDHP still managed to distinguish stopwords among the vocabulary, while classifying them by language. We explain this result by the facts that our embeddings work with contexts, and that French words are much more likely to appear within French documents. Despite its accidental origin, the result is interesting, as multilingual topic modeling and text mining is still an open issue [27,31]. To confirm our intuition about multilingual topic modeling, we take benefit from our model's ability to generate embeddings. In particular, we extracted the 50 nearest neighbors of our French stopwords. Table 4 below shows an example. As expected, most of the neighbors are also French words, including non-stopwords. However, as we get lower in the ranking, non-French words start to appear (in bold). We hypothesize that in a multilingual corpus, some words do appear in several languages, especially event-related nouns, for instance. We think these words can act as pivots to link words from other languages, thus potentially enabling both supervised and unsupervised cross-lingual topic modeling with no extra adaptations.

Table 3. Complete list of topics extracted from HADR by the EHDP.

Topic	Word list
India & Bangladesh	tongi, manu, gorai, rly, storey, serjganj, kanaighat
United Nations	gva, dhagva, metzner, masayo, pbp, spaak, pos
Weather	tpc, nws, knhc, outward, forecaster, accumulations, ast
Floods & landslides	floods, landslides, padang, flooding, rain, mudslides, sichuan
Africa	drc, lusaka, monuc, burundi, darfur, congolese, amis
Economic development	development, financing, macroeconomic, management, reduction, usaid, sustainable
Politics & diplomacy	paragraph, decides, resolution, pursuant, vii, welcomes, stresses
French stopwords	les, qui, de, que, à , une, des

Table 4. 50 nearest neighbors of the "les" french stopword in decreasing order. Non-French words appear in bold.

les, par, à, que, », rapport, pas, «, lui, ses, nt, entre, fin, qui, cet, aux, ou, gouvernement, fui, bien, han, ces, haïtien, deux, manger, **gomez**, unies, rdc, ya, sont, ne, une, notre, ont, locales, avril, sur, première, dans, cours, santé, **una**, croix, **zona**, **santander**, vie, **sailing**, mais, milliers, **casos**

6 Conclusion and Future Work

In this paper, we developed two nonparametric models: the Embedded Dirichlet Process and its hierarchical version, the Embedded Hierarchical Dirichlet Process. We found that EHDP outperforms other state of the art algorithms in most configurations, and also shows signs of increased robustness to adapt to the dataset at hand. Besides, we found that EHDP can handle stopwords and make regroupments in a multilingual environment. As for its summarization capabilities, we noticed that the algorithm tend to combine several human-annotated topics into a single one. Future work will aim at exploring EHDP's multilingual capabilities and at including a nested Chinese Restaurant Process (nCRP) in its design to get more precise topics with a hierarchy. Last but not least, and as a third research direction, we will investigate alternatives to the NPMI to use with word embedding-enabled models.

References

1. Batmanghelich, K., et al.: Nonparametric spherical topic modeling with word embeddings, pp. 537–542 (2016)
2. Blei, D., et al.: Latent Dirichlet allocation. J. Mach. Learn. Res. **3**, 993–1022 (2003)
3. Blei, D.M.: Build, compute, critique, repeat: data analysis with latent variable models. Ann. Rev. Stat. Appl. **1**(1), 203–232 (2014)
4. Blei, D.M., Lafferty, J.D.: A correlated topic model of Science. Ann. Appl. Stat. **1**(1), 17–35 (2007). arXiv: 0708.3601
5. Blei, D.M., et al.: Variational inference: a review for statisticians. J. Am. Stat. Assoc. **112**(518), 859–877 (2017). arXiv: 1601.00670
6. Boyd-Graber, J., et al.: Applications of topic models. Found. Trends® Inf. Retrieval **11**, 143–296 (2017)
7. Chang, J., Blei, D.: Relational topic models for document networks. J. Mach. Learn. Res. Proc. Track **5**, 81–88 (2009)
8. Chang, J., et al.: Reading tea leaves: how humans interpret topic models. In: Bengio, Y., Schuurmans, D., Lafferty, J., Williams, C., Culotta, A. (eds.) Advances in Neural Information Processing Systems, vol. 22. Curran Associates, Inc. (2009)
9. Das, R., et al.: Gaussian LDA for topic models with word embeddings. In: Association for Computational Linguistics, Beijing, China, pp. 795–804, July 2015
10. Dieng, A.B., et al.: TopicRNN: a recurrent neural network with long-range semantic dependency. arXiv:1611.01702 (2017)
11. Dieng, A.B., et al.: Topic modeling in embedding spaces. arXiv:1907.04907 [cs, stat], July 2019

12. Figurnov, M., et al.: Implicit reparameterization gradients. arXiv: 1805.08498 (2019)
13. Griffiths, T.L., Steyvers, M.: Finding scientific topics. Proc. Nat. Acad. Sci. **101**(Supplement 1), 5228–5235 (2004)
14. Griffiths, T.L., et al.: Hierarchical topic models and the nested Chinese restaurant process, p. 8 (2003)
15. Horwood, G.V.: reliefweb_corpus_raw_20160331.json. In: Humanitarian Assistance and Disaster Relief (HA/DR) Articles and Lexicon. Harvard Dataverse (2017). version Number: V1
16. Hu, Y., et al.: Interactive topic modeling. Mach. Learn. **95**(3), 423–469 (2014)
17. Ishwaran, H., James, L.F.: Gibbs sampling methods for stick-breaking priors. J. Am. Stat. Assoc. **96**, 161–173 (2001)
18. Kingma, D.P., Welling, M.: Auto-encoding variational bayes, May. arXiv:1312.6114 (2014)
19. Miao, Y., et al.: Neural variational inference for text processing. arXiv:1511.06038 [cs, stat], Jun 2016
20. Miao, Y., et al.: Discovering Discrete latent topics with neural variational inference. arXiv:1706.00359 [cs], May 2018
21. Mikolov, T., et al.: Efficient estimation of word representations in vector space. arXiv:1301.3781 [cs], September 2013
22. Nalisnick, E., Smyth, P.: Stick-breaking variational autoencoders. arXiv:1605.06197 (2017)
23. Ning, X., et al.: A Bayesian nonparametric topic model with variational auto-encoders (2018)
24. Rezende, D.J., et al.: Stochastic Backpropagation and Approximate Inference in Deep Generative Models. **arXiv**, 1401.4082 (2014)
25. Srivastava, A., Sutton, C.: Autoencoding variational inference for topic models. arXiv:1703.01488 (2017)
26. Teh, Y.W., et al.: Hierarchical Dirichlet processes. J. Am. Stat. Assoc. **101**(476), 1566–1581 (2006)
27. Vulić, I., De Smet, W., Tang, J., Moens, M.F.: Probabilistic topic modeling in multilingual settings: an overview of its methodology and applications. Inf. Process. Manage. **51**(1), 111–147 (2015)
28. Wallach, H.M., et al.: Evaluation methods for topic models. In: ICML 2009, pp. 1105–1112. Association for Computing Machinery, New York (2009)
29. Xun, G., et al.: A correlated topic model using word embeddings. In: Proceedings of the Twenty-Sixth International Joint Conference on Artificial Intelligence, IJCAI 2017, pp. 4207–4213 (2017)
30. Yan, X., et al.: A biterm topic model for short texts. In: WWW 2013 - Proceedings of the 22nd International Conference on World Wide Web, pp. 1445–1456 (2013)
31. Yang, W., et al.: A multilingual topic model for learning weighted topic links across corpora with low comparability. In: Association for Computational Linguistics, Hong Kong, China, pp. 1243–1248, November 2019

Density-Based Evaluation Metrics in Unsupervised Anomaly Detection Contexts

Rui Maia(✉) and Cláudia Antunes

Instituto Superior Técnico, Avenida Rovisco Pais 1, 1049-001 Lisbon, Portugal
{rui.maia,claudia.antunes}@tecnico.ulisboa.pt

Abstract. From cybersecurity to life sciences, anomaly detection is considered crucial as it often enables the identification of relevant semantic information that can help to prevent and detect events such as cyber attacks or patients heart-attacks. Although anomaly detection is a prominent research area it still encompasses several challenges, namely regarding results evaluation in real-world unlabelled and imbalanced datasets. This work contributes to understand and compare the behaviour of different evaluation metrics, namely classic metrics based on positive and negative rates, and density based metrics without classes information. We experiment five state-of-art anomaly detection approaches over two datasets with contrasting characteristics regarding dimensionality or contamination. Each metrics' ability to give trustful results is analysed regarding different datasets or approaches properties focusing on the possibility of evaluating real-world unsupervised learning models using density metrics.

Keywords: Anomaly · Detection and learning · Evaluation · Metric

1 Introduction

Anomalies in data series have been earlier described as occasional observations that deviate from the most common behaviour in the series [12]. Real world data series are frequently characterized by being dynamic, noisy and irregular, having both categorical and real valued dimensions. Seasonality and context are also relevant factors. While anomaly detection approaches strongly depend on dataset properties [1,5], other complexities sustain challenges in this research area.

Although most real-world datasets are unlabelled, meaning that there is no information on which observations are regular or abnormal, research works frequently experiment over labelled datasets in order to facilitate the evaluation process. Multiple authors [7,9,18] underline that there is no correct evaluation metric for unsupervised anomaly detection.

This work focuses on this evaluation challenge, by comparing the behaviour of different metrics when applied to the anomaly detection task in very different datasets and approaches. We compare two types of metrics: (a) based on true

V. Torra and Y. Narukawa (Eds.): MDAI 2021, LNAI 12898, pp. 311–322, 2021.
https://doi.org/10.1007/978-3-030-85529-1_25

and false positive and negative rates such as Receiver Operator Characteristics and Precision-Recall curves, and (b) Excess-Mass and Mass-Volume curves, both density based metrics that do not use classes information.

Our experiments indicate that the studied density metrics frequently present high or very high correlation with reference metrics. Because they do not depend on labelled information, rather they do on class probabilities function, they can be considered when anomaly detection tasks target unlabelled datasets.

In the following sections of this paper we present the fundamental concepts and related work in Sect. 2. Section 3 introduces the experimented evaluation metrics while the experimental process and results analysis is done in Sect. 4. Finally conclusions, are drawn in Sect. 5.

2 Background Knowledge

The definition of anomaly (also found as *outliers, exceptions, faults or discordant observations*) has been revisited since 1969, when Frank Grubbs [12] identified it as unusual behaviours of isolated samples or groups of samples. Chandola et al. [5] defined *Anomaly* as a pattern that do not follows an expected behaviour. While this generic definition can be intuitively illustrated onto three dimensions, higher dimensional spaces will escape human visualization capabilities resulting in human intuition loss. Most anomaly detection approaches deal with non temporal univariate series as recognized in different overviews [1,5,13,15].

Previous research has been published regarding the comparison of anomaly detection approaches, some focusing neighbourhood based methods [4] by Campos et al. or time-series focused detection by Lavin and Ahmad [17]. Goldstein et al. [11] published a broader analysis focusing density and neighbourhood methods, but included a support vector machine and a dimensionality reduction based approach.

Although some updated relevant work in anomaly detection has been published [2,16,20,21], there is still no coverage regarding the comparison of this methods or the analysis of their application in unsupervised contexts.

Regarding performance evaluation, the use of simple statistical metrics such as accuracy are invalidated by imbalanced datasets, where a naive classifier predicting always the majority class would have an accuracy of 99% if the dataset contamination was 1%. Research works use Receiver Operator Characteristic curve as a standard performance measure in supervised or semi-supervised contexts, by calculating rates using classes information (i.e. labelled datasets). Coming from statistics research area, Excess-Mass [10] and Mass-Volume [6] are density based evaluation techniques already experimented in the anomaly detection task. Both assume the possibility of ranking observations level of abnormality without any knowledge about classes (i.e. unlabelled datasets).

3 Evaluation Metrics

Most anomaly detection problems are characterized by very imbalanced classes, invalidating the use of some evaluation methods such as Accuracy. Other measures such as Precision and Recall (PR) or Receiver Operator Characteristic (ROC) have the preference of researchers due to their better behaviour when dealing with imbalanced classes. The ROC curve is created by calculating the True Positive Rate (TPR) (see Precision Eq. 2) and False Positive Rate (FPR) as defined by Eq. 1, in a varying threshold. PR curve, is also calculated by iteratively calculating Precision and Recall values for different thresholds (see Eqs. 2 and 3).

$$FPR = \frac{False\ Positives}{False\ Positives + True\ Negatives} \tag{1}$$

$$Precision = \frac{True\ Positives}{True\ Positives + False\ Positives} \tag{2}$$

$$Recall = \frac{True\ Positives}{True\ Positives + False\ Negatives} \tag{3}$$

Clémençon and Jakubowicz [6] proposed to leverage Excess-Mass [18] as an evaluation method applicable both in supervised or unsupervised approaches. Nicolas Goix et al. [9,10] further developed this work by testing Excess-Mass (EM) as an abnormality ranking metric, comparing its results against Mass-Volume, Receiver Operator Characteristic and Precision-Recall curves. Both Excess-Mass and Mass-Volume metrics are supported by aligning a probability density function and the classifier scoring function assuming data follows a normal distribution F and anomalies occur in the tail of F. The density function s is defined w.r.t. the Lebesgue measure of dimensions $\mathbb{R}^n$, which means defined in terms of the n-dimensional volume sets of the distribution (see the example in Fig. 1).

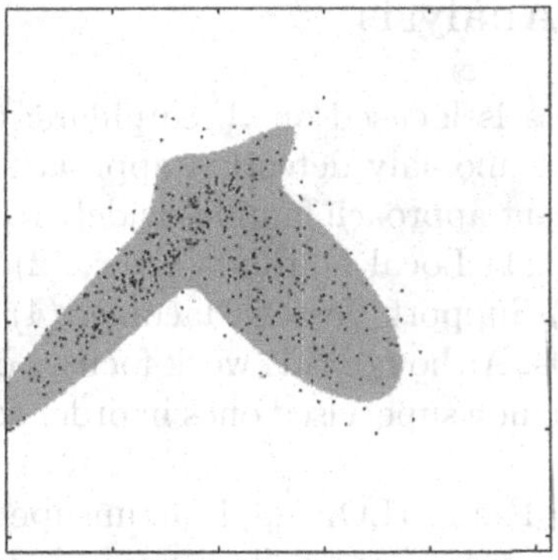

Fig. 1. Volume sets of a gaussian mixture distribution in two-dimensional space.

EM and MV have an inverted correlation, where high EM scores will correspond to low MV scores, by varying the threshold t of t_α as illustrated by Fig. 2.

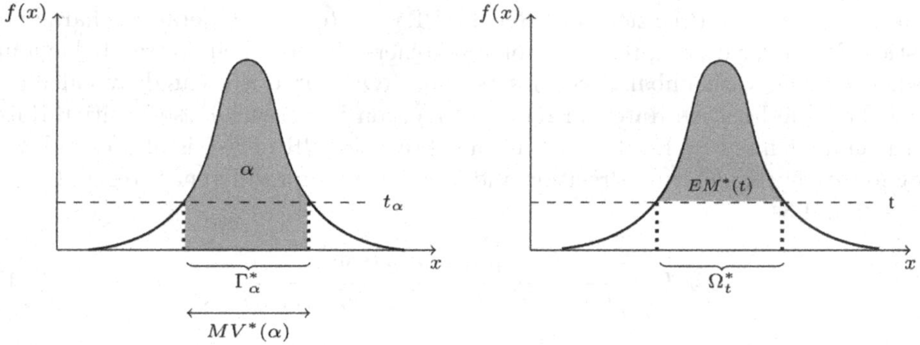

Fig. 2. Excess-Mass (EM) and Mass-Volume (MV) curves.

Considering $s \in S$ as a scoring function, $t > 0$ and $\alpha \in (0, 1)$, both Eqs. 5 and 4 depend on the classifier scoring function $f(x)$. that approximates the original distribution with t close to 0 (see [7,9,10] for detailed definitions).

$$MV_s(\alpha) = \inf_{u \geq 0} Leb(s \geq u) \ s.t. \ \mathbb{P}(s(X) \geq u) \ \geq \alpha \tag{4}$$

$$EM_s(t) = \sup_{u \geq 0} \mathbb{P}(s(X) \geq u) - tLeb(s \geq u) \tag{5}$$

Since our experimental work focuses on binary classification in imbalanced datasets we also included Precision (Eq. 2) and Recall (Eq. 3) metrics for results comparison.

4 Experimental Analysis

Our experimental process is focused on three pillars: the metrics, the selected datasets, and state-of-art anomaly detection approaches. We choose to include some of the most prominent approach families widely recognized in different classification tasks, namely: (1) Local Outlier Factor, (2) Long Short-Term Memory Neural Network, (3) Support Vector Machine, (4) Isolation Forest and (5) Gaussian Mixture models. Although this work focuses on unsupervised anomaly detection, we chose to include supervised ones in order to have a broader problem analysis.

The (1) Local Outlier Factor (LOF) [3] is an unsupervised density based approach that calculates the degree of an observation being abnormal considering the k nearest neighbours. The calculated degree is based on the Local Reachability Density (LRD) of each sample and it is calculated based on distance calculations against n neighbours. The final LOF score for each observation is obtained by comparing k neighbours LRD. (2) A Long Short-Term Memory Neural Network [14] is a Recurrent Neural Network type of architecture that provides feedback connections between cells, enabling the network to keep memory over arbitrary time intervals. (3) A Support Vector Machine [8] aims at finding the decision boundary that best separates the different classes of observations. The boundary is obtained by defining an hyper-plane. (4) An Isolation Forest (IF) is an unsupervised approach consisting in several decision trees that try to isolate abnormal samples. Each tree randomly selects a subset of features and data subsets for each calculation in order to avoid over fitting. The recursive process of splitting observations in a tree structure generates shorter paths, potentially associated to abnormal observations. Finally, the (5) Gaussian based approach [19], or Elliptic Envelop model, is an unsupervised method that assumes data is normally distributed. It leverages the Minimum Covariance Determinant to find the observations having lowest determinant.

4.1 Approaches Hyper-Parameters Values Definition

To train the LSTM Neural Network based approach we tested every combination of the following hyper-parameters: *epochs* [10, 50, 100], *batch_size* [10, 50, 100], *dropout_rate* [0.1, 0.5, 0.75, 0.9], *neurons* [1, 10, 25, 50, 100], *init_mode* ['uniform', 'normal', 'zero', 'glorot_normal'] and *activation* ['softmax', 'relu', 'sigmoid', 'linear']. The network is has one LSTM layer with 50 memory units and a Dense output layer with a single neuron and a sigmoid activation function returning binary predictions.

The LOF approach was tested by varying the number of k neighbours and with different distance metrics: *Number of Neighbours (k)* [5, 20, 50, 100, 200] and *metric* ['cosine', 'mahalanobis']. The metrics were chose regarding the different datasets dimensionality characteristics.

Isolation Forest trees were fit with random subsets of the training data (using sample-with-replacement) and tested against all combinations of the following hyper-parameters: *maximum samples* [250, 500, 'auto'], *n_estimators* [50, 100, 250, 500] and *max_features* [0.3, 1, 5].

The Gaussian or Elliptic Envelop approach was tested considering two different assumptions, namely, if the gaussian distribution was assumed to be centred or not, and finally the SVM was instantiated using: *kernel* ['rbf', 'poly'], *gamma* [1e−3, 1e−4] and *nu* [0.001, 0.01, 0.1, 1].

4.2 Datasets: Harvard Dataverse and Numenta Anomaly Benchmark

For the experimental phase of this work we used two open access datasets, (1) the Harvard Dataverse, a multivariate non temporal dataset with different levels of contamination, and (2) Numenta Anomaly Benchmark dataset, a collection of highly imbalanced univariate temporal series. These were selected because they cover a wide range of data series characteristics and are frequently used by the research community. Dataverse series contamination ranges from 0.149%

Table 1. Harvard Dataverse multivariate series

Name	N. samples	N. features	Inliers	Outliers	% contamination
aloi-unsupervised-ad	50000	27	48492	1508	3.016
annthyroid-unsupervised-ad	6916	21	6666	250	3.615
breast-cancer-unsupervised-ad	367	30	357	10	2.725
kdd99-unsupervised-ad	50000	29	50000	0	0.000
letter-unsupervised-ad	1600	32	1500	100	6.250
pen-global-unsupervised-ad	809	16	719	90	11.125
pen-local-unsupervised-ad	6724	16	6714	10	0.149
satellite-unsupervised-ad	5100	36	5025	75	1.471
shuttle-unsupervised-ad	46464	9	45586	878	1.890
speech-unsupervised-ad	3686	400	3625	61	1.655

to 11.125% while NAB univariate temporal series have 0.011% to 0.355% contamination. Tables 1 and 2 describe the Harvard Dataverse and NAB datasets respectively. Figure 3 illustrates the comparison between both datasets regarding, size, dimensionaly and contamination.

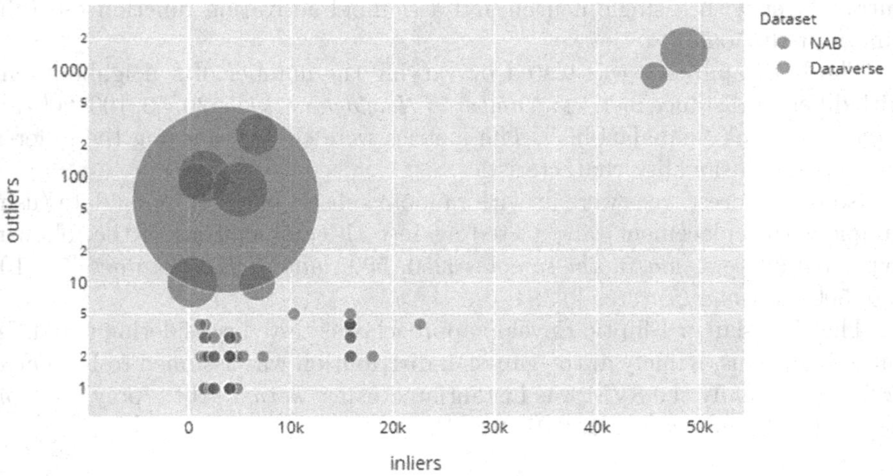

Fig. 3. Numenta Anomaly Benchmark and Harvard Dataverse datasets. Figure shows the number of inliers and outliers as axis in log scale while the circle radius represent the number of features.

Table 2. Numenta Anomaly Benchmark univariate series

Name	N. samples	N. features	Inliers	Outliers	% contamination
art daily flatmiddle	4032	2	4031	1	0.025
art daily jumpsdown	4032	2	4031	1	0.025
art daily jumpsup	4032	2	4031	1	0.025
art daily nojump	4032	2	4031	1	0.025
art increase spike density	4032	2	4031	1	0.025
art load balancer spikes	4032	2	4031	1	0.025
ec2 cpu utilization 24ae8d	4032	2	4030	2	0.050
ec2 cpu utilization 53ea38	4032	2	4030	2	0.050
ec2 cpu utilization 5f5533	4032	2	4030	2	0.050
ec2 cpu utilization 77c1ca	4032	2	4031	1	0.025

4.3 Experimental Setup

Each approach was tested in grid search parameter evaluation that covered a wide hyper-parameters configuration range. Our goal was to have a solid setup for each approach where in every execution the complete dataset was used for training. Some notes should be underlined to better understand the evaluation process:

- The goal was not to compare the performance of approaches, rather they were used as tools to support a wide perspective of metrics behaviour in different and realistic perspectives: temporal datasets vs non-temporal ones, multiple anomaly detection approaches, varying data series sizes and a wide range of contamination and dimensionality.
- Commonly strategies such as k-fold cross-validation require observations from all classes to be present in training. For supervised approaches such as SVM and LSTM this can be difficult without artificially introducing anomalies or removing normal observations (changing the dataset characteristics).
- Regarding series are ordered sequences of observations, all training and test folds were created for the k-fold process following observations order.
- Each generated k training fold was always superset of the previous *k-1* fold, with the exception of $k = 1$.
- Some approaches were supervised, receiving the target class information, other received either the contamination factor or the class weights.
- All models were trained and tested with complete data series.

For both datasets, and for each data series, results were kept from the best hyper-parameters set for each model. The results were calculated using the Area Under Curve for curves (1) Receiver Operator Characteristic, (2) Precision-Recall, (3) Mass Volume and (4) Excess-Mass. Precision and Recall were also calculated only for the outlier class.

4.4 Empirical Evaluation

Outlier classification results were got from all the tests: (NAB Dataset Series + Dataverse Dataset Series) * (Support Vector Machine, Long Short-Term Memory Neural Network, Elliptic/Gaussian Envelop, Isolation Forest, Local Outlier Factor). These results - classification predictions and scores - were then used to get the final metrics results, namely: Area Under Curve for Excess-Mass, Mass-Volume, Receiver Operator Characteristic and Precision-Recall, and Anomaly Class Detection metrics, specifically, Precision and Recall.

Figure 4 shows the distribution of values got for each metric in each approach considering both all series from both datasets. Excess-Mass seems to polarize its results depending on the dataset. This may be explained by different factors such as: temporal data series (NAB) or non temporal (Dataverse), and the very different contamination rate (see datasets description). Precision-Recall Area Under Curve suggest a more clear look over the methods results although having an inverse polarization of results when compared with AUC (ROC) at least for three of the studied approaches.

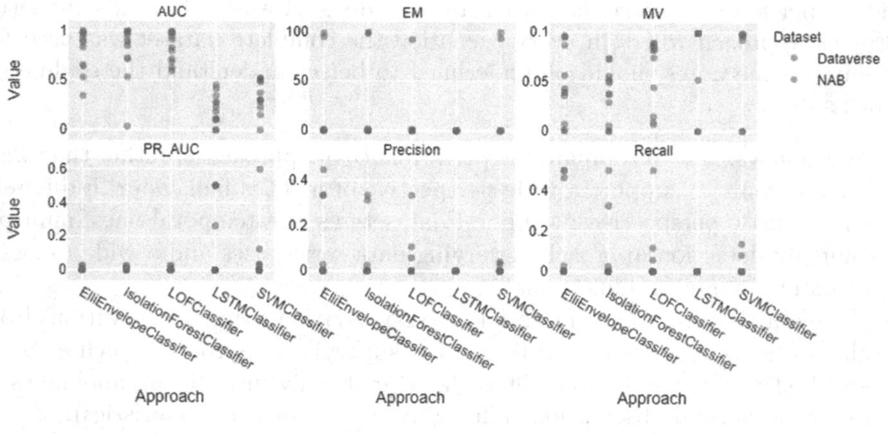

Fig. 4. Approaches and Metrics. Area Under Curve calculated metrics: Receiver Operating Characteristics Area Under Curve (AUC), Excess-Mass (EM), Mass-Volume (MV) and Precision-Recall Area Under Curve (PR_AUC). Anomaly class calculated metrics: Precision and Recall.

Using the result vector of all results - all series from both datasets - of each metric, we calculated the correlation between pairs of metrics using Pearson's coefficient. The correlation was interpreted assuming the following levels: very high correlation when the coefficient value lies between ± 0.90 and ± 1, high correlation when it lies between ± 0.70 and ± 0.9, moderate correlation when the coefficient locates between ± 0.5 and ± 0.7, low correlation between 0.3 and ± 0.5 and finally a negligible correlation when the value fall between 0.0 and ± 0.3.

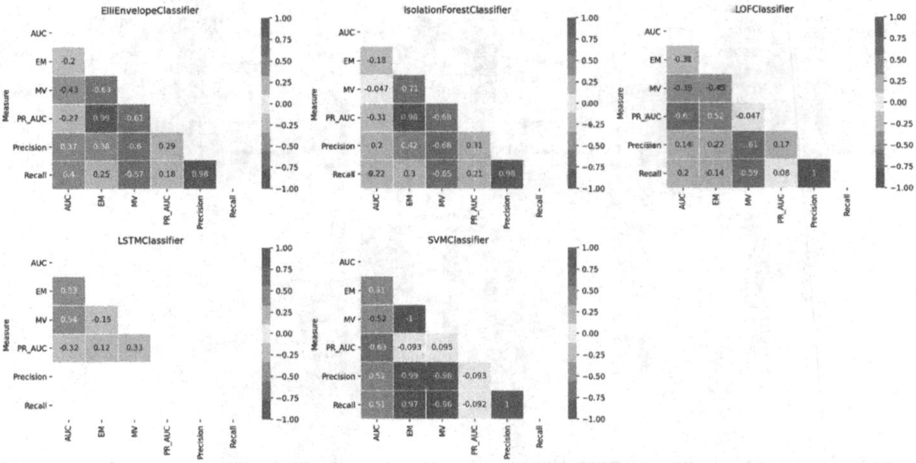

Fig. 5. Heatmap of the pairwise Pearson's correlation of metrics in each of the studied approaches.

Figure 5 underline different metrics behaviours depending on approach results. Most correlations are near negligible, with two exceptions: (1) the inverted correlation between Excess-Mass (EM) and Mass-Volume (MV), which confirms the expected behaviour since EM returns lower values for better anomaly detection results while MV does the opposite, i.e. returns bigger values for better classification results, and (2) EM shows a very-high correlation with Precision-Recall curve or with Outlier Precision and Recall rates in most approaches evaluation. It is hard to take any conclusion on LSTM results since it was experimented only against the NAB dataset temporal series. MV shows a moderate inverted correlation with Precision-Recall, and both Outlier Precision and Recall.

Figure 6 shows high or very high correlation between Excess-Mass and datasets features in four of five studied approaches. This also happens with (outlier) Precision-Recall results in three of five approaches.

It is worth to mention the considerable alignment between the behaviour of EM and PR curve, specially considering PR curve is defined by True and False, Positive and Negative Rates, while EM is supported by each classifier probability distribution function.

When analysing the extremely imbalanced NAB dataset (see the bottom row of plots on Fig. 5), there is clearly less correlations between metrics behaviour. This is due not to better or worst classification performance but to different approaches results rankings, which in turn is the result of less clear classes probabilities distributions.

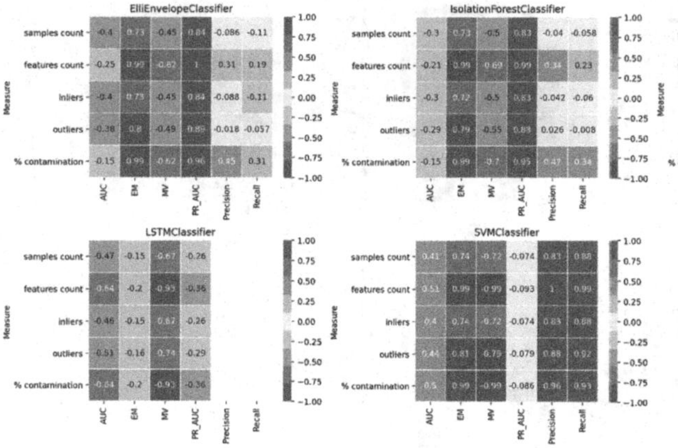

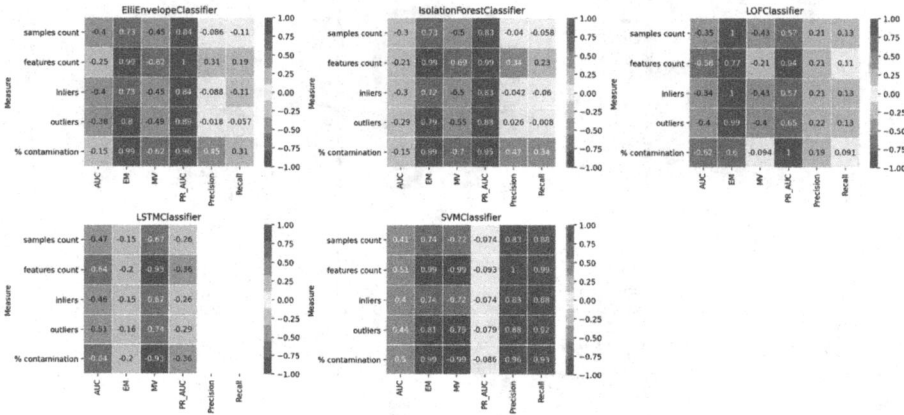

Fig. 6. Heatmap of the pairwise Pearson's correlation of metrics in each of the studied approaches and datasets features.

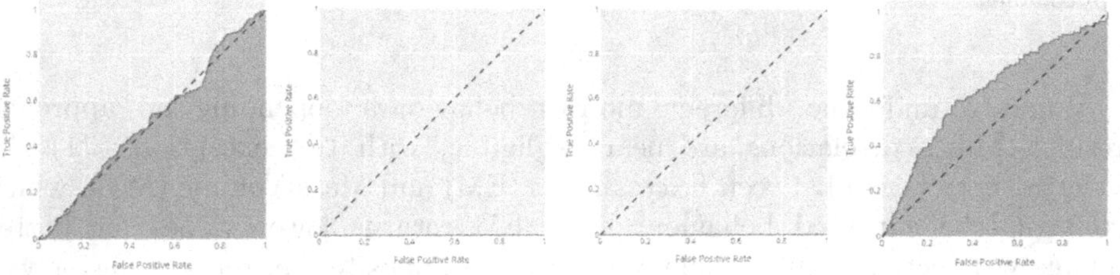

(a) Fold 1 (b) Fold 2 (c) Fold 3 (d) Fold 4

Fig. 7. Receiver Operator Characteristic curve in 4 folds of K-fold cross validation. Example got from LSTM Neural Network experiments.

Finally, in multiple folds of k-fold cross-validation process, there was no possibility of correctly calculating the area under curve for multiple metrics due to inexistent anomalies, as illustrated in Fig. 7. This fact can severely skew the results got by Precision (Eq. 2) and Recall (Eq. 3), both depending on the positive class, which in the presented context is the abnormal class.

5 Conclusions

This study focused on the evaluation process and metrics of anomaly detection tasks contributing to better understand the challenges associated to unsupervised contexts. Specifically on the evaluation process this work's experiments showed how extremely low contaminated datasets can affect not only each metric's scores but also the evaluation process. We underlined how the most common training and hyper-parameter definition approach - the k-fold cross validation - is affected by low contamination contexts and discussed the possibility of building folds that respect not only the temporal alignment of series but also the balance of classes and their dynamics through under or oversampling.

Using all available data for training in each experiment we calculated the contamination parameter (or class weights) used in unsupervised training, an advantage not available in real-world problems. This, although, did not solve the unbalanced distribution of anomalies in low contamination series and the folds representation of the data. In some of our experiments training results were improved by incrementing or decrementing fold counts, depending on the data series and even on the contamination location.

Regarding evaluation metrics Mass-Volume and Excess-Mass density based metrics showed relevant coherence when compared to the classic Precision-Recall curve, Outlier Precision and Recall rates, suggesting that these can be taken into account in unsupervised anomaly detection problems. Moreover, these density based metrics also seem to present an expressive relation with datasets features, indicating they are sensible to the dimensionality, size or contamination factors. Experiments also showed that all metrics present different behaviour in the experimented datasets, which may be due to the fact that one dataset is an univariate temporal series dataset, with extremely low contamination factor, while the other is a multivariate non temporal dataset, presenting varied conditions.

This research unveiled the potential of using density based metrics in the evaluation of unsupervised anomaly detection approaches. Nonetheless we observed different correlations with classical reference measures. These differences should be deeper investigated namely by investigating at what level the input data distributions - considering the multivariate case - justify the change of behaviour of Excess-Mass or Mass-Volume.

Our work will continue to focus on the unsupervised anomaly detection evaluation task in real-world scenarios where classes are highly imbalanced aiming to contribute for the development of new measures and evaluation strategies that finally can help to improve trained models and their performance.

References

1. Aggarwal, C.C.: Outlier Analysis. Springer, New York (2013). https://doi.org/10.1007/978-1-4614-6396-2
2. Audibert, J., Michiardi, P., Guyard, F., Marti, S., Zuluaga, M.A.: USAD: unsupervised anomaly detection on multivariate time series. In: Proceedings of the 26th ACM SIGKDD International Conference on Knowledge Discovery & Data Mining, pp. 3395–3404 (2020)
3. Breunig, M.M., Kriegel, H.P., Ng, R.T., Sander, J.: LOF: identifying density-based local outliers. In: ACM SIGMOD Record, vol. 29, pp. 93–104. ACM (2000)
4. Campos, G.O., et al.: On the evaluation of unsupervised outlier detection: measures, datasets, and an empirical study. Data Mining Knowl. Discov. **30**(4), 891–927 (2016). https://doi.org/10.1007/s10618-015-0444-8
5. Chandola, V., Banerjee, A., Kumar, V.: Anomaly detection: a survey. ACM Comput. Surv. (CSUR) **41**(3), 15 (2009)
6. Clémençon, S., Jakubowicz, J.: Scoring anomalies: a M-estimation formulation. In: Artificial Intelligence and Statistics, pp. 659–667 (2013)

7. Clémençon, S., Thomas, A.: Mass volume curves and anomaly ranking. arXiv preprint arXiv:1705.01305 (2017)
8. Cortes, C., Vapnik, V.: Support-vector networks. Mach. Learn. **20**(3), 273–297 (1995). https://doi.org/10.1007/BF00994018
9. Goix, N.: How to evaluate the quality of unsupervised anomaly detection algorithms? arXiv:1607.01152 [cs, stat], July 2016
10. Goix, N., Sabourin, A., Clémençon, S.: On anomaly ranking and excess-mass curves. In: Artificial Intelligence and Statistics, pp. 287–295 (2015)
11. Goldstein, M., Uchida, S.: A comparative evaluation of unsupervised anomaly detection algorithms for multivariate data. PLoS ONE **11**(4), e0152173 (2016). https://doi.org/10.1371/journal.pone.0152173. https://journals.plos.org/plosone/article?id=10.1371/journal.pone.0152173
12. Grubbs, F.E.: Procedures for detecting outlying observations in samples. Technometrics **11**(1), 1–21 (1969)
13. Gupta, M., Gao, J., Aggarwal, C.C., Han, J.: Outlier detection for temporal data: a survey. IEEE Trans. Knowl. Data Eng. **26**(9), 2250–2267 (2014)
14. Hochreiter, S., Schmidhuber, J.: Long short-term memory. Neural Comput. **9**, 1735–80 (1997). https://doi.org/10.1162/neco.1997.9.8.1735
15. Hodge, V., Austin, J.: A survey of outlier detection methodologies. Artif. Intell. Rev . **22**(2), 85–126 (2004)
16. Ismail Fawaz, H., Forestier, G., Weber, J., Idoumghar, L., Muller, P.A.: Deep learning for time series classification: a review. Data Mining Knowl. Discov. (2019). https://doi.org/10.1007/s10618-019-00619-1
17. Lavin, A., Ahmad, S.: Evaluating real-time anomaly detection algorithms-the Numenta Anomaly Benchmark. In: 2015 IEEE 14th International Conference on Machine Learning and Applications (ICMLA), pp. 38–44. IEEE (2015)
18. Müller, D.W., Sawitzki, G.: Excess mass estimates and tests for multimodality. J. Ame. Stat. Assoc. **86**(415), 738–746 (1991)
19. Rousseeuw, P.J., Driessen, K.V.: A fast algorithm for the minimum covariance determinant estimator. Technometrics **41**(3), 212–223 (1999)
20. Su, Y., Zhao, Y., Niu, C., Liu, R., Sun, W., Pei, D.: Robust anomaly detection for multivariate time series through stochastic recurrent neural network. In: Proceedings of the 25th ACM SIGKDD International Conference on Knowledge Discovery & Data Mining, pp. 2828–2837 (2019)
21. Zhao, H., et al.: Multivariate time-series anomaly detection via graph attention network. arXiv preprint arXiv:2009.02040 (2020)

Explaining Image Misclassification in Deep Learning via Adversarial Examples

Rami Haffar(✉), Najeeb Moharram Jebreel, Josep Domingo-Ferrer, and David Sánchez

Department of Computer Engineering and Mathematics, CYBERCAT-Center for Cybersecurity Research of Catalonia, UNESCO Chair in Data Privacy, Universitat Rovira i Virgili, Av. Països Catalans 26, 43007 Tarragona, Catalonia
{rami.haffar,najeebmoharramsalim.jebreel,josep.domingo, david.sanchez}@urv.cat

Abstract. With the increasing use of convolutional neural networks (CNNs) for computer vision and other artificial intelligence tasks, the need arises to interpret their predictions. In this work, we tackle the problem of explaining CNN misclassification of images. We propose to construct adversarial examples that allow identifying the regions of the input images that had the largest impact on the CNN wrong predictions. More specifically, for each image that was incorrectly classified by the CNN, we implemented an inverted adversarial attack consisting on modifying the input image as little as possible so that it becomes correctly classified. The changes made to the image to fix classification errors *explain* the causes of misclassification and allow adjusting the model and the data set to obtain more accurate models. We present two methods, of which the first one employs the gradients from the CNN itself to create the adversarial examples and is meant for model developers. However, end users only have access to the CNN model as a black box. Our second method is intended for end users and employs a surrogate model to estimate the gradients of the original CNN model, which are then used to create the adversarial examples. In our experiments, the first method achieved 99.67% success rate at finding the misclassification explanations and needed on average 1.96 queries per misclassified image to build the corresponding adversarial example. The second method achieved 73.08% success rate at finding the explanations with 8.73 queries per image on average.

Keywords: Explainability · Deep learning · Image classification · Adversarial examples · Convolutional neural networks

1 Introduction

The use of deep learning, and of convolutional neural networks (CNNs) in particular, has brought great advances in computer vision [11] and many other artificial

V. Torra and Y. Narukawa (Eds.): MDAI 2021, LNAI 12898, pp. 323–334, 2021.
https://doi.org/10.1007/978-3-030-85529-1_26

intelligence (AI) endeavors. Although CNNs can achieve high accuracy in classification, detection, and segmentation tasks, they are black-box models. This means that their predictions do not come with explanations or justifications and, therefore, it is not possible for humans to understand how decisions were made. To avoid blind algorithm-based decisions, AI models should be explainable [2]. Explainability is not only an ethical principle but also a legal requirement set out in the European General Data Protection Regulation (GDPR) [6]. The lack of explanations about the decisions made by CNNs is a problem both for developers who train such networks and for the citizens affected by their decisions:

- Developers want to know how a decision is made, to ensure that the AI model takes into account the correct features of the input during the training phase. In some cases, it may happen that wrong features are used to make decisions, such as in the well-known example of [14] where a dog-like animal is classified as a wolf if the image has a snow background and as a husky dog if the image has a grass background because in the training pictures all the wolves were displayed in a snowy landscape and the huskies were not.
- Citizens are affected by a growing number of automated decisions: credit granting, insurance premiums, medical diagnoses, etc. For that reason, legal regulations [6] and ethics guidelines [4,16] have appeared that assert the citizen's right to an explanation on every automated decision affecting her. Lacking such explanations, even *de iure* democracies risk becoming *de facto* AI-driven authoritarian societies.

In computer vision, the interpretations of CNN predictions are usually presented as saliency maps. Those maps suggest specific regions in the images that are most important in the decision made by the CNN [7].

Contributions and Plan of This Paper

For the generation of explanations to be scalable and efficient, it must be automated. In this work, we present two methods that explain CNN-based image classification by identifying the features that were most influential in the CNN predictions. The first method assumes access to the gradients of the CNN and is meant for model developers. The second method treats the model as a black box and, therefore, assumes that the party generating the explanations, such as a model end user, only has access to the model predictions.

Both methods leverage adversarial examples [13] to generate explanations. While the first methods computes adversarial examples by directly employing the CNN gradients, the second approach builds a simpler surrogate model to estimate the gradients of the original model, and then uses these estimated gradients to obtain adversarial examples. More specifically, for each image that was incorrectly classified by the CNN, we implemented an inverted adversarial attack consisting in modifying the input image as little as possible so that it becomes correctly classified. The changes made to the image to fix classification errors highlight the regions that had the highest influence in the decisions and thus *explain* the causes of model misclassification. By identifying the causes of wrong predictions, one may tailor the model or the training data to improve the classification accuracy.

The remainder of this paper is organized as follows. Section 2 discusses related works devoted to explaining CNNs. Section 3 describes our methods for explaining CNN-based image misclassification from adversarial examples. Experimental results are reported in Sect. 4. Finally, in Sect. 5 we gather conclusions and sketch future research lines.

2 Related Work

Several methods have been proposed to interpret CNN predictions in image classification. They attempt to link inputs to outputs to identify the regions in the image that have the highest impact on the classification decision.

One of the most commonly used approaches to generate explanations is to calculate the influence of the image features by back-propagating the decision score across all layers of the network. Works that follow this approach are XRAI [10], Guided Backprop [20], Gradient Input [17], SmoothGrad [19] and GradCAM [15]. This approach is fast and the methods following it usually require a fixed number of queries to the black-box model. However, they need access to all the internal layers of the model. Also, the computational cost of those methods is high because a second back-propagation is necessary.

It is also possible to locally train a simpler model, a.k.a. surrogate model, to approximate the black-box behavior and obtain an explanation of the black-box model [14]. This approach requires that the surrogate model be simple and understandable to humans. Unfortunately, the surrogate model is often much inferior to the black-box model in terms of accuracy, and hence the explanations provided by the former are not very reliable [8].

Another approach, known as perturbation-based, is to modify the input images and measure the effect of this change on the black-box prediction [3,5,22]. Perturbation-based methods allow directly detecting the regions of the images that had the highest impact on the predictions. Typically, these methods require multiple queries to the black box in order to interpret its predictions, which makes them slow [23].

The methods we propose in this paper fall in the perturbation-based category. By using gradient-based adversarial examples to add the perturbation to the original image, we are able to minimize the required number of queries to the model and we keep the computational cost at a minimum. Also, thanks to a surrogate model with reasonable complexity and accuracy compared to the black-box model, we can generate explanations just from the black-box prediction, without knowing any details on the black box's internal layers.

3 Our Proposals

We focus on generating explanations for the wrong predictions made by CNNs by identifying the regions of the input images that had the highest influence on those predictions. To this end, we need a way to modify the input images towards the correct classification while keeping the number of required queries and the computational cost at a minimum. Our choice is to use gradient-based adversarial examples [21]. Specifically, we add minimal perturbations to incorrectly classified images to create correctly classified adversarial examples. Then, by comparing the original image with the modified image, we can find the regions that had the highest impact on the wrong black-box predictions.

3.1 Adversarial Examples

An adversarial example is a sample from the same distribution as the original data in which small, intentional perturbations of its features cause an AI model to change its prediction [12]. Adversarial examples can be used to alter predictions of a variety of machine learning models, including state-of-the-art neural networks [21]. Even though adversarial examples are usually employed to cause the AI models to produce wrong predictions, in this work we use them the other way around: to correct wrongly classified samples.

To create adversarial examples we used the gradient-based optimization approach proposed in [21], in which we set the target to be the correct label for the wrongly predicted samples. The adversarial examples are created by minimizing the following function with respect to r:

$$loss(f(x + r), l) + \epsilon \cdot |r|, \tag{1}$$

where f is the AI classifier, x is the original image, r is the perturbation added to the pixels of x to create the perturbed image that constitutes the adversarial example, l is the target class label and ϵ is used to balance the distance between images and the distance between predictions. The smaller ϵ, the more similar is the created perturbed image to the original image. To minimize the loss function in Eq. (1), the party that computes it needs access to the model gradients.

3.2 Explaining Model Predictions on the Developer's Side

To explain the wrong predictions made by the model on the developer's side, we consider that the developer has full access to the CNN and, more specifically, to the gradients of the model.

As shown in Algorithm 1, first the developer splits the input images into training and testing sets and trains the model with the training images. Then, in the testing phase the developer keeps track of all the wrongly classified images. For each of these images, she tries to find the closest adversarial example that is correctly classified. The developer does this in the following way: i) she calculates the value of the loss function between the model prediction and the correct prediction; ii) she calculates the gradients of the model according to the image and the loss value; iii) she modifies the image according to the gradients and the perturbation ratio ϵ. These steps are repeated until the adversarial example is obtained or ϵ exceeds the α value signaling the termination condition (in the latter case the image misclassification cannot be explained). The final step consists in comparing each original image with its corresponding adversarial example. To draw a saliency map that identifies the features that caused wrong predictions, Algorithm 2 prescribes that pixels in perturbations with values smaller than $q3 + iqr \cdot \tau$, where $q3$ is the third quartile of perturbations, iqr is their interquartile range and $\tau > 0$ is relaxation parameter, are neglected because they do not identify regions of interest, whereas the remaining pixels are multiplied by $\beta > 1$ to boost them in the saliency map.

Algorithm 1. Explaining the model predictions on the developer's side

```
input: Data set X, CNN model model
Train_X, Test_X ← Split_Train_Test(X)
model ← Train_Model(Train_X)
perturbations ← {}
for i in Test_X do
    model_prediction ← model.predict(Test_X[i])
    ϵ ← 0.1
    while model_prediction ≠ correct_prediction OR ϵ < α do
        loss ← loss_function(model_prediction, correct_prediction)
        gradients ← get_model_gradients(Test_X[i], loss)
        perturbed_image ← Test_X[i] − ϵ · gradients
        model_prediction ← model.predict(perturbed_image)
        ϵ ← ϵ + 0.1
    end while
    if model.prediction = correct_prediction then
        perturbations[i] ← perturbed_image − Test_X[i]
    else
        perturbations[i] ← NIL
    end if
end for
return perturbations
```

3.3 Explaining Model Predictions on the User's Side

End users should have the right to obtain explanations about predictions made by the AI models that concern them. However, for end users, the model is a black box and they only have access to the model predictions. Therefore, they must create their own local explanations. In our work, we considered that the user who wants to generate explanations of an AI model must have enough data to train a simpler CNN model, a.k.a. a surrogate model. It is shown in [9] that knowledge of one or more models can be compressed into another, less complex model, which allows us to estimate the gradations of the original model using a surrogate model.

Algorithm 2. Drawing the saliency maps

```
input: perturbations
q1, q3 = get_quartiles_of_non_NIL_perturbations(perturbations)
iqr ← q3 − q1
for i such that perturbations[i] ≠ NIL do
    for pixel in perturbations[i] do
        if perturbations[i][pixel] < q3 + iqr · τ then
            perturbations[i][pixel] ← 0
        else
            perturbations[i][pixel] ← perturbations[i][pixel] · β
        end if
    end for
    Draw(perturbations[i])
end for
```

The method we propose is formalized in Algorithm 3. First, the user splits the data she has into training and testing data sets. Then she builds a surrogate model by using the local training data set. Afterwards, she uses the test data set to identify the wrong predictions to be explained. Finally, she generates the adversarial examples in a similar way as in Algorithm 1. The only difference is that the gradients from the surrogate model will be used instead of the gradients from the original model.

4 Experimental Results

We tested the two proposed methods introduced above on the gender classification data set[1] from the Kaggle website. This data set consists of cropped RGB images of male and female faces. The training data set contains 23,200 female images and 23,800 male images. The validation data set contains 5,800 images in each class. The images are rectangular, but not all of them are of the same size. Thus, we first resized all the images to 100 × 100 pixels.

4.1 Explanations for the Developer

To test Algorithm 1 we used the CNN shown in Fig. 1 with four Conv blocks followed by six fully connected layers. Each Conv block contained two convolutional layers and a max-pooling layer. The output depths for Conv blocks were, respectively, 64, 128, 256 and 512. The numbers of nodes of fully connected layers were, respectively, 2048, 1024, 512, 128, 32 and 2. We trained the model for 20 epochs, with a batch size 64 and a learning rate 0.001. The test accuracy was 96.3%.

[1] https://www.kaggle.com/cashutosh/gender-classification-dataset.

Algorithm 3. Explaining the model predictions on the user's side

```
input: local data X_local, black-box model black_box, surrogate model local_surrogate
Train_X_local, Test_X_local ← Split_Train_Test(X_local)
local_surrogate ← Train_Black_Box(Train_X_local)
for i in Test_X_local do
    black_box_prediction ← black_box.predict(Test_X_local[i])
    ε ← 0.1
    while black_box_prediction ≠ correct_prediction OR ε < α do
        local_prediction ← local_surrogate.predict(Test_X_local[i])
        loss ← loss_function(local_prediction, correct_prediction)
        gradients ← get_surrogate_gradients(Test_X_local[i], loss)
        perturbed_image ← Test_X_local[i] − ε · gradients
        black_box_prediction ← black_box.predict(perturbed_image)
        ε ← ε + 0.1
    end while
    if black_box_prediction = correct_prediction then
        perturbations[i] ← perturbed_image − Test_X_local[i]
    else
        perturbations[i] ← NIL
    end if
end for
return perturbations
```

The number of misclassified images in the test data set was 301, for which our method created adversarial examples. 300 of the 301 adversarial examples were classified into the correct labels, which corresponds to a success rate 99.67%, with an average number 1.96 queries per image. Figure 2 shows four examples of the explanations created by Algorithm 1, in the form of saliency maps highlighting the differences between original images and adversarial examples.

In Fig. 2, we can see that perturbations added to the original images to create the adversarial examples are not noticeable to the naked eye. Nevertheless, the explanations resulting from Algorithm 1 in the form of saliency maps shed light on the most important regions that caused the wrong predictions. For example, in Image 1 the important pixels were those around the eyes and the edge of the nose. In Image 2 the causes of wrong classification were also found in the eyes and the nose in addition to the left cheek. In Image 3, the causes of misclassification were mainly the left eye and the edge of the right eye in addition to parts of the covered forehead. The most relevant regions for Image 4 were the left eye, the edge of the nose and the left cheek.

4.2 Explanations for the User

Testing the performance of Algorithm 3 tells whether the user is capable of creating model explanations locally. We assumed the user's local data consisted of a random 10% sample of the training data described in the previous section. With

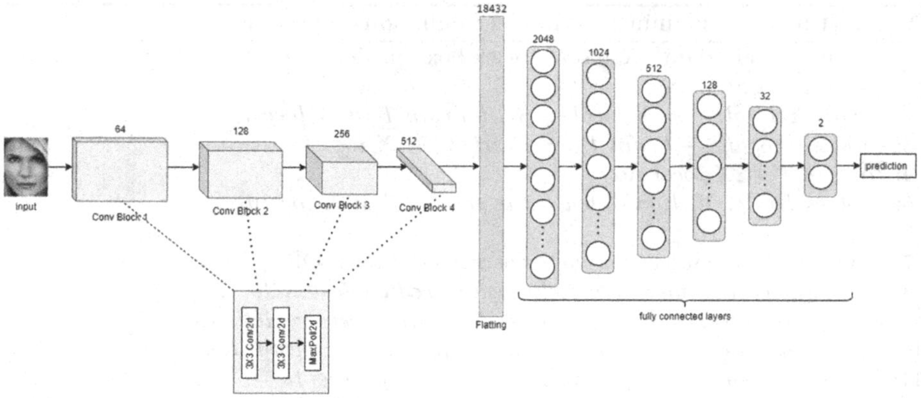

Fig. 1. Architecture of the CNN used as original model in the experiments

these local data, the user trained her surrogate model. The black-box model was the same CNN described in the previous section, whereas the local surrogate model built by the user consisted of a CNN with three Conv blocks followed by four fully connected layers. Each Conv block contained two convolutional layers and a max-pooling layer. The output depths for Conv blocks were, respectively, 64, 256 and 512. The numbers of nodes of fully connected layers were, respectively, 1024, 256, 32 and 2. We trained the model for 50 epochs, with a batch size 64 and a learning rate 0.001. The test accuracy for the surrogate model was 87.78%.

The complete test data set was used to generate explanations. Therefore, we got the same 301 misclassified images. Following Algorithm 3, explanations were obtained as follows: i) obtain the prediction and gradients of the local surrogate model; ii) create the adversarial example using the gradients of the surrogate model; iii) test whether the adversarial example was correctly predicted by using the original black-box model; iv) draw the saliency map. Out of the 301 adversarial examples, the original black-box model correctly classified 220 images, which corresponds to a 73.08% success rate. The average number of queries to the original CNN model per image required to create the adversarial example was 8.73.

Figure 3 shows the same four samples of Fig. 2 but with saliency maps that were locally generated using the gradients of the surrogate model. As in the previous test, the differences between the adversarial examples and the original images are not noticeable to the naked eye. However, the most relevant regions of the images are similar to those obtained with Algorithm 1: in the four images the same regions highlighted by Algorithm 1 are also highlighted here, even though in a less focused way due to the less accurate surrogate model.

The number of queries per image needed to create adversarial examples with Algorithm 1 was lower than with Algorithm 3: 1.96 vs 8.73. The reason is that the former algorithm uses the gradients from the original model, whereas the

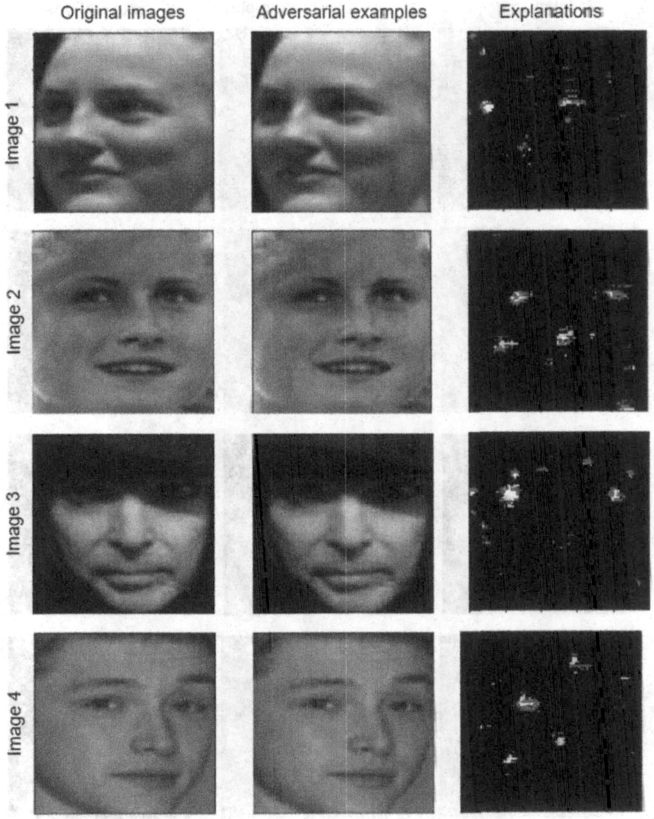

Fig. 2. Four examples of explanations generated on the developer's side

latter uses the gradients from the surrogate model. Hence, the generation of adversarial examples is less accurate with the second algorithm. However, both algorithms were successful in generating explanations for the wrong predictions of the original model and both highlighted the same regions as important.

The explanations provided by our methods can help model developers to identify the weaknesses of the data sets used to train the model. Specifically, for the gender classification data set, the eye regions are highlighted in most saliency maps as important regions. This suggests that classification accuracy may be improved by training on images where the eye region is clear. For model users, it is important to know which features influenced the predictions since some of the black boxes may be artificially biased or employ features that may discriminate some minorities [1]. Beyond face classification, an even more crucial application could be to help understand medical diagnoses [18].

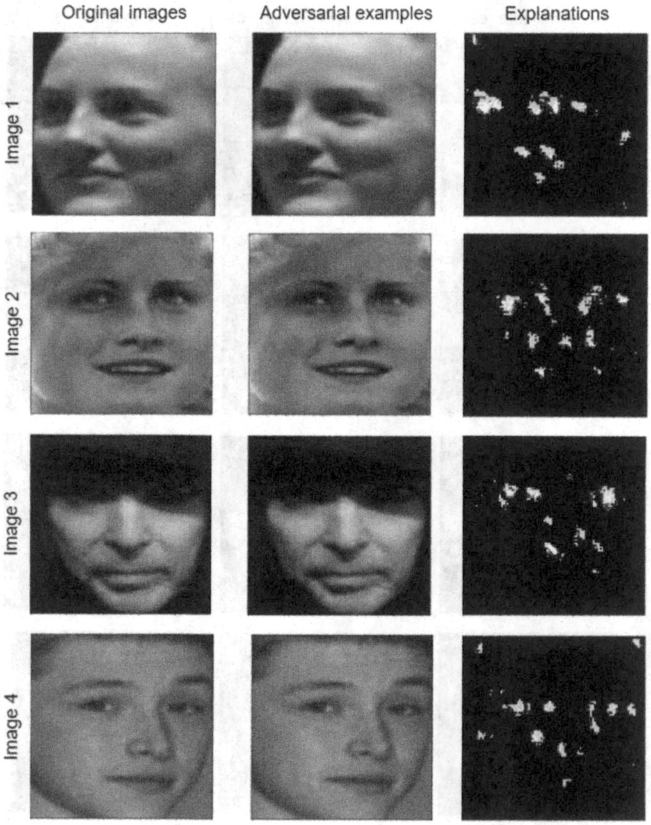

Fig. 3. Four examples of explanations generated on the user's side

5 Conclusions and Future Research

We have presented two methods employing gradient-based adversarial examples to obtain explanations of the predictions of CNNs in image classification.

We have reduced the number of queries needed to create the adversarial examples by adding targeted perturbations to change the predictions for each image. In our experimental work, developer-side Algorithm 1 required only 1.96 queries per image, whereas user-side Algorithm 3 needed 8.73 queries per image.

The two proposed algorithms showed promising results to explain misclassification by CNNs. Both produced similar explanations on the same samples. Algorithm 1 had a higher success rate (99.67%) thanks to using the gradients of the original model, whereas Algorithm 3 had a lower success rate (73.08%) due to using a surrogate.

As future work, we plan to test the performance of our approach on data that are not identically and independently distributed. We also plan to tailor the generation of adversarial examples to highlight regions of interest in correctly classified images.

Acknowledgments. We acknowledge support from the European Commission (projects H2020-871042 "SoBigData++" and H2020-101006879 "MobiDataLab"), the Government of Catalonia (ICREA Acadèmia Prizes to J. Domingo-Ferrer and D. Sánchez, and grant 2017 SGR 705) and from the Spanish Government (projects RTI2018-095094-B-C21 "Consent" and TIN2016-80250-R "Sec-MCloud"). The authors are with the UNESCO Chair in Data Privacy, but the views in this paper are their own and are not necessarily shared by UNESCO.

References

1. Azad, R., Fayjie, A.R., Kauffmann, C., Ben Ayed, I., Pedersoli, M., Dolz, J.: On the texture bias for few-shot CNN segmentation. In: Proceedings of the IEEE/CVF Winter Conference on Applications of Computer Vision, pp. 2674–2683 (2021)
2. Blanco-Justicia, A., Domingo-Ferrer, J., Martínez, S., Sánchez, D.: Machine learning explainability via microaggregation and shallow decision trees. Knowl. Based Syst. **194**, 105532 (2020)
3. Carter, B., Mueller, J., Jain, S., Gifford, D.: What made you do this? Understanding black-box decisions with sufficient input subsets. In: The 22nd International Conference on Artificial Intelligence and Statistics, pp. 567–576. PMLR (2019)
4. European Commission's High-Level Expert Group on Artificial Intelligence: Draft Ethics Guidelines for Trustworthy AI (2019). https://ec.europa.eu/futurium/en/ai-alliance-consultation
5. Fong, R.C., Vedaldi, A.: Interpretable explanations of black boxes by meaningful perturbation. In: Proceedings of the IEEE International Conference on Computer Vision, pp. 3429–3437 (2017)
6. GDPR: General Data Protection Regulation, Regulation EU 2016/679 of the European Parliament and of the Council of 27 April 2016. Official Journal of the European Union (2016). http://ec.europa.eu/justice/data-protection/reform/files/regulation_oj_en.pdf. Accessed 20 Sept 2017
7. Ghorbani, A., Abid, A., Zou, J.: Interpretation of neural networks is fragile. In: Proceedings of the AAAI Conference on Artificial Intelligence, vol. 33, pp. 3681–3688 (2019)
8. Haffar, R., Domingo-Ferrer, J., Sánchez, D.: Explaining misclassification and attacks in deep learning via random forests. In: Torra, V., Narukawa, Y., Nin, J., Agell, N. (eds.) MDAI 2020. LNCS (LNAI), vol. 12256, pp. 273–285. Springer, Cham (2020). https://doi.org/10.1007/978-3-030-57524-3_23
9. Hinton, G., Vinyals, O., Dean, J.: Distilling the knowledge in a neural network. arXiv preprint arXiv:1503.02531 (2015)
10. Kapishnikov, A., Bolukbasi, T., Viégas, F., Terry, M.: XRAI: better attributions through regions. In: Proceedings of the IEEE/CVF International Conference on Computer Vision, pp. 4948–4957 (2019)
11. Krizhevsky, A., Sutskever, I., Hinton, G.E.: ImageNet classification with deep convolutional neural networks. In: Advances in Neural Information Processing Systems, vol. 25, pp. 1097–1105 (2012)
12. Molnar, C.: Interpretable machine learning (2020). Lulu.com

13. Nguyen, A., Yosinski, J., Clune, J.: Deep neural networks are easily fooled: high confidence predictions for unrecognizable images. In: Proceedings of the IEEE Conference on Computer Vision and Pattern Recognition, pp. 427–436 (2015)
14. Ribeiro, M.T., Singh, S., Guestrin, C.: "why should i trust you?" explaining the predictions of any classifier. In: Proceedings of the 22nd ACM SIGKDD International Conference on Knowledge Discovery and Data Mining, pp. 1135–1144 (2016)
15. Selvaraju, R.R., Cogswell, M., Das, A., Vedantam, R., Parikh, D., Batra, D.: Grad-cam: Visual explanations from deep networks via gradient-based localization. In: Proceedings of the IEEE International Conference on Computer Vision, pp. 618–626 (2017)
16. Shahriari, K., Shahriari, M.: IEEE standard review. ethically aligned design: a vision for prioritizing human wellbeing with artificial intelligence and autonomous systems. In: 2017 IEEE Canada International Humanitarian Technology Conference (IHTC), pp. 197–201. IEEE (2017)
17. Shrikumar, A., Greenside, P., Shcherbina, A., Kundaje, A.: Not just a black box: learning important features through propagating activation differences. arXiv preprint arXiv:1605.01713 (2016)
18. Singh, V.K., et al.: Mass detection in mammograms using a robust deep learning model. In: CCIA, pp. 365–372 (2019)
19. Smilkov, D., Thorat, N., Kim, B., Viégas, F., Wattenberg, M.: SmoothGrad: removing noise by adding noise. arXiv preprint arXiv:1706.03825 (2017)
20. Sundararajan, M., Taly, A., Yan, Q.: Axiomatic attribution for deep networks. In: International Conference on Machine Learning, pp. 3319–3328. PMLR (2017)
21. Szegedy, C., et al.: Intriguing properties of neural networks. arXiv preprint arXiv:1312.6199 (2013)
22. Zeiler, M.D., Fergus, R.: Visualizing and understanding convolutional networks. In: Fleet, D., Pajdla, T., Schiele, B., Tuytelaars, T. (eds.) ECCV 2014. LNCS, vol. 8689, pp. 818–833. Springer, Cham (2014). https://doi.org/10.1007/978-3-319-10590-1_53
23. Zintgraf, L.M., Cohen, T.S., Adel, T., Welling, M.: Visualizing deep neural network decisions: Prediction difference analysis. arXiv preprint arXiv:1702.04595 (2017)

Towards Machine Learning-Assisted Output Checking for Statistical Disclosure Control

Josep Domingo-Ferrer(✉) and Alberto Blanco-Justicia

Department of Computer Engineering and Mathematics, CYBERCAT-Center for Cybersecurity Research of Catalonia, UNESCO Chair in Data Privacy, Universitat Rovira i Virgili, Av. Països Catalans 26, 43007 Tarragona, Catalonia
{josep.domingo,alberto.blanco}@urv.cat

Abstract. There is an increasing demand by researchers to access the microdata (data on individual persons or enterprises) collected by national statistical institutes or other data controllers. If microdata are personally identifiable information, the most usual way for data controllers to share them in a way compliant with the privacy legislation (notably the EU General Data Protection Regulation) is to release anonymized microdata. Yet, data analysts often need access to the original microdata in order to avoid the information loss caused by anonymization. To answer that need, safe access centers (on physical premises or on-line) have been set up by several national statistical institutes. In these centers, users can run their analyses on original data using the controller's software, and the controller checks the outputs of the users' analyses before returning those outputs to them, in order to make sure users do not take home any result that might leak the confidential microdata on which it has been computed. Output checking is currently implemented with human checkers, which is expensive and slow, especially because checkers need to have specific statistical expertise. In this work, we explore the use of machine learning to partially automate output checking. We follow the rule-based approach and our empirical results show that our system can generalize the rules it is trained on. In conclusion, output checking assisted by machine learning seems to work well and should be trialed in safe access centers and decentralized data marketplaces.

Keywords: Output checking · Statistical disclosure control · Safe data access centers · Privacy · Machine learning

1 Introduction

Researchers want data that are as accurate as possible to reach meaningful and trustworthy conclusions. Microdata, that is, data at the level of individual persons or enterprises, are in increasing demand. Privacy legislation, epitomized by the European Union's General Data Protection Regulation [3], prevents data controllers from sharing for secondary use microdata that contain

V. Torra and Y. Narukawa (Eds.): MDAI 2021, LNAI 12898, pp. 335–345, 2021.
https://doi.org/10.1007/978-3-030-85529-1_27

personally identifiable information (PII). The most usual solution is for the controller to anonymize microdata before releasing them for secondary use [2,5]. Yet, anonymization entails information loss and hence the analyses on anonymized data may not be entirely trustworthy. For this reason, researchers often require access to the original microdata.

Some data controllers, such as those involved in the nascent decentralized data markeplaces, such as Ocean [6], intend to sell not only anonymized data (data-as-a-service) but also the possibility of running computations on the original data (compute-to-data). Yet, they offer no solution to avert possible data leakages associated with the results of computations.

Other data controllers, especially national statistical institutes and data archives, have set up safe access centers as an alternative for those situations in which resarchers cannot use anonymized data. A safe access center may be a physical facility to which the researcher must travel or an on-line service that the researcher can remotely access. Whatever the case, it is a controlled environment in which the researcher runs her analyses using software provided by the controller and is under monitoring by the controller's staff during her entire work session.

A salient feature of safe access centers is that any output of the researcher's analysis is checked by the data controller's staff before returning it to the researcher [1,4]. The purpose of output checking is to make sure that the researcher will not take home any result that might leak the confidential microdata on which it has been computed.

Highly expert output checkers can follow the so-called principles-based model [1,4]. In this model, no output is ruled in or out in advance. Rather, checkers collaborate with researchers and take the entire context of the analysis into account to make a decision on whether an output is safe enough to be returned or not. Although this model is quite costly, it minimizes the probabilities of false positives (labeling an output as safe when in fact it leaks sensitive information) and false negatives (labeling as unsafe an output that actually leaks no confidential information).

An easier alternative that requires less interaction and expertise on the checker's side is the rule-based model. In this case, the checker uses simple rules of thumb to label an output as safe or unsafe. The price paid is a higher probability of false positives and false negatives.

Contribution and Plan of This Paper

Output checking currently relies on human checkers. Even if they guide themselves by rules rather than principles, checking is time-consuming and hence expensive and slow. Besides, it is not easy for data controllers to appoint dedicated output checkers: staff with the required statistical expertise are difficult to recruit and output checking is often not regarded as a core task.

We propose to relieve some of the burden of output checking by (partially) automating it via a machine learning approach. This can be useful to all kinds of controllers, from national statistical institutes to decentralized data marketplaces.

The principles-based model is definitely very difficult to automate, because it requires contextual input to be obtained from the interaction between checkers and researchers. In contrast, the rule-based approach is more amenable to automation, as rules can easily be learned using machine learning.

Taking as a starting point the rules set forth in [1], we create synthetic output checking log files based on different subsets of rules. Then we train deep learning models on each synthetic log file, and we examine how well the rules used to generate the log file have been learned and, more importantly, how the rules *not* used to generate the log file have also been learned. Our results show that our deep learning approach can generalize the rules embedded in the training data, and hence captures the general flavor of safe and unsafe outputs. Admittedly, our system does not completely eliminate the need for human checking, but it can be used to reduce the human workload to filtering out any false positives, that is, outputs labeled as safe by our system which turn out to be unsafe under a more sophisticated checking.

The rest of this paper is organized as follows. Section 2 rewrites the checking rules proposed in [1] in view of using them to create synthetic output checking logs. Section 3 describes how to generate synthetic training data and test data from the rewritten rules. Section 4 reports experimental work and assesses how the deep learning models learned can generalize the rules embedded in the training data. Conclusions and future work suggestions are summarized in Sect. 5.

2 Rewriting Checking Rules for Synthetic Log Generation

In [1,4], rules of thumb are proposed to decide whether an output can be safely returned to the researcher. Both documents propose similar rules based on similar rationales.

For the sake of concreteness, we take the rules proposed in [1], because they are easier to automate than those in [4]. We rewrite the rules in terms of the following attributes: *AnalysisType*, *Output*, *Confidential*, *Context* and *Decision*. In this way, we get:

RULE 1

AnalysisType: FrequencyTable.
Output: Number of units in each cell.
Confidential: YES/NO (YES means the data on which the frequency table is computed are confidential).
Decision: YES/NO.

The decision is NO, that is, the output is not returned if data are confidential AND {some cell contains less than 10 units OR a single cell contains more than 90% of the total number of units in a row or column}.

RULE 2

AnalysisType: MagnitudeTable.
Output: Magnitudes in each cell (average or total).
Confidential: YES/NO.

Context: Number of units in each cell, and percentage of cell total represented by the maximum contribution to the cell.
Decision: YES/NO.
The decision is NO if data are confidential AND {some cell contains less than 10 units OR a single cell contains more than 90% of the units in a row or column OR in some cell the largest contributor contributes more than 50% of cell total}.

RULES 3a/3b/3c
AnalysisType: Maximum/Minimum/Percentile.
Output: Value of Maximum/Minimum/Percentile.
Confidential: YES/NO.
Decision: YES/NO.
The decision is NO if data are confidential.

RULE 4
AnalysisType: Mode.
Output: Modal value.
Confidential: YES/NO.
Context: Sample size.
Decision: YES/NO.
The decision is NO if {data are confidential AND the frequency of the modal value is more than 90% of the sample size}.

RULES 5a/5b/5c/5d
AnalysisType: Mean/Index/Ratio/Indicator.
Output: Value of the statistic.
Confidential: YES/NO.
Context: Sample size, percentage of sample total represented by the largest value in the sample.
Decision: YES/NO
The decision is NO if {data are confidential AND {sample size $<$ 10 OR a single contribution accounts for more than 50% of the sample total}}.

RULE 6
AnalysisType: ConcentrationRatio.
Output: Value of the ratio.
Confidential: YES/NO.
Context: Sample size, percentage of sample total represented by the largest value in the sample.
Decision: YES/NO.
The decision NO if {data are confidential AND {sample size $<$ 10 OR a single contribution accounts for more than 90% of the sample total}}.

RULES 7a/7b/7c
AnalysisType: Variance/Skewness/Kurtosis.
Output: Value of the statistic.
Confidential: YES/NO.
Context: Sample size.
Decision: YES/NO.
The decision is NO if {data are confidential AND sample size $<$ 10}.

RULE 8

AnalysisType: Graph.
Output: Graph.
Confidential: YES/NO .
Decision: YES/NO.
The decision is NO if data are confidential.

RULES 9a/9b

AnalysisType: LinearRegressionCoefficients/ NonLinearRegressionCoefficients.
Output: Value of coefficients.
Confidential: YES/NO.
Context: Intercept is to be returned?
Decision: YES/NO.
The decision is NO if {data are confidential AND intercept is one of the coefficients to be returned}.

RULE 10

AnalysisType: RegressionResiduals/RegressionResidualsPlot.
Output: Values of residuals/Plot of residuals.
Confidential: YES/NO.
Decision: YES/NO.
The decision is NO if data are confidential.

RULES 11a/11b

AnalysisType: TestStatistic_t/TestStatistic_F.
Output: Value of statistic.
Confidential: YES/NO.
Context: Degrees of freedom.
Decision: YES/NO.
The decision NO if {data are confidential AND degrees of freedom < 10}.

RULE 12

AnalysisType: FactorAnalysis.
Output: Factor scores.
Decision: YES.

RULE 13

AnalysisType: Correlations.
Output: Matrix of correlation coefficients.
Confidential: YES/NO.
Context: Number of units contributing to each correlation coefficient.
Decision: YES/NO.
If data are confidential, then the decision is NO for those coefficients that are -1, 0, -1 OR that have been computed on less than 10 units.

RULE 14

AnalysisType: CorrespondenceAnalysis.
Output: Loadings of factors.
Confidential: YES/NO.
Context: Number of variables, sample size.
Decision: YES/NO.
The decision is NO if {data are confidential AND {number of variables < 2 OR sample size < 10}}.

3 Generation of Synthetic Training and Test Data

In this section we discuss how to generate synthetic data from the above rules that can be used to train and test a deep learning model. On the one side, we will generate training data from a subset of the rules and, on the other side, we will generate test data from *all* the rules. The purpose of the test data is to enable an assessment of how well the model trained with the training data has been able to learn not only the rules embedded in the training data but also the rules that were not embedded (by generalizing the former).

We realize that records in a training data set should be of fixed length, which allows them to be fed to the inputs of, say, a neural network model. To that end, we need to derive a record schema of fixed length that can describe the decisions made by all the above rules.

However, there are some outputs in the above rules that have a variable number of components: frequency tables (Rule 1), magnitude tables (Rule 2), linear regression coefficients (Rule 9a), non-linear regression coefficients (Rule 9b), regression residuals (Rule 10), factor scores (Rule 12), correlation coefficients (Rule 13) and loadings of factors (Rule 14). To deal with that problem, we will split those rules into rules that *separately apply to each single output component*, *e.g.* to each single table cell, regression coefficient, regression residual, factor score, correlation coefficient or factor loading. Then we can always create a post-processing rule whereby the entire output is only returned to the researcher if all its components are labeled as safe. In what follows we will denote by Rule i_s, the *split* version of Rule i for a single output component.

With the above arrangement, we have a single output component for all split rules. Yet, the number of context components for split rules increases with respect to the original rules, because more context on a split output (*e.g.* a cell) within the total output (*e.g.* a table) is required to make a decision on the split output. Let us assess the number of context attributes required for each rule:

- For Rule 1_s, we need a context attribute *PercentageRows* that contains the maximum percentage the cell total represents over all the totals of rows and columns the cell is part of.
- For Rule 2_s, we need a context attribute *CellUnits* that stores the number of units in the cell, a second context attribute *PercentageRows* that contains the maximum percentage the cell total represents over all the totals of rows and columns the cell is part of, and a third context attribute *PercentageCellTotal* that contains the percentage of the cell total represented by the maximum contribution to the cell.
- For Rules 3a, 3b, 3c and 8, no context attributes are needed.
- For Rule 4, one context attribute *SampleSize* is needed to store the sample size.
- For Rules 5a, 5b, 5c, 5d and 6 a context attribute *SampleSize* is needed to store the sample size and another context attribute *PercentageSampleTotal* is needed to store the percentage of the sample total represented by the largest value in the sample.

- For Rules 7a, 7b and 7c, a context attribute *SampleSize* is needed to store the sample size.
- For Rules $9a_s$ and $9b_s$, one context attribute *Intercept* is needed that stores YES if the corresponding coefficient is the intercept and NO otherwise.
- For Rules 10_s (split rule for each residual) and 12_s (split rule for each factor score) no context attributes are needed.
- For Rule 11a, one context attribute *DegreesOfFreedom* is needed to store the degrees of freedom for the t test statistic.
- For Rule 11b, two context attributes *DegreesOfFreedom* and *DegreesOfFreedom2* are needed to store the degrees of freedom of the F test statistic.
- For Rule 13_s (split rule for each correlation coefficient), one context attribute *CellUnits* is needed that contains the number of units contributing to the correlation coefficient.
- For Rule 14_s (split rule for each factor loading), a context attribute *NumberOfVariables* is needed for the number of variables and another attribute *SampleSize* is needed for the sample size.

Therefore, the schema that can describe the decisions made by all rules is formed by the following superset of attributes: *AnalysisType*, *Output*, *Confidential*, *CellUnits*, *PercentageRows*, *PercentageCellTotal*, *SampleSize*, *PercentageSampleTotal*, *Intercept*, *DegreesOfFreedom*, *DegreesOfFreedom2*, *NumberOfVariables* and *Decision*.

Given the above schema, a synthetic record to describe an instance of a certain Rule i can be generated as follows:

1. Initialize *AnalysisType* to the analysis corresponding to Rule i.
2. Randomly choose an output that is compatible with the analysis type. *E.g.* in Rule 13_s the output is a correlation coefficient and hence it must lie in the interval $[-1, 1]$.
3. Randomly set *Confidential* to YES or NO.
4. Randomly choose context attributes that fit the expected semantics for the analysis type. *E.g.* in Rules $9a_s$ and $9b_s$, *Intercept* must be YES/NO, whereas the other context attributes must be blank; in Rules 11a and 11b, the context attributes must be natural numbers expressing degrees of freedom for the t and F distributions, respectively.
5. Finally, compute *Decision* according to the decision algorithm for the rule.

Given any subset of rules, the above procedure can be applied for each rule in the subset as many times as desired. The result can be used as synthetic training data for a deep learning model to learn the rules in the set.

To obtain test data, the above procedure should be used for the entire set of rules. In this way, one can test how well the rules in the training data have been learnt, and how well the deep learning model can generalize to capture the rules not present in the training data.

4 Experimental Work

We took the rules identified in the previous section, and we unified similar rules having the same decision algorithm. That is, we merged Rules 3a, 3b, 3c into a Rule 3*, Rules 5a, 5b, 5c, 5d into a Rule 5*, Rules 7a, 7b, 7c into a Rule 7*, Rules $9a_s$, $9b_s$ into a Rule $9*_s$, and Rules 11a, 11b into a Rule 11*. This left us with 14 total rules.

Then, following the procedure in Sect. 3, we generated a synthetic data set with 200,000 training samples, with each of the 14 rules contributing approximately 14,700 samples, half of which with positive decisions (the analysis results can be released) and half with negative decisions. As an example, in Table 1 we list a few generated records.

Table 1. Sample of generated records

AnalysisType	Output	Confidential	CellUnits	PercentageRows	PercentageCellTotal	SampleSize	PercentageSampleTotal	Intercept	DegreesOfFreedom	DegreesOfFreedom2	NumberOfVariables	Decision
2	0.658	0	0	0	0	508	0	0	0	0	0	True
3	0.613	1	0	0	0	358	0	0	0	0	0	True
11	0.568	1	0	0	0	9	0	0	0	0	0	False
7	0.564	0	0	0	0	955	68.5	0	0	0	0	True
9	0.337	1	0	0	0	763	5.8	0	0	0	0	False

As a machine learning model, we chose a neural network because deep neural networks have demonstrated the capacity to generalize well, and several frameworks allow for easy prototyping and deployment of neural models. Other machine learning models, such as gradient boosting, have also shown good performance on structured data, such as microdata sets. Thus, we trained feed-forward neural network with 2 hidden layers of 64 units each and obtained a 94.08% accuracy, a 4.2% false positive rate and a 7.4% false negative rate. Note that false positives indicate outputs that should not be released but whose decision is YES (release) and false negatives indicate outputs that could be released without privacy risk but whose decision is NO (do not release). We are mainly interested in a low false positive rate (FPR), since false positives are those that are dangerous for the privacy of the respondents on whose data the outputs are computed.

Next, we conducted a series of experiments to find out if a neural network can generalize when exposed to samples generated from rules it has not been exposed to during training. First, we generated a testing data set that contains samples generated using the 14 rules. This testing data set was used throughout all experiments. Then, from a number n of rules ranging from 1 to 14 we generated 100 training data sets using random subsets of n rules. That is, we built 100 data sets with samples generated from random subsets of one rule, 100 data sets with samples generated from random subsets of 2 rules, and so on, which yielded 1,400 data sets with 200,000 training samples each. We trained a neural network like the one described above for each of the training data sets and tested it against the previously described single testing data set whose samples were generated using all 14 rules. The source code and the results of our experiments are available in GitHub[1].

Figure 1 displays the distributions of obtained accuracies with respect to the number n of rules used to generate the training data sets. The figure shows how the accuracy of the trained models increases with the number n of rules. For a single rule ($n = 1$), however, the figure indicates that some data sets result in accuracies over 80%, although the median accuracy sits below 70% and the mode is below 60%. From $n = 6$ rules or more, most of the generated training data sets result in accuracies over 80%.

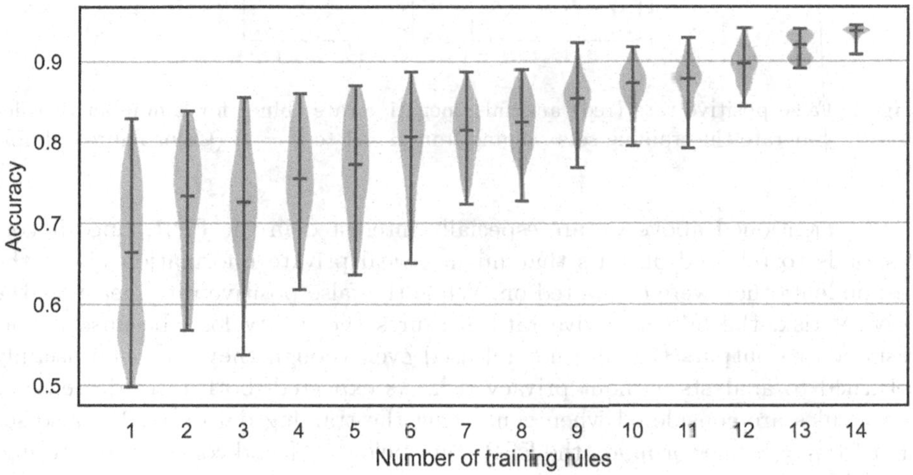

Fig. 1. Accuracy of the models with respect to the number n of rules used to generate the training sets

Figure 2 displays the false positive rate (red) and the false negative rate (blue) for a number of rules used to generate the training data sets ranging from $n = 1$ to $n = 14$.

[1] https://github.com/ablancoj/output-checking.

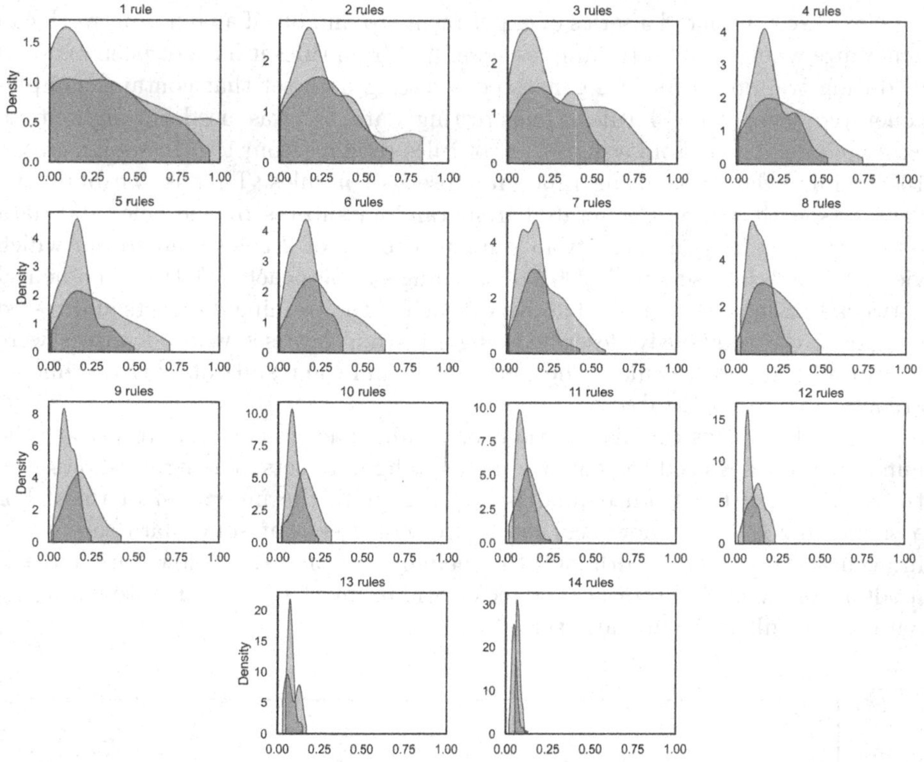

Fig. 2. False positive rate (red) and false negative rate (blue) for a number of rules used to generate the training sets ranging from $n = 1$ to $n = 14$ (Color figure online)

As mentioned above we are especially interested in the FPR, since it corresponds to released outputs that might reveal private information about the respondents they were computed on. While the false positive rate measures the privacy risk, the false negative rate measures the utility loss, because it corresponds to outputs that are not released even though they could be usefully returned to analysts without privacy risk. As expected, both rates decrease as more rules are considered when generating the training data sets. We also see that for $n = 8$ rules or more, the FPR stays below 50% and concentrates around 15–20%.

5 Conclusions and Future Research

We have presented an approach that leverages machine learning to assist human experts in output checking at safe data access centers. Our system follows the rule-based model, and we have shown that it can generalize the rules it is trained on. In our opinion, automating output checking is a pressing need for safe access centers and decentralized data marketplaces to take off.

A limitation of the presented research is that it does not use real log files obtained by the current manual output checking services, because such data are scarce and not public. Future research will strive to gather such data to further validate our approach. We also aim at increasing the level of automation of the entire process. Ideally, given the code of the analysis submitted by the analyst, it should be possible to automatically derive all the inputs required to make rule-based decisions. Optimizing the set of rules for maximum coverage and minimum overlap is another possible direction. Finally, extending automation to the principles-based model is also an important and daunting challenge.

Acknowledgments. Partial support is acknowledged from the European Commission (projects H2020-871042 "SoBigData++" and H2020-101006879 "MobiDataLab"), the Government of Catalonia (ICREA Acadèmia Prize to J. Domingo-Ferrer and grant 2017 SGR 705), and the Spanish Government (project RTI2018-095094-B-C21 "Consent"). The authors are with the UNESCO Chair in Data Privacy, but the views in this paper are their own and are not necessarily shared by UNESCO.

References

1. Bond, S., Brandt, M., de Wolf, P.-P.: Guidelines for the checking of output based on microdata research. Deliverable D11.8, project FP7-262608 "DwB: Data without Boundaries" (2015). https://ec.europa.eu/eurostat/cros/system/files/dwb_standalone-document_output-checking-guidelines.pdf
2. Domingo-Ferrer, J., Sánchez, D., Soria-Comas, J.: Database Anonymization: Privacy Models, Data Utility, and Microaggregation-Based Inter-Model Connections. Morgan & Claypool (2016)
3. General Data Protection Regulation. Regulation (EU) 2016/679. https://gdpr-info.eu
4. Griffiths, E., et al.: Handbook on Statistical Disclosure Control for Outputs (version 1.0). Safe Data Access Professionals Working Group (2019). https://ukdataservice.ac.uk/media/622521/thf_datareport_aw_web.pdf
5. Hundepool, A., et al.: Statistical Disclosure Control. Wiley, Hoboken (2012)
6. Ocean Protocol. https://oceanprotocol.com. Accessed 28 July 2021

Author Index